KU-110-545

# Apoptosis in Toxicology

# Apoptosis in Toxicology

Edited by
Ruth Roberts
AstraZeneca Central Toxicology Laboratory
Alderley Park
Macclesfield, UK

First published 2000 by Taylor & Francis Limited
11 New Fetter Lane, London EC4P 4EE

Simultaneously published in the USA and Canada by Taylor & Francis
29 West 35th Street, New York, NY 10001

© 2000 Taylor & Francis

*Taylor & Francis is an imprint of the Taylor & Francis Group*

Typeset in Stone Serif by Graphicraft Ltd, Hong Kong
Printed and bound in Great Britain by TJ International Ltd, Padstow, Cornwall

All rights reserved. No part of this book may be reprinted or reproduced or utilized in any form or by any electronic, mechanical, or other means, now known or hereafter invented, including photocopying and recording, or in any information storage or retrieval system, without permission in writing from the publishers.

Every effort has been made to ensure that the advice and information in this book is true and accurate at the time of going to press. However, neither the publisher nor the authors can accept any legal responsibility or liability for any errors or omissions that may be made. In the case of drug administration, any medical procedure or the use of technical equipment mentioned within this book, you are strongly advised to consult the manufacturer's guidelines.

*British Library Cataloguing in Publication Data*
A catalogue record for this book is available from the British Library

*Library of Congress Cataloging in Publication Data are available*

ISBN 0-7484-0815-0

# Apoptosis in Toxicology

# Foreword

For many years it was thought that chemical-induced cell injury and death occurred primarily by necrosis. Now, however, it is recognized that cell death may also be the result of another major mechanism, namely apoptosis. It is realized that apoptosis is of fundamental importance in many areas of cell biology, including pathology, immunology, developmental biology, carcinogenesis and, more recently, pharmacology and toxicology.

The early pioneering work of Andrew Wyllie and his colleagues described the importance of apoptosis in various tumours and in certain forms of chemical-induced injury. Subsequent studies by Horvitz and colleagues, in understanding programmed cell death in the nematode *Caenorhabditis elegans*, recognized the importance of three genes, namely CED-3, CED-4 and CED-9, as being critically important for developmental cell death. During normal development, 131 cells of the 1090 cells generated die by apoptosis. Two genes, CED-3 and CED-4, are vital for cell death in *C. elegans*, while the CED-9 gene antagonizes their function and prevents cell death. Over the past few years, the mammalian homologues of these genes have been identified: the BCL2 family members (CED-9), the caspases (CED-3) and Apaf-1 (CED-4). The identification of these mammalian homologues has led to a rapid increase in our knowledge of the fundamental mechanism(s) of apoptosis, so enabling us to begin to define its role in many different toxicological scenarios.

The importance of apoptosis in toxicology has long been underestimated, largely because of the difficulty in identifying apoptotic cells in the intact organism, due to their rapid phagocytosis and disposal. New methods providing better identification and quantitation of apoptotic cells have aided in the recognition of the importance of apoptosis in various areas of toxicology. While most of the methods are applicable largely to *in vitro* systems, some newer methods are being developed to aid in the early recognition of apoptotic cells *in vivo* in tissues exposed to toxicants.

The present book deals with basic mechanisms of apoptosis and their relevance to toxicology, and several chapters highlight the importance of apoptosis in various tissues in response to toxic agents. This volume is an important contribution to understanding the role of apoptosis in toxicology and will be of particular value to those working on fundamental mechanisms of toxicity and cell death.

Professor Gerry Cohen
MRC Toxicology Unit
University of Leicester, UK

# Apoptosis: Basic Mechanisms and Relevance to Toxicology

JASON H. GILL AND CAROLINE DIVE

School of Biological Sciences, University of Manchester, Manchester, M13 9PT, UK

## Contents

## 1.1 INTRODUCTION

The death of a cell was thought for many years to be an uncontrolled, degenerative and catastrophic failure of homeostasis in response to cellular injury and was thus of little scientific interest. However, the discovery over the past two decades that all cells within multicellular organisms have the ability to activate a cell death programme termed apoptosis has challenged this idea and resulted in a re-evaluation of all aspects of biology including immunology, development, carcinogenesis, and more recently, toxicology. Toxicologists have known for many years that cells can be killed by a variety of toxicants and pathological conditions. In general, when the toxicant is at high enough concentration to cause gross cellular injury or a major perturbation in the cellular environment the cell dies by necrosis or 'cell murder'. However, it is more common for the affected cells to be deleted by apoptosis or 'cell suicide'. It is now believed that apoptosis is the major form of pathophysiological cell death and that necrosis is much rarer, occurring only in circumstances of gross cell injury (reviewed in Raffray and Cohen, 1997).

Apoptosis is desirable to the organism as a whole since it provides a mechanism for the disposal of cells damaged by mutagenic chemicals or irradiation without perturbing the homeostatic balance of its environment. In addition to a role during a cellular response to toxicants, apoptosis occurs throughout development and is vitally important in the immune system (Jacobson *et al.*, 1997). Examples include the removal of interdigital cells from a solid limb paddle during digit formation (Hammar and Mottet, 1971) and deletion of potentially lethal self-reactive B- and T-lymphocytes in the immune system (MacDonald and Lees, 1990). Moreover, apoptosis is of major importance in the pathogenesis of several diseases (Thompson, 1995) such as cancer (Dive and Wyllie, 1992, Lee and Bernstein, 1995, Reed, 1995), AIDS (Gougeon and Montagnier, 1993) and neurological disorders such as Alzheimer's and Parkinson's disease (Carson and Ribeiro, 1993, Bredesen, 1995). Therefore, a basic understanding of the molecular mechanisms that regulate apoptosis will have widespread implications for the prevention and treatment of these diseases. Apoptosis is of major interest to molecular toxicologists since it is central to the action of many toxicants and, as such, is important in the study of toxicological potential.

## 1.2 MORPHOLOGY OF APOPTOSIS

The execution phase of apoptosis has striking morphological characteristics (Alison and Sarraf, 1995, Raffray and Cohen, 1997, Zamzami *et al.*, 1998) although, at present, the latent phases of apoptosis are undetectable. The morphology of the end stages of apoptosis is remarkably similar across a wide variety of tissues; an apoptotic body from the liver may be indistinguishable from an apoptotic body from the crypt of the small intestine (Alison and Sarraf, 1995). These morphological changes are described in detail and illustrated in Chapter 10. Briefly, they include loss of cell contact with neighbouring viable cells, chromatin condensation to form dense compact masses and breakdown of the cytoskeletal scaffolding. The cell surface membrane, which is already contracted, begins to ruffle and bleb. This is followed by breakdown of the cell into membrane-bound fragments or 'apoptotic bodies'. In contrast to necrosis, the plasma membrane and cytoplasmic organelles of the apoptotic cell remain intact. The apoptotic cell is swiftly recognized and engulfed then degraded by professional or amateur phagocytes (macrophages and neighbouring cells, respectively) (Savill *et al.*, 1993). *In vitro*, in the absence of phagocytosis, a secondary non-specific degeneration occurs which results in the uptake of vital dyes such as trypan blue and is commonly mistaken for necrosis. This process is often referred to as secondary necrosis (see Chapter 10). Importantly, the whole apoptotic process occurs without the development of an inflammatory response and thus facilitates the deletion of cells with minimal disruption to the surrounding cellular environment and tissue architecture.

## 1.3 BIOCHEMICAL EVENTS DURING APOPTOSIS INDUCED BY A TOXICANT



As with proliferation and differentiation, apoptosis has conserved, regulated stages. There also appear to be as yet incompletely defined biochemical checkpoints in the pathway to apoptosis where the cellular decision to live or die is made (Figure 1.1). The biochemical stages of apoptosis explained in this chapter are conserved throughout most of the tissues discussed in this book. There are events which occur prior to cellular commitment to apoptosis and others which are post commitment events. However, the events which actually commit a cell to apoptosis remain elusive. The idea of commitment, however, is an important one especially when considering sites for therapeutic intervention and molecular targeting in diseases where normal control of apoptosis has been lost or when considering antidotes to be applied after exposure of cells to a toxicant.

The biochemical stages of apoptosis induced by a toxicant are illustrated in Figure 1.1 and can generally be summarized as:

(1) the *imposition* of damage by the toxicant
(2) *sensing* the damage imposed by the toxicant
(3) *coupling* the damage to the engagement of apoptosis and
(4) *execution* of the cell and disposal of the corpse.

Each of these stages will be described in turn below. In addition to apoptosis induced by a toxic substance, apoptosis can be induced by the absence of necessary survival signals (see section 1.4.1) or via transmission of a lethal signal via activation of members of a family of death receptors including Fas (CD95/APO-1) (see section 1.3.2).

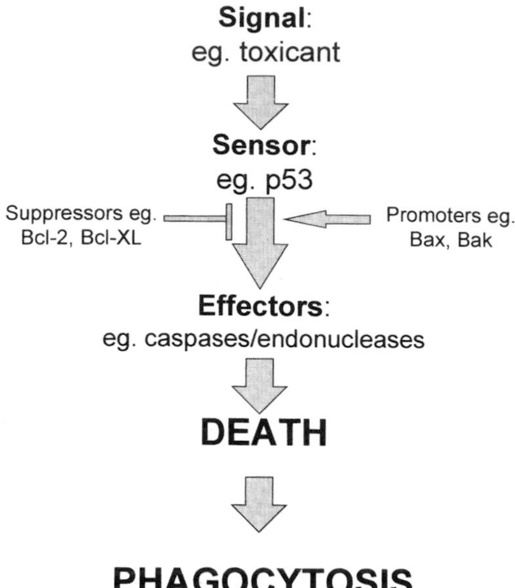


**Figure 1.1** The stages of apoptosis: signal, sensor and execution.

### 1.3.1 Imposition of damage by a toxicant

Apoptosis can be induced by a plethora of toxicants which have different mechanisms of action including agents which damage DNA and nongenotoxic agents.

### 1.3.2 Sensing the damage imposed by the toxicant

How do cells monitor toxicant-induced damage? This important question has yet to be fully answered particularly for nongenotoxic agents. However, more is known about responses to DNA damage. Since DNA contains all the 'instructions' for the function of the cell and the future of further cell populations, its integrity must be of the highest order. Any mutations, adducts or loss of integrity of the DNA may lead to the loss of function of the cell or the development of a neoplastic potential. Therefore, a tissue disposes of cells with irreparable DNA lesions or a mutagenic potential by apoptosis. DNA damage may be on either one or both strands of the DNA, and may be caused by many agents such as ionizing radiation or the DNA alkylating nitrogen mustards (by direct DNA damage), or as a result of agents such as the topoisomerase II inhibitor etoposide (by damaging DNA indirectly). There are many candidate proteins which are known to respond to DNA damage such as poly (ADP-ribose) polymerase (PARP), DNA protein kinase (DNA PK), ATM, the c-abl tyrosine kinase and the tumour suppressor p53, but apart perhaps from p53, their roles in sensing DNA damage and coupling it to repair, arrest or apoptosis have yet to be well characterized.

### 1.3.2.1 The tumour suppressor gene p53

The p53 gene encodes a nuclear phosphoprotein which acts as a transcription factor and a 'sensor' of DNA damage (Kastan, 1996). Normally, p53 protein is kept at a low basal cellular concentration due to its relatively short half-life. In addition, p53 probably exists in a latent form, inactive for transcription (Levine, 1997). Exposure of cells to DNA-damaging agents leads to an increase in p53 protein levels which is thought to result from a combination of both an increase in p53 protein half-life by post-translational stabilization and an increase in p53 protein translation (Levine, 1997). This increase in the levels of p53 in the cell results in the activation of p53 as a transcription factor. The activation of wild-type p53 leads to either growth arrest in G1 (Kastan *et al.*, 1991) and/or the induction of apoptosis (Yonish-Rouach *et al.*, 1991). Therefore, p53 functions as a cell cycle checkpoint between G1 and S-phase preventing the progression of damaged or potentially tumourigenic cells. Growth arrest allows time to repair the damage before replication, or if the damage is too great to commit to apoptosis (Hale *et al.*, 1996). The function of p53 as a tumour suppressor is exemplified by p53-null mice which develop normally but are prone to a variety of spontaneous tumours by 6 months of age (Donehower *et al.*, 1992). The coupling between DNA damage, p53 induction and apoptosis is cell type specific both *in vivo* and *in vitro*. This is demonstrated by studies of the effects of whole body irradiation on the individual tissues of the mouse (Midgley *et al.*, 1995, MacCallum *et al.*, 1996); although the whole body received a DNA damaging dose of radiation, apoptosis was observed in the small intestine, thymus and spleen, but not the lung, muscle or liver. Lack of apoptosis in the thymus and spleen following irradiation of p53-null mice confirms the requirement for p53 (Komarova *et al.*, 1997). Apoptosis following DNA damage was initially believed to be completely p53 dependent in small intestinal epithelia (Merritt *et al.*, 1994), but further investigations using p53-null mice demonstrated a delayed initiation of apoptosis compared to wild-type mice, indicating the presence of a p53-independent mechanism (Merritt *et al.*, 1997 and see below).

In most experimental systems, p53-dependent apoptosis is dependent on p53-mediated transcription, although this remains controversial (reviewed in Levine, 1997). Proteins which have been shown to be regulated by p53 include members of the Bcl-2 family (Miyashita *et al.*, 1994, Selvakumaran *et al.*, 1994); it has been shown to up-regulate the synthesis of the pro-apoptotic protein, Bax, and to down-regulate levels of the anti-apoptotic member Bcl-2 (Miyashita *et al.*, 1994). Since the intracellular balance of the pro-apoptotic Bcl-2 family members to the anti-apoptotic members has been demonstrated to regulate apoptosis in a variety of cell types, p53 would be predicted to induce cell death by tipping the Bcl-2 family rheostat towards apoptosis (Korsmeyer *et al.*, 1993).

Another gene which has been recently associated with DNA damage-induced, p53-dependent apoptosis is *fas* (CD95/APO-1). Fas is a member of the tumour necrosis factor receptor (TNF-R) superfamily (Cleveland and Ihle, 1995). Upon binding of its ligand (FasL)

Fas trimerizes and an intracellular receptor bound complex is assembled termed a death-inducing signalling complex (DISC) which in turn activates a caspase cascade (Chapter 2) leading to the death of the cell (Kischkel *et al.*, 1995). Increased transcriptional activity of p53 appeared to play a role in up-regulation of Fas in doxorubicin-treated hepatoblastoma cells (Muller *et al.*, 1997). In addition, there are temperature sensitive mutants of p53 which can up-regulate Fas (Owen-Schaub *et al.*, 1995). More recently, the TRAIL (TNF-related apoptosis inducing ligand) receptor-5 (also called KILLER/DR5), another member of the TNFR superfamily, has been shown to be induced in a p53-dependent manner following exposure of a variety of wild-type p53 containing cells to DNA damaging agents (Wu *et al.*, 1997).

p53 also transcriptionally up-regulates p21WAF1/CIP1 (El-Diery *et al.*, 1993) which acts as a potent inhibitor of cyclin dependent kinases (CDKs) causing cell cycle arrest at the GI/S border (El-Diery *et al.*, 1994). p21WAF1/CIP1 itself may also play a role in the regulation of apoptosis although this may be via a p53 independent mechanisms and is also cell context specific (El-Diery *et al.*, 1994, Polyak *et al.*, 1996). p53 also results in the up-regulation of the MDM2 protein (Hainaut, 1995, Hale *et al.*, 1996) which acts in a regulatory negative feedback loop by binding to p53 and blocking its transactivation (Chen *et al.*, 1994). This would lead to down-regulation of p53-dependent genes, such as those involved in apoptosis.

After DNA damage has caused a p53-mediated G1 arrest, DNA damaged cell can progress down three possible pathways: remain in a G1 arrest pending repair, repair the DNA and return to the cell cycle or die via apoptosis. The decision to undergo cell cycle arrest, repair or apoptosis is dependent upon cell type, the level of DNA damage received and the trophic environment. For instance, p53 is present in the liver but does not appear to accumulate following DNA damage induced by irradiation (Midgley *et al.*, 1995) or following administration of the cytotoxic drug etoposide (Gill *et al.*, 1998). Also, if mice are irradiated, the p53 protein is strongly expressed in the small intestinal crypts but not in the large bowel crypts (Merritt *et al.*, 1994). This may explain why tumours are more common in humans in the large bowel compared with the small bowel (Alison and Sarraf, 1995).

The role of p53 may be of major importance to toxicologists with reference to risk assessment. For instance, at high doses of a genotoxic chemical, cells may engage the apoptotic pathway. At lower doses, below the threshold level, cells may growth arrest, survive and thereby increase its chance of mutation. However the existence of p53 independent pathways linking DNA damage to apoptosis must also be considered.

### 1.3.2.2 p53 independent mechanisms for sensing DNA damage

p53 independent mechanisms to 'sense' DNA damage certainly exist. For example, in mice null for p53, apoptosis occurred following irradiation, but at a later time-point to that observed in wild-type mice (Merritt *et al.*, 1997). One molecule implicated in

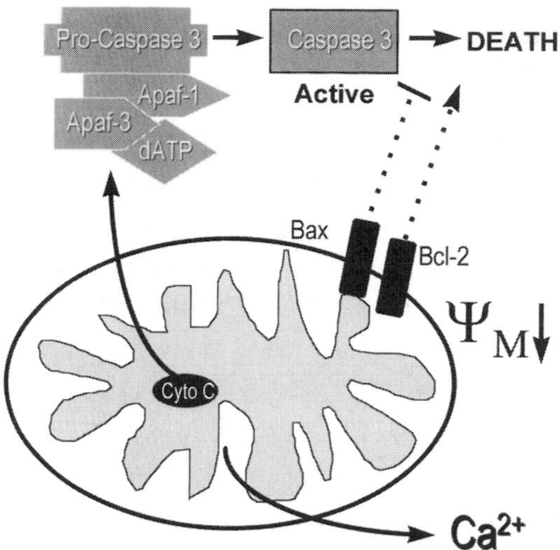


**Figure 1.2** The principal molecular regulators of apoptosis.



p53-independent DNA damage response pathways is interferon regulatory factor 1 (IRF-1) (Nozaki *et al.*, 1997) since G1 arrest and apoptosis were lost in DNA-damaged IRF-1 null fibroblasts and in IRF-1 null spleen-derived mature T-lymphocytes, respectively. DNA damaging agents activate the nuclear tyrosine kinase c-Abl, considered to be a *sensor* of DNA damage involved in coupling DNA damage to apoptosis (Kharbanda *et al.*, 1995). Cells which are null for c-Abl are more resistant to DNA damage-induced apoptosis than their wild-type counterparts (Yuan *et al.*, 1997). c-Abl binds to p53 at least *in vitro* but appears to have a pro-apoptotic function in the presence or absence of p53.

## 1.3.3 Coupling cellular damage to the engagement of apoptosis

After cellular damage has been *sensed*, this information has to be processed and coupled to a cellular response to the damage, and the decision to survive or to die must be taken. There is very little known about how these cellular processes might occur but the molecules and mechanisms implicated are discussed below.

### 1.3.3.1 Mitochondrial events

Mitochondria play a central role in mediating apoptosis and appear to be recipients of damage-induced signals (reviewed by Kroemer *et al.*, 1995). There is a loss of electrical potential across the inner mitochondrial membrane ($\Psi$m) early in the apoptosis pathway (Figure 1.2) which is termed the mitochondrial permeability transition (PT) (Kroemer *et al.*, 1995, Zamzami *et al.*, 1998). Cytochrome c release from mitochondria has also been shown to be a central event in the regulation of apoptosis (Liu *et al.*, 1996, Kluck *et al.*,

1997, Yang *et al.*, 1997). Once released from the mitochondria, cytochrome c (also termed Apaf-2, apoptosis protease activating factor-2) binds to an adaptor protein Apaf-1. This is the mammalian homolog of the nematode death gene *ced-4* which acts upstream of the worm caspase *ced-3* and downstream of the worm homolog of Bcl-2, *ced-9* (see Chapter 2, section 2.2.2). Apaf-1 has also been demonstrated to bind to dATP and caspase 9 (Apaf-3) (Vaux, 1997, Zou *et al.*, 1997). This complex of Apaf-l,-2,-3 and dATP is thought to trigger the cleavage of pro-caspase 3 which in turn initiates a caspase cascade, the executionery machinery of apoptotic cell death (discussed in Chapter 2, section 2.2.2). Exactly how cytochrome c is released from mitochondria into the cytosol is unclear but one hypothesis is that the outer mitochondrial membrane ruptures permitting the release of cytochrome c from the inter-membrane space (Zamzami *et al.*, 1998). The rupture of the outer mitochondrial membrane is believed to occur due to a PT-mediated influx of ions and water into the mitochondrial matrix. Since the inner mitochondrial membrane has a larger surface area than the outer membrane, the former could expand until it physically ruptures the latter (reviewed in Zamzami *et al.*, 1998). The relationship between PT and the release of apoptogenic substances such as cytochrome c and apoptosis inducing factor (AIF) from the mitochondria into the cytosol is currently unclear. The PT pore is not large enough to allow passage of cytochrome c. The release of cytochrome c from mitochondria can be regulated by members of the Bcl-2 family (1.1.3.2 and Chapter 2, section 2.2.2). A perplexing question remains unanswered: how does damage induced by a diverse array of toxicants with widely different cellular targets result in the induction of the mitochondrial events described above?

### 1.3.3.2 Intracellular calcium

One signalling molecule which could conceivably link diverse types of cellular damage to the mitochondrial events involved in mediating apoptosis is intracellular calcium, $[Ca^{2+}]_i$. Early studies suggested that alterations in $[Ca^{2+}]_i$ homeostasis were frequently observed in cells undergoing apoptosis (McConkey and Orrenius, 1997). However, it was never clear whether this was a cause or consequence of the apoptotic process. A role for $[Ca^{2+}]_i$ in regulating apoptosis was supported by the observation that $[Ca^{2+}]_i$ buffering agents, $Ca^{2+}$ chelators and inhibitors of the endoplasmic reticular (ER) $Ca^{2+}$-ATPase can inhibit caspase activation, DNA fragmentation, and cell death (McConkey and Orrenius, 1997). Conversely, calcium ionophores, which increase $[Ca^{2+}]_i$ trigger apoptosis in several cell types (McConkey and Orrenius, 1997). Further support for the involvement of $Ca^{2+}$ influx in the triggering of apoptosis comes from studies with specific $Ca^{2+}$ channel blockers, which abrogate apoptosis in the regressing prostate following testosterone withdrawal and in pancreatic β-cells treated with serum from patients with type I diabetes (McConkey and Orrenius, 1997). Recently, improved techniques and probes have re-established a role for calcium in apoptosis signalling (He *et al.*, 1997) but the mechanism remains unclear. Outflow from intracellular stores and influx of $Ca^{2+}$ across the plasma membrane could result in a sustained $Ca^{2+}$ increase that acts as a signal for apoptosis. Depletion of $[Ca^{2+}]_i$ from the ER, thought to be an early event in the apoptotic pathway, has been shown to be sufficient to induce apoptosis by causing the subsequent depletion of $Ca^{2+}$ from mitochondrial stores (Baffy *et al.*, 1993, He *et al.*, 1997). Anti-apoptotic members of the Bcl-2 family (discussed in section 1.3.3.4) have been shown to regulate $Ca^{2+}$ compartmentalization in both ER and mitochondria (Baffy *et al.*, 1993, He *et al.*, 1997). Constitutive levels of $C^{2+}$ in mitochondria were shown to be significantly lower in a Bcl-2-expressing haemopoietic cell line (32D) compared to vector control transfectants, supporting the idea that Bcl-2 regulates $Ca^{2+}$ in mitochondria (Baffy *et al.*, 1993). The exact mechanism by which Bcl-2 regulates $Ca^{2+}$ is currently unclear. One possible model is that Bcl-2 prevents the opening of Bax-induced channels, preventing $Ca^{2+}$ efflux from the mitochondria.

A growing family of apoptotic proteases, the caspases are generally considered to be the executioners of apoptosis. The caspases are discussed in depth in Chapter 2. Once the caspases have enzymatically broken down the intercellular components, the cell fragments form apoptotic bodies and are destroyed by phagocytosis.

### 1.3.3.3 Signalling for apoptosis by ceramide

Ceramide is a secondary messenger liberated by hydrolysis of membrane sphingolipids (Ballou *et al.*, 1996, Hale *et al.*, 1996). The generation of ceramide following an apoptotic stimulus results in the activation of a protein kinase cascade (Anderson, 1997). A wide range of apoptotic stimuli have been demonstrated to lead to the generation of ceramide. Activation of TNFα and Fas activate sphingomyelinases resulting in the production of ceramide and apoptotic cell death (Ballou *et al.*, 1996). Apoptosis induced by γ-irradiation was also demonstrated to involve ceramide signalling using an endothelial cell line. Accumulation of ceramide was observed within minutes of irradiation and cell permeable ceramides were shown to be capable of causing apoptosis in these cells (reviewed in Ballou *et al.*, 1996). In contrast to TNFα, Fas and γ-irradiation, apoptosis induced by the DNA damaging agent daunorubicin was demonstrated to induce ceramide but without sphingomyelinase activation. Daunorubicin-induced apoptosis was shown to involve *de novo* synthesis of ceramide via stimulation of ceramide synthase, which was further supported by the use of inhibitors of ceramide synthase that blocked daunorubicin induced apoptosis (reviewed in Ballou *et al.*, 1996). A role for ceramide in apoptosis was reinforced by the observation that resistance to certain types of apoptosis was associated with insufficient production of ceramide (Gottschalk *et al.*, 1995).

The downstream events of ceramide production and their link to the signalling of apoptosis are relatively unknown. Ceramide is demonstrated to activate both the MAP kinase (mitogen-activated protein kinase) and the SAP kinase (stress-activated protein kinase) pathways (Canman and Kastan, 1996, Anderson, 1997). The MAP kinase pathway is thought to be involved in cell survival and growth whereas the SAP kinase pathway is thought to have the opposing effect and has been associated with cell death (Canman and

Kastan, 1996, Anderson, 1997). Therefore, the apoptotic response of ceramide may be as a result of the perturbation of these two opposing pathways in favour of the SAP kinase pathway, culminating in the apoptotic death of the cell (Canman and Kastan, 1996, Anderson, 1997).

#### 1.3.3.4 The Bcl-2 family

Bcl-2, the acronym for the B-cell lymphoma/leukaemia-2 protein, was first discovered because it is overexpressed in follicular lymphoma and other B-cell malignancies (Tsujimoto *et al.*, 1985). Following a seminal report showing that Bcl-2 could synergize with c-myc to transform cells (Vaux *et al.*, 1988), Bcl-2 became the focus of much study (Reed, 1997). Bcl-2 is the founding member of a rapidly expanding family which include death antagonists (Bcl-2, Bcl-XL, Bcl-w, Bfl-I, Brag-1, Mcl-1 and AI) and death agonists (Bax, Bak, Bcl-xs, Bad, Bid, Bik and Hrk) (Reed, 1994, Kroemer, 1997). We therefore review the Bcl-2 family in its entirety here, but refer again to the protective effects of the Bcl-2-like members of the family in section 1.3.

The involvement of the Bcl-2 family in apoptosis is supported by studies of mice deficient for a particular Bcl-2 family member. Mice which are null for the anti-apoptotic Bcl-2 protein can be observed to be lymphopenic, azotemic and hypopigmented; all of these characteristics resulting from increased cell death (Veis *et al.*, 1993, Knudson and Korsmeyer, 1997). Bcl-XL deficient mice are embryonic lethal, displaying increased apoptosis in both haematopoietic and neuronal lineages (reviewed in Knudson and Korsmeyer, 1997). In contrast, mice deficient for the pro-apoptotic protein Bax demonstrate an excess of lymphocytes, granulosa cells and spermatogonia and have some neurons that display marked resistance to neurotrophic factor deprivation (reviewed in Knudson and Korsmeyer, 1997). Mice which have the genes for both *bcl*-2 and *bax* knocked out had equivalent numbers of thymocytes compared to the control mice, illustrating that the lymphocyte loss observed in Bcl-2 null mice depends on Bax (reviewed in Knudson and Korsmeyer, 1997).

The family possess variable amounts of Bcl-2 homology (BH) regions (BH1 to BH4), governing their capacity to interact with each other or with other unrelated proteins (reviewed in Kroemer, 1997). The ratio of death antagonists to agonists may determine whether a cell will respond to an apoptotic signal. This life–death balancing act is believed to be mediated, at least in part, by competitive dimerization between selective pairs of antagonists and agonists (reviewed in Kroemer, 1997). Whether one group can neutralize the functions of the other is not clear since the family members may act independently of one another (Knudson and Korsmeyer, 1997).

Most members of the Bcl-2 family possess a carboxy-terminal transmembrane region which acts to tether them into internal membranes of the ER, mitochondria and nucleus (Krajewski *et al.*, 1993, Kroemer, 1997). However some family members such as Bid and Bad do not possess this hydrophobic region and are located in the cytoplasm. The localization of Bcl-2 to membranes other than the mitochondria may neutralize the pro-apoptotic Bcl-2 family members, thereby preventing them from acting on the mitochondria (Kroemer, 1997). This theory is supported by the observation that ER-targeted Bcl-2 interacts with Bax, which itself must act on mitochondria to kill cells (reviewed in Kroemer, 1997). Alternatively, the effects of the Bcl-2 family may be mediated through the ER and nuclear membranes and not just through the mitochondria (reviewed in Kroemer, 1997). This theory is supported by the observation that ER-targeted Bcl-2 is capable of suppressing apoptosis induced by c-myc overexpression and serum withdrawal but not that induced by etoposide. In contrast, mitochondria-targeted Bcl-2 is an efficient inhibitor of apoptosis induced by all these stimuli, suggesting that Bcl-2 is a more efficient inhibitor of apoptosis when targeted to the mitochondria (reviewed in Zamzami *et al.*, 1998).

The anti-apoptotic proteins Bcl-2 and Bcl-XL prevent release of cytochrome c from mitochondria and the subsequent activation of caspase-3 (Yang *et al.*, 1997, Hengartner, 1998) (Chapter 2, Figure 2.1). Overexpression of Bax has been shown to result in its localization to the mitochondria and to induce the release of cytochrome c (Rosse *et al.*, 1998). In cells over-expressing Bcl-2, microinjection of cytochrome c no longer induces apoptosis (Zhivotovsky *et al.*, 1998). Taken together, these data suggest that Bcl-2 can regulate cytochrome c release from mitochondria but can also interfere with Bax mediated cell killing downstream of and independently of cytochrome c release (Rosse *et al.*, 1998). Several hypotheses have been put forward to explain these observations: Bcl-2 may bind directly to the Apaf complex or to cytochrome c in the cytosol to quench its effects (Zhivotovsky *et al.*, 1998).

Although Bcl-2 family proteins fail to show significant amino-acid sequence homology with other proteins, the three-dimensional structure of Bcl-XL appears to be similar to the pore-forming domains of certain bacterial toxins that act as channels for ions or proteins, such as diptheria toxin (Muchmore *et al.*, 1996). Thus, the Bcl-2 family of proteins may function as channels for ions, proteins or both (reviewed in Reed, 1997) and both Bcl-XL and Bcl-2 possess ion-channel activity in synthetic lipid membranes (Minn *et al.*, 1997, Reed, 1997). However, the pore formation is only observed at low pH (pH 4.0 or 5.5 for Bcl-XL and Bcl-2, respectively), suggesting that protonation of these proteins is required for pore formation. Recently, it has been demonstrated that Bax also exhibits channel activity *in vitro* (Zamzaini *et al.*, 1998) but the existence and function of pore formation *in vivo* at physiological pH remains unclear (reviewed in Zamzami *et al.*, 1998). The anti- and pro-apoptotic pores may have differential selectivity, providing either conduit, charge or direction specific transport of ions or molecules that regulate apoptosis (reviewed in Reed, 1997). Circumstantial evidence suggests that overexpression of Bcl-2 inhibits the mitochondrial permeability transition whereas Bax overexpression induced it (reviewed in Reed, 1997).

As well as forming pores, Bcl-2 and Bcl-XL can bind to a vast range of proteins including the protein kinase Raf-1, the protein phosphatase calcineurin, the GTPases R-Ras and

H-Ras, Nip-1, -2, -3 and a protein called BAG-I (reviewed in Reed, 1997). The binding of these signalling proteins provides a second function for the Bcl-2 family by acting as signal adapters or docking proteins.

The involvement of Bcl-2 in the Fas pathway is controversial and appears to depend on the experimental model (Itoh *et al.*, 1993, Chiu *et al.*, 1995, Lacronique *et al.*, 1996). Mice expressing human Bcl-2 in hepatocytes were protected against apoptosis induced by agonistic anti-Fas antibody (Lacronique *et al.*, 1996) and Bcl-2 down-regulation may be the first step in Fas-mediated apoptotic process in the male mouse reproductive tract following gonadectomy (Suzuki *et al.*, 1996).

Given the functional importance and central role of Bcl-2 related proteins in apoptosis control, they constitute prime targets for therapeutic interventions on numerous disease states mediated by dysregulated cell death. Knowledge of the mode of action of Bcl-2 related proteins may provide us with information for the development of new treatments and toxicological tests, as well as for the induction of apoptosis in tumour cells.

## 1.4 SUPPRESSION OF APOPTOSIS INDUCED BY TOXICANTS

### 1.4.1 Survival signals

As originally suggested by Raff (1992), cells must receive survival signals to avoid apoptosis. Survival signals comprise receptor binding of soluble growth factors such as interleukin-3 (IL3), cell-cell communications such as gap junctions and interaction with extracellular matrix components. It is not clear whether the various survival stimuli share common downstream signalling elements to suppress the default apoptotic pathway or whether they signal survival independently.

The mechanism by which growth factors permit cell survival is an area of great interest. It was found that signalling through Ras, Raf, and Src pathways, although mitogenic, was not sufficient to permit cell survival in fibroblasts (Kennedy *et al.*, 1997). One molecule which was found to be involved in the signalling of survival factors mediated by a wide range of growth factors is phosphoinositide 3 kinase (PI3K). Blockade of PI3K activity can suppress the ability of trophic factors to promote survival; conversely, in many cell types PI3K activity is sufficient to promote survival (Datta *et al.*, 1997) (see also Chapter 8, section 8.4.4.2). One target of PI3K found to be central to the delivery of survival signals is the serine-threonine kinase c-Akt, also known as protein kinase B (PKB). Activation of Akt is known to deliver a survival signal that inhibits apoptosis induced by a variety of stimuli, including growth factor withdrawal, cell-cycle discordance, loss of cell adhesion, and DNA damage (reviewed in Datta *et al.*, 1997). Akt activation results in the inhibition of caspase activity to protect from apoptotic cell death (Del Peso *et al.*, 1997). One possible mechanism by which PI3K/Akt suppresses apoptosis and promotes cell survival is via modulation of members of the Bcl-2 family. This was supported by the observations that activation of Akt in cells resulted in the phosphorylation and inactivation of the pro-apoptotic Bcl-2 family member, Bad (Datta *et al.*, 1997, Del Peso *et al.*, 1997). Activation of PI3K/Akt does not result in changes in either Bcl-2, Bcl-XL or Bax (Kennedy *et al.*, 1997). Therefore, Bad phosphorylation by Akt is one of many possible mechanisms by which survival signals may result in the suppression of apoptosis.

In addition to soluble factors, cell-cell contact and microenvironment plays an important role in cell survival (Bates *et al.*, 1994). Within the germinal centre, ligation of the cell surface antigen CD40 sends a signal to maturing B cells that is necessary to prevent their death by apoptosis (Wang *et al.*, 1997). The PI3K/Akt pathway and the Bcl-2 family have been implicated in the survival pathways. Interestingly, CD40 signalling also suppresses drug-induced apoptosis in B lymphoma cells (Walker *et al.*, 1997) increasing the probability of relapse after chemotherapy. Extracellular matrix also determines cell survival since epithelial cells from the mammary gland (Pullan *et al.*, 1996) survive only when attached to the correct extracellular matrix; inappropriate or inadequate matrices resulted in the death of the cells, even in the presence of cell-cell contacts and soluble trophic factors. In summary, survival factors and the extracellular environment is of major importance with respect to the control of apoptosis and the outcome of exposure to toxicants.



### 1.4.2 The Bcl-2 family

There have been a multitude of reports showing that enforced overexpression of Bcl-2 or of Bcl-XL acts to delay the onset of apoptosis induced by toxicants (Fisher *et al.*, 1993, Schott *et al.*, 1995). The vast majority of these reports concentrate on apoptosis induced by DNA damaging agents in cancer cell lines and conclude that Bcl-2 or Bcl-XL renders the cells drug resistant. However, most of these studies assess short-term survival and overlook subsequent clonogenicity, questioning their relevance to long-term cell survival. In the few studies that consider clonogenicity, Bcl-2 overexpression can give a clonogenic advantage after drug induced damage (Walker *et al.*, 1997), allowing cells with DNA damage to survive and proliferate.

## 1.5 SUMMARY

The maintenance of tissue homeostasis is a fine balance between cell survival and cell death. Since apoptosis appears a common response to many toxicants, it has become a particularly hot topic for toxicologists. Recent research has identified key molecules involved in the regulation of apoptosis and these (and others yet to be identified on the apoptotic pathways) are now major sites for investigation by toxicologists with a view to therapeutic intervention.

## ACKNOWLEDGEMENTS

The authors would like to thank Dr Sian Taylor for her contribution to Figure 1.2, and Dr Rosemary Gibson, Dr Danniela Riccardi and Professor John Hickman for their helpful discussions. Dr Caroline Dive is a Lister Institute Research Fellow.

The editor would like to thank Dr Sabina Cosulich for expert critical appraisal of this chapter and Zana Dye for invaluable editorial assistance.

## REFERENCES

ALISON, M.R. and SARRAF, C.E., 1995, Apoptosis: regulation and relevance to toxicology, *Human and Experimental Toxicology*, **14**, 234–247

ANDERSON, P., 1997, Kinase cascades regulating entry into apoptosis, *Microbiology and Molecular Biology Reviews*, **61**, 33–46

BAFFY, G., MIYASHITA, T., WILLIAMSON, J.R. and REED, J.C., 1993, Apoptosis induced by withdrawal of interleukin-3 (IL-3) from an IL-3-dependent hematopoietic cell line is associated with repartitioning of intracellular calcium and is blocked by enforced Bcl-2 oncoprotein production, *Journal of Biological Chemistry*, **268**, 6511–6519

BALLOU, L.R., LAULEDERKIND, S.J.F., ROSLONIEC, E.F. and RAGHOW, R., 1996, Ceramide signalling and the immune response, *Biochimica et Biophysica Acta*, **1301**, 273–287

BATES, R.C., BURET, A., VAN HELDEN, D.F., HORTON, M.A. and BUMS, G.F., 1994, Apoptosis induced by inhibition of intercellular contact, *Journal of Cell Biology*, **125**, 403–415

BREDESEN, D.E., 1995, Neural apoptosis, *Annals of Neurology*, **38**, 839–851

CANMAN, C.E. and KASTAN, M.B., 1996, Three paths to stress relief, *Nature*, **384**, 213–214

CARSON, D.A. and RIBEIRO, J.M., 1993, Apoptosis and disease, *The Lancet*, **341**, 1251–1254

CHEN, C.Y., OLINER, J.D., ZHAN, Q., FOMACE, A.J., VOGELSTEIN, B. and KASTAN, M.B., 1994, Interactions between p53 and MDM2 in a mammalian cell cycle checkpoint pathway, *Proceedings of the National Academy of Science (USA)*, **91**, 2684–2688

CHIU, V.K., WALSH, C.M., LIU, C.C., REED, J.C. and CLARK, W.R., 1995, Bcl-2 blocks degranulation but not Fas-based cell-mediated cytotoxicity, *Journal of Immunology*, **154**, 2023–2032

CLEVELAND, J.L. and IHLE, J.N., 1995, Contenders in Fas/'INF death signalling, *Cell*, **81**, 479–482

DATTA, S.R., DUDEK, H., TAO, X., MASTERS, S., FU, H., GOTOH, Y. and GREENBERG, M.E., 1997, Akt phosphoiylation of BAD couples survival signals to the cell-intrinsic death machinery, *Cell*, **91**, 231–241

DEL PESO, L., GONZALEZ-GARCIA, M., PAGE, C., HERRERA, R. and NUNEZ, G., 1997, Interleukin-3induced phosphorylation of BAD through the protein kinase Akt, *Science*, **278**, 687–689

DIVE, C. and WYLLIE, A.H., 1992, Apoptosis and cancer chemotherapy, in HICKMAN, J.A. and TRITTON, T.R. (eds), *Frontiers in Pharmacology and Therapeutics*, Oxford: Blackwell Scientific, 21–56

DONEHOWER, L.A., HARVEY, M., SLAGLE, B.L., MCARTHUR, M.J., MONTGOMERY, C.A., BUTEL, J.S. and BRADLEY, A., 1992, Mice deficient for p53 are developmentally normal but susceptible to spontaneous tumours, *Nature*, **356**, 215–221

EL-DIERY, W.S., TOKINO, T., VELCULESCU, V.E., LEVY, D.B., PARSONS, R., TRENT, J.M., *et al.*, 1993, WAFI, a potential mediator of p53 tumour suppression, *Cell*, **75**, 817–825

EL-DIERY, W.S., WADE HARPER, J., O'CONNOR, P.M., VELCULESCU, V.E., CANMAN, C.E., JACKMAN, J., *et al.*, 1994, WAFI/CIPI is induced in p53-mediated GI arrest and apoptosis, *Cancer Research*, **54**, 1169–1174

FISHER, T.C., MILNER, A.E., GREGORY, C.D., JACKMAN, A.L., AHEME, G.W., HARTLEY, J.A., *et al.*, 1993, Bcl-2-modulation of apoptosis induced by anticancer drugs: resistance to thymidylate stress is independent of classical resistance pathways, *Cancer Research*, **53**, 3321–3326

GILL, J.H., ROBERTS, R.A. and DIVE, C., 1998, The nongenotoxic hepatocarcinogen nafenopin suppresses rodent hepatocyte apoptosis induced by TGFβ1, DNA damage and Fas, *Carcinogenesis*, **19**, 299–304

GOTTSCHALK, A.R., MCSHAN, C.L., KILKUS, J., DAWSON, G. and QUINTANS, J., 1995, Resistance to anti-lgm induced apoptosis in a WEHI-231 subline is due to insufficient production of ceramide, *European Journal of Immunology*, **25**, 1032–1038

GOUGEON, M.L. and MONTAGNIER, L., 1993, Apoptosis in AIDS, *Science*, **260**, 1269–1270

HAINAUT, P., 1995, The tumor suppressor protein p53: a receptor to genotoxic stress that controls cell growth and survival, *Current Opinion in Oncology*, **7**, 76–82

HALE, A.J., SMITH, C.A., SUTHERLAND, L.C., STONEMAN, V.E.A., LONGTHOME, V.L., CULHANE, A.C. and WILLIAMS, G.T., 1996, Apoptosis: molecular regulation of cell death, *European Journal of Biochemistry*, **236**, 1–26

HAMMAR, S.P. and MOTTET, N.K., 1971, Tetrazolium salt and electron-microscopic study of cellular degeneration and necrosis in the interdigital areas of the developing limb, *Journal of Cell Science*, **8**, 229–251

HE, H., LAM, M., MCCONNICK, T.S. and DISTELHORST, C.W., 1997, Maintenance of calcium homeostasis in the endoplasmic reticulum by Bel-2, *Journal of Cell Biology*, **138**, 1219–1228

HENGARTNER, M.O., 1998, Death cycle and swiss army knives, *Nature*, **391**, 441–442

ITOH, N., TSUJIMOTO, Y. and NAGATA, S., 1993, Effect of Bcl-2 on Fas antigen-mediated cell death, *Journal of Immunology*, **151**, 621–627

JACOBSON, M.D., WEIL, M. and RAFF, M.C., 1997, Programmed cell death in animal development, *Cell*, **88**, 347–354

KASTAN, M.B., 1996, Signalling to p53: where does it all start? *BioEssays*, **18**, 617–619

KASTAN, M.B., ONYEKWERE, O., SIDRANSKY, D., VOGELSTEIN, B. and CRAIG, R.W., 1991, Participation of p53 protein in the cellular response to DNA damage, *Cancer Research*, **51**, 6304–6311

KENNEDY, S.G., WAGNER, A.J., CONZEN, S.D., JORDAN, J., BELLACOSA, A., TSICHLIS, P.N. and HAY, N., 1997, The PI 3 -kinase/Akt signalling pathway delivers an anti-apoptotic signal, *Genes and Development*, **11**, 701–713

KHARBANDA, S., REN, R., PANDEY, P., SHAFMAN, T.D., FELLER, S.M., WEICHSELBAUM, R.R. and KUFE, D.W., 1995, Activation of the c-Abl tyrosine kinase in the stress response to DNA damaging agents, *Nature*, **376**, 785–788

KISCHKEL, F.C., HELLBARDT, S., BEHRMANN, I., GERMER, M., PAWLITA, M., KRAMMER, P.H. and PETER, M.E., 1995, Cytotoxicity-dependent APO- I (Fas/CD95)-associated proteins form a death-inducing signalling complex (DISC) with the receptor, *EMBO Journal*, **14**, 5579–5588

KLUCK, R.M., BOSSY-WETZEL, E., GREEN, D.R. and NEWMEYER, D.D., 1997, The release of cytochrome c from mitochondria: a primary site for Bel-2 regulation of apoptosis, *Science*, **275**, 1132–1136

KNUDSON, C.M. and KORSMEYER, S.J., 1997, Bcl-2 and Bax function independently to regulate cell death, *Nature Genetics*, **16**, 358–363

KOMAROVA, E.A., CHEMOV, M., FRANKS, R., WANG, K., ANNIN, G., ZELNIK, C.R., *et al.*, 1997, Transgenic mice with p53-responsive IACZ: p53 activity varies dramatically during normal development and determines radiation and drug sensitivity *in vivo*, *EMBO Journal*, **16**, 1391–1400

KORSMEYER, S.J., SHUTTER, J.R., VEIS, D.J., MERRY, D.E. and OLTVAI, Z.N., 1993, Bcl-2/Bax: a rheostat that regulates an anti-oxidant pathway and cell death, *Seminars in Cancer Biology*, **4**, 327–332

KRAJEWSKI, S., TANAKA, S., TAKAYARNA, S., SCHIBLER, M.J., FENTON, W. and REED, J.C., 1993, Investigation of the subcellular distribution of the Bcl-2 oncoprotein. Residence in the nucleur envelope, endoplasmic reticulw-n, and outer mitochondrial membranes, *Cancer Research*, **53**, 4701–4714

KROEMER, G., 1997, The proto-oncogene Bcl-2 and its role in regulating apoptosis, *Nature Medicine*, **3**, 614–620

KROEMER, G., PETIT, P., ZARNZAMI, N., VAYSSIERE, J.-L. and MIGNOTTE, B., 1995, The biochemistry of programmed cell death, *FASEB Journal*, **9**, 1277–1287

LACRONIQUE, V., MIGNON, A., FABRE, M., VIOLLET, B., ROUQUET, N., MOLINA, T., *et al.*, 1996, Bcl-2 protects from lethal hepatic apoptosis induced by an anti-Fas antibody in mice, *Nature Medicine*, **2**, 80–86

LEE, J.M. and BERNSTEIN, A., 1995, Apoptosis, cancer and the p53 tumour suppressor gene, *Cancer Metastasis Reviews*, **14**, 149–161

LEVINE, A.J., 1997, p53, the cellular gatekeeper for growth and division, *Cell*, **88**, 323–331

LIU, X., KIM, C.N., YANG, J., JEMMERSON, R. and WANG, X., 1996, Induction of apoptotic program in cell-free extracts: requirement for DATP and cytochrome c, *Cell*, **86**, 147–157

MACCALLUIN, D.E., HUPP, T.R., MIDGLEY, C.A., STUART, D., CAMPBELL, S.J., HARPER, A., *et al.*, 1996, The p53 response to ionising radiation in adult and developing murine tissues. *Oncogene*, **13**, 2575–2587

MACDONALD, H.R. and LEES, R.K., 1990, Programmed death of autoreactive thymocytes, *Nature*, **343**, 642–644

MCCONKEY, D.J. and ORRENIUS, S., 1997, The role of calcium in the regulation of apoptosis, *Biochemical Biophysical Research Communications*, **239**, 357–366

MERRITT, A.J., ALLEN, T.D., POTTEN, C.S. and HICKMAN, J.A., 1997, Apoptosis in small intestinal epithelia from p53-null mice: evidence for a delayed, p53-independent G2/M-associated cell death after gamma-irradiation, *Oncogene*, **14**, 2759–2766

MERRITT, A.J., POTTEN, C.S., KEMP, C.J., HICKINAN, J.A., BALMAIN, A., LANE, D.P. and HALL, P.A., 1994, The role of p53 in spontaneous and radiation-induced apoptosis in the gastrointestinal tract of normal and p53-deficient mice, *Cancer Research*, **54**, 614–617

MIDGLEY, C.A., OWENS, B., BRISCOE, C.V., THOMAS, D.B., LANE, D.P. and HALL, P.A., 1995, Coupling between gamma irradiation, p53 induction and the apoptotic response depends upon cell type *in vivo*, *Journal of Cell Science*, **108**, 1843–1848

MINN, A.J., VELEZ, P., SCHENDEL, S.L., LIANG, H., MUCHMORE, S.W., FESIK, S.W., *et al.*, 1997, Bcl-XL forms an ion channel in synthetic lipid membranes, *Nature*, **385**, 353–357

MIYASHITA, T., KRAJEWSKI, S., KRAJEWSKA, M., WANG, H.G., LIN, H.K., LIEBENNANN, D.A., *et al.*, 1994, Tumor suppressor p53 is a regulator of Bcl-2 and bax gene expression *in vitro* and *in vivo*, *Oncogene*, **9**, 1799–1805

MUCHMORE, S.W., SATTLER, M., LIANG, H., MEADOWS, R.P., HARLAN, J.E., YOON, H.S., *et al.*, 1996, X-ray and NMR structure of human Bcl-XL, an inhibitor of programmed cell death, *Nature*, **381**, 335–341

MULLER, M., STRAND, S., HUG, H., HEINEMANN, E.-M., WALCZAK, H., HOFINANN, W.J., *et al.*, 1997, Drug-induced apoptosis in hepatoma cells is mediated by the CD95 (APO- I /Fas) receptor/ligand system and involves activation of VAld-type p53, *Journal of Clinical Investigation*, **99**, 403–413

Nozaki, T., Masutani, M., Sugimura, T., Takato, T. and Wakabayashi, K., 1997, Abrogation of GI arrest after DNA damage is associated with constitutive overexpression of mdm2, Cdk4, and lrfl mRNAs in the BALB/c 3 T3 variant 1–1 clone, *Biochemical Biophysical Research Communications*, **233**, 216–220

Owen-Schaub, L.B., Zhang, W., Cusack, J.C., Angelo, L.S., Santee, S.M., Fujiwara, T., *et al.*, 1995, Wild-type human p53 and a temperature-sensitive mutant induce Fas/APO-1 expression, *Molecular and Cellular Biology*, **15**, 3032–3040

Polyak, K., Waldman, T., He, T.-C., Kinzler, K.W. and Vogelstein, B., 1996, Genetic determinants of p53-induced apoptosis and growth arrest, *Genes and Development*, **10**, 1945–1952

Pullan, S., Wilson, J., Metcalfe, A., Edwards, G.M., Goberdhan, N., Tilly, J., *et al.*, 1996, Requirement of basement membrane for the suppression of programmed cell death in mammary epithelium, *Journal of Cell Science*, **109**, 631–642

Raff, M.C., 1992, Social controls on cell survival and cell death, *Nature,* **356**, 397–400

Raffray, M. and Cohen, G.M., 1997, Apoptosis and necrosis in toxicology: a continuum or distinct modes of cell death, *Pharmacology & Therapeutics*, **75**, 153–177

Reed, J.C., 1994, Bcl-2 and the regulation of programmed cell death, *Journal of Cell Biology*, **124**, 1–6

Reed, J.C., 1995, Regulation of apoptosis by Bcl-2 family proteins and their role in cancer and chemoresistance, *Current Opinion in Oncology*, **7**, 541

Reed, J.C., 1997, Double identity for proteins of the Bcl-2 family, *Nature*, **387**, 773–776

Rosse, T., Olivier, R., Monney, L., Rager, M., Conus, S., Fellay, I., *et al.*, 1998, Bcl-2 prolongs cell survival after Bax-induced release of cytochrome c, *Nature*, **391**, 496–499

Savill, J., Fadok, V., Henson, P. and Haslett, C., 1993, Phagocyte recognition of cells undergoing apoptosis, *Immunology Today*, **14**, 131–136

Schott, A.F., Apel, I.J., Nunez, G. and Clarke, M.F., 1995, Bcl-xl protects cancer cells from p53 mediated apoptosis, *Oncogene*, **11**, 1389–1394

Selvakumaran, M., Lin, H.K., Miyashita, T., Wang, H.G., Krajewski, S., Reed, J.C., *et al.*, 1994, Immediate early up-regulation of bax expression by p53 but not TGF-P a paradigm for distinct apoptotic pathways, *Oncogene*, **9**, 1791–1798

Suzuki, A., Matsuzawa, A. and Lguchi, T., 1996, Down regulation of Bcl-2 is the first step on Fas-mediated apoptosis of male reproductive tract, *Oncogene* **13**, 31–37

Tsujimoto, Y., Cossman, J., Jaffe, E. and Croce, C.M., 1985, Involvement of the Bcl-2 oncogene in human follicular lymphoma, *Science*, **228**, 1440–1443

VAUX, D.L., 1997, CED-4. The third horseman of apoptosis, *Cell*, **90**, 389–390

VAUX, D.L., CORY, S. and ADAMS, J.M., 1988, Bcl-2 gene promotes haemopoietic cell survival and co-operates with c-myc to immortalize pre-B cells, *Nature*, **335**, 440–442

VEIS, D.J., SORENSON, C.M., SHUTTER, J.R. and KORSMEYER, S.J., 1993, Bcl-2 deficient mice demonstrate fulminant lymphoid apoptosis, polycyctic kidneys and hypoppigmented hair, *Cell*, **75**, 229–240

WALKER, A., TAYLOR, S.T., HICKMAN, J.A. and DIVE, C., 1997, Germinal centre-derived signals act with Bcl-2 to decrease apoptosis and increase clonogenicity of drug-treated human B lymphoma cells, *Cancer Research*, **57**, 1939–1945

WANG, D., FREEMAN, G.J., LEVINE, H., RITZ, J., ROBERTSON, M.J., 1997, Role of the CD40 and CD95 (Apo- I /Fas) antigens in the apoptosis of human B-cell malignancies, *British Journal of Haematology*, **97**, 409–417

WU, G.S., BUMS, T.F., MCDONALD, E.R., JIANG, W., MENG, R., KRANTZ, I.D., *et al.*, 1997, KILLER/DR5 is a DNA damage-inducible p53-regulated death receptor gene, *Nature Genetics*, **17**, 141–143

YANG, J., LIU, X., BHAILA, K., KIM, C.N., IBRADO, A.M., CAI, J., *et al.*, 1997, Prevention of apoptosis by Bcl-2: Release of cytochrome c from mitochondria blocked, *Science*, **275**, 1129–1136

YONISH-ROUACH, E., LOTEM, J., SACHS, L., KIMICHI, A. and OREN, M., 1991, Wild-type p53 induces apoptosis of myeloid leukaemic cells that is inhibited by interleukin 6, *Nature*, **352**, 345–347

YUAN, Z.-M., HUANG, Y., ISHIKO, T., KHARBANDA, S., WEICHSELBAUM, R. and KUFE, D., 1997, Regulation of DNA damage-induced apoptosis by the c-Abl tyrosine kinase, *Proceedings of the National Academy of Science (USA)*, **94**, 1437–1440

ZAMZAMI, N., BRENNER, C., MARZO, I., SUSIN, S.A. and KROEMER, G., 1998, Subcellular and submitochondrial mode of action of Bcl-2 like oncoproteins, *Oncogene*, **16**, 2265–2282

ZHIVOTOVSKY, B., ORRENIUS, S., BRUSTUGUN, O.T. and DOSKELAND, S.O., 1998, Injected cytochrome c induces apoptosis, *Nature*, **391**, 449–450

ZOU, H., HENZEL, W.J., LIU, X., LUTSCHG, A. and WANG, X., 1997, Apaf-l, a human protein homologous to *C. elegans* CED-4, participates in cytochrome c-dependent activation of caspase-3, *Cell*, **90**, 405–413

CHAPTER

2

# *Toxicological Consequences of Caspase Inhibition and Activation*

**KELVIN CAIN**

MRC Toxicology Unit, Hodgkin Building, University of Leicester, Lancaster Road, Leicester, LE1 9HN, UK

## *Contents*

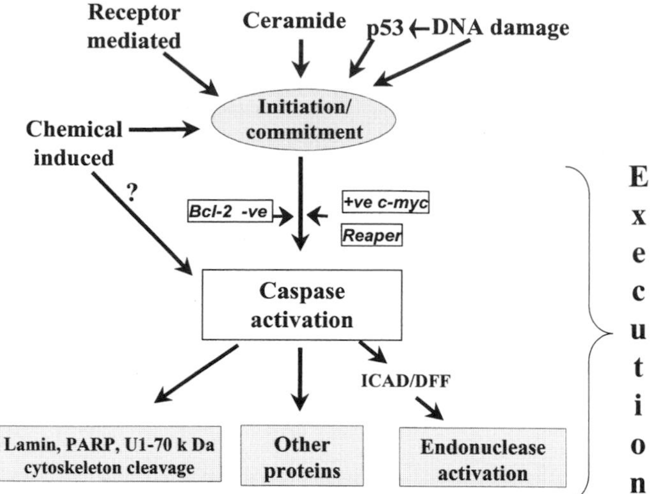


**Figure 2.1** A schematic diagram depicting the pathways of apoptosis, showing the pivotal role played by caspases in the execution of cell death.

## 2.1 INTRODUCTION

Since the original finding by Kerr *et al.* (1972) which described the morphology of apoptotic cell death, considerable progress has been made in unravelling the biochemistry and molecular biology of the process (reviewed in Hale *et al.*, 1996, Adams and Cory, 1998, Ashkenazi and Dixit, 1998, Evan and Littlewood, 1998, Green, 1998, Green and Reed, 1998, Thornberry and Lazebnik, 1998). These pathways have been described in detail in Chapter 1 and are summarized in Figure 2.1. Irrespective of how cell death is initiated, death signalling proceeds to an execution stage that involves caspase activation and the subsequent proteolytic dismembering of the cell. The toxicological consequences and mechanisms of this caspase activation and its inhibition are considered in this chapter.

The activation of the caspases seems to represent an all or nothing, one way process concluding in life or death of the cell. Thus, the possible toxicological consequences of interference with caspase activity are great and could be derived from either inappropriate activation or inhibition of the caspases. To address this, the enzymology of the caspase enzymes is discussed in the next section followed by a final section describing modulation of caspase activity and the possible consequences of activating or inhibiting caspases.

## 2.2 ENZYMOLOGY OF CASPASES

### 2.2.1 General properties

Caspases were first implicated in apoptosis when CED-3, a protein required for programmed cell death in *Caenorhabditis elegans*, was found to have close homology with the

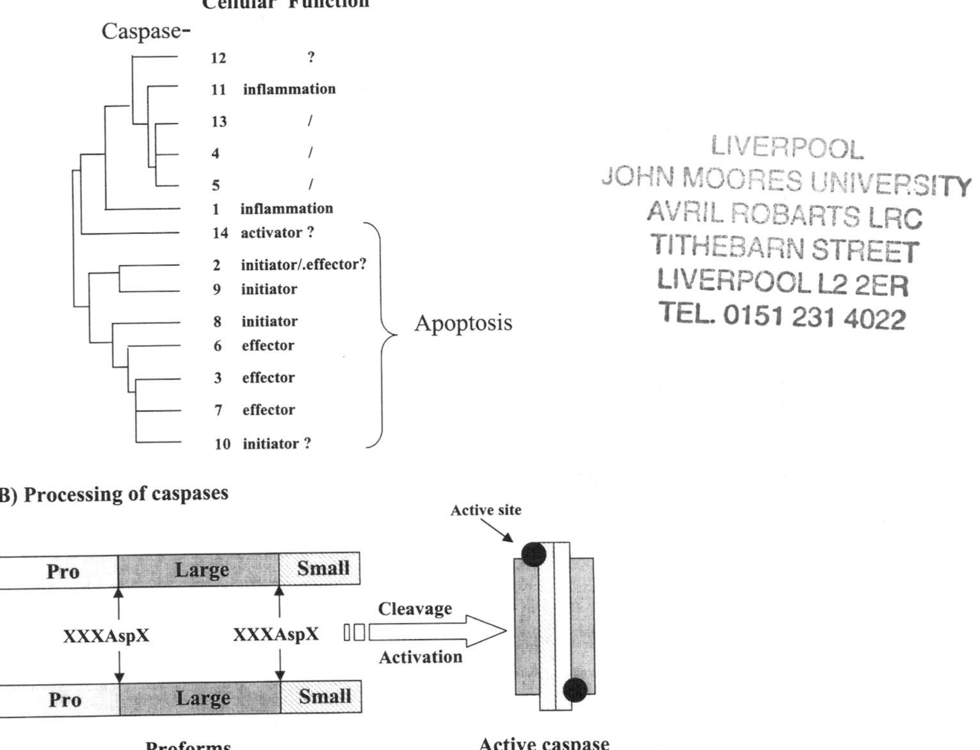


LIVERPOOL
JOHN MOORES UNIVERSITY
AVRIL ROBARTS LRC
TITHEBARN STREET
LIVERPOOL L2 2ER
TEL. 0151 231 4022



**Figure 2.2** (A) Nomenclature and sequence homology for the caspase family. (B) Schematic diagram depicting the processing of caspases from a pro-form to the active form.

mammalian interleukin-1β-converting enzyme (ICE or caspase-1) and that overexpression of ICE-induced apoptosis (Miura *et al.*, 1993, Yuan *et al.*, 1993). Since this initial discovery there has been an avalanche of papers describing new CED-3 homologues (reviewed in Cohen, 1997). This rush of papers inevitably led to a plethora of acronyms for the new homologues, often with duplication. In order to rationalize this confusing picture, a uniform nomenclature was proposed and now seems to be universally accepted (Alnemri *et al.*, 1996). Thus, the generic name for all family members is caspase with the *c* denoting a cysteine protease and *aspase* referring to the aspartate specific cleaving ability of these enzymes. The individual members are then numbered according to their chronological order of publication and thus ICE became caspase-1. Additional descriptives for the main functions are now quite commonly used such as initiator/activator or effector caspases.

To date 14 caspases have been identified (12 human and 2 murine) and are listed in Figure 2.2, which depicts the phylogenetic and probable cellular functions (Thornberry and Lazebnik, 1998). Caspases have marked similarities in amino acid sequence, structure and substrate specificity (Cohen, 1997, Thornberry and Lazebnik, 1998). They are synthesized as pro-enzymes with molecular weights ranging from 30–50 kDa and contain three domains: a N-terminal pro-domain which varies considerably between family members, a large subunit (~20 kDa) and small subunit (~10 kDa) domain which show considerable homology between family members. The pro-caspases are believed to be essentially, catalytically inactive (Thornberry and Lazebnik, 1998, but see later) and must be cleaved at specific XXXAsp↓X sites to yield the large and small subunits which assemble as the active 2 X heterodimer. The active caspases cleave proteins at unique tetrapeptide motifs ($NH_2$—$P_4$-$P_3$-$P_2$-$P_1$—COOH) and have an absolute specificity for aspartic acid in the $P_1$ position. Crystallographic studies on caspase-1 and caspase-3 (Wilson *et al.*, 1994, Mittl, 1996) show that the tetrameric form is a sandwich-like structure in which the two small subunits are in association with one another and bounded by the large subunits (Figure 2.2B). Each large/small subunit pair shares amino residues to form an active site, which contains the critical catalytic cysteine (Cys-238), required for caspase activity. All the caspases contain an active site pentapeptide motif QACXG (where X can be R, Q or G) which contains the catalytic cysteine. In the case of caspase-1 the Asp pocket ($P_1$) is formed from residues Arg-179, Gln-238, Arg-341 and Ser-347.

Caspase are relatively specific proteases and a positional-scanning combinatorial substrate approach (Thornberry *et al.*, 1997) has been used to characterize the caspases on the basis of their substrate specificities into three functional subgroups (Figure 2.2). Group I caspases (-1, -4 and -5) cleave at W/LEHD tetrapeptide motifs. The function of the group-1 caspases in apoptosis is unclear (reviewed in Cohen, 1997) and it is more likely that there main cellular function is in the production of the inflammatory response. The effector caspases (Group II) which include -2, -3 and -7 (and also CED-3) cleave the DEXD tetrapeptide motif which is found in many proteins associated with apoptosis (Table 2.1). Group III (activator) caspases include caspases-6, -8 and -9 as well as granzyme B and cleave at (I, V, L) EXD tetrapeptide sequences. The latter group of caspases are so-called because they are believed to be responsible for processing and activating the effector caspases. While this provides a rational approach to classifying the caspases, it cannot be regarded as an absolute definition of caspase functionality. Caspase-6, for example, on the basis of its substrate specificity would be predicted to be an activator caspase (Thornberry *et al.*, 1997). However, there is good evidence (Takahashi *et al.*, 1996) to suggest that this is the caspase responsible for lamin cleavage and the dismantling of the nuclear envelope (Table 2.1) and as such should be regarded as an effector caspase.

### 2.2.2 Activation and initiation of the caspase cascade

The activation of the caspase cascade, whether it be during receptor or non-receptor (chemical) induced-apoptosis, requires that the activator caspases are processed and

**TABLE 2.1**
Proteins cleaved by effector caspases and their functions

| Protein target | Cleavage motif(s) | Caspase(s) | Biochemical activity of target |
|---|---|---|---|
| PARP* | DEVD↓G | 3, 7 | DNA repair |
| U1-70 kDa | DGPD↓G | 3 | RNA splicing |
| DNA-PK$_{cs}$ | DEVD↓N | 3 | DNA double strand-break repair |
| Lamin-A | VEID↓N | 6 | Essential for nuclear membrane integrity |
| ICAD/DFF | DETD↓S;DAVD↓T | 3 | Activation of DNAse |
| Gas2 | SRVD↓G | ? | Element of microfilament system |
| Protein kinase C δ | DMQD↓N | 3 | Activated during apoptosis |
| SREBP-1 | DEPD↓S | 3, 7 | Sterol regulatory element binding proteins |
| SREBP-2 | | | |
| Huntingtin | DXXD↓S | 3 | Huntington disease gene product |
| Fodrin | DETD↓S | ? | Cytoskeletal protein |

* poly (ADP ribose) polymerase



activated. For an activator caspase to process an effector caspase, it must first be activated itself. The key question is how is this achieved? In the case of receptor mediated apoptosis (see Ashkenazi and Dixit, 1998 for review) there is considerable evidence to suggest that during Fas/Apo1/CD 95 induced apoptosis the apical caspase-8 is activated by association with the receptor itself. The Fas receptor which is a member of the tumour necrosis factor (TNF) receptor gene superfamily, is characterized by a cysteine-rich extracellular region, a transmembrane domain and a homologous cytosolic region known as the death domain (DD). Trimerization of the receptors occurs when the Fas ligand engages with its receptor which then binds an adaptor protein, FADD (Fas-associated death domain) which also has a similar death domain to the death domains of the trimerized receptor (Figure 2.3). FADD also contains a death effector domain (DED) sequence which binds to an analogous (one of a tandem repeat) region on pro-caspase-8. Both pro-caspase-8 and pro-caspase-10 contain tandem DED sequences, which are basically protein–protein interaction domains and are similar, but unrelated to the caspase recruitment domains (CARDS) of caspases-1, -2, -4, -5, and -9 (Takahashi *et al.*, 1996). The recruitment of pro-caspase-8 to the receptor is believed to result in oligomerization and autocatalytic processing of caspase-8 to its active form (Muzio *et al.*, 1998, Yang *et al.*, 1998a). This induced proximity or oligomerization model of caspase activation provides, if correct, an elegant explanation for activator caspase activation. As discussed by Thornberry and Lazebnik (1998) this hypothesis is based on the fact that pro-caspase-8 has low (1–2% of the processed enzyme) proteolytic activity and that when artificially forced to oligomerize, it auto/transprocesses into its active form and causes cell death (Muzio *et al.*, 1998, Yang *et al.*, 1998a). However, there is as yet no

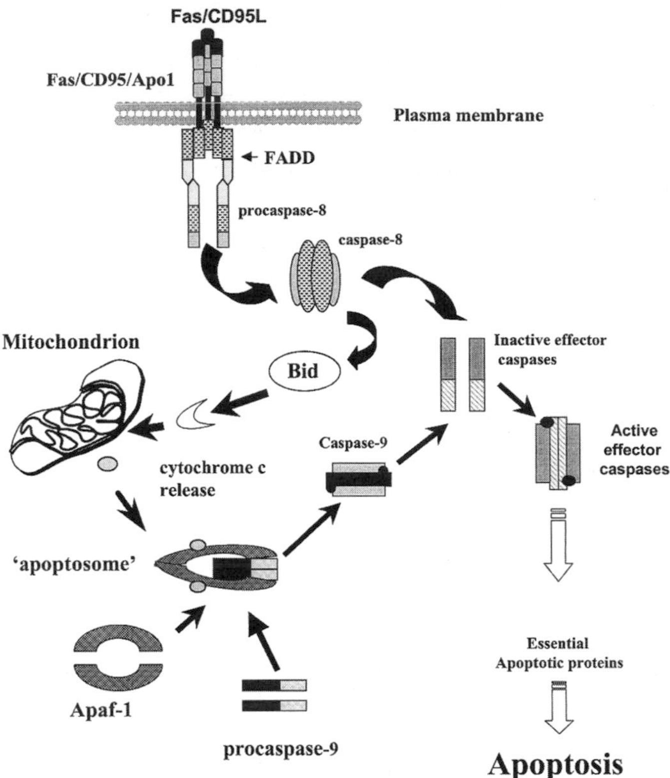


**Figure 2.3** Caspase activation during receptor and non-receptor mediated apoptosis.

evidence that pro-caspases exist as monomers and also it is unclear as to how the caspase catalytic active site could be formed in the intact pro-enzyme or after oligomerization occurs. Validation of this model of caspase activation thus requires the structure of the pro-caspase molecule to be solved.

In the case of non-receptor (chemically) mediated caspase activation, there is considerable evidence to suggest that additional co-factors are needed to activate the caspase cascade. Three apoptotic protease-activating factors have been identified which are required for caspase-3 activation (Liu *et al.*, 1996, Li *et al.*, 1997, Zou *et al.*, 1997). Apaf-1 is the first identified mammalian homologue of CED-4. It is a large (Mr = 130 kDa) soluble protein with a short N-terminal CARD domain, a central CED-4 homology domain followed by twelve WD-40 repeats, which are probably involved in protein–protein interactions. Apaf-1 in the presence of dATP, complexes with Apaf-2 (cytochrome c) and Apaf-3 (pro-caspase-9) to form an 'apoptosome' (Hengartner, 1997) in which caspase-9 is processed to its active form (Figure 2.3). The activated caspase then processes caspase-3 and initiates the caspase cascade. As caspase-9 has a DQLD↓A cleavage site (Srinivasula *et al.*, 1996) which can be cleaved by caspase-3, it is possible that activation of caspase-3 leads to further cleavage of caspase-9, thereby amplifying the caspase activation event. Interestingly, caspase-9 also has a granzyme-B cleavage site (PEPD↓A) which suggests that caspase-9 can be activated in an Apaf-1 independent manner, during T-cell mediated apoptosis (Duan *et al.*, 1996). The mechanism by which the apoptosome complex activates caspase-9 remains to be elucidated, although evidence has been presented for a model in which Apaf-1 binds caspase-9 (through their respective CARD interaction domains) and then oligomerizes via the CED-4 homology domains, resulting in procaspase aggregation and activation (Srinivasula, 1998). Removal of the WD-40 repeats eliminates the requirement for cytochrome c, indicating that cytochrome c activates Apaf-1 by binding to the WD-40 repeat domain. Support for oligomerization of Apaf-1 as an intrinsic mechanism for caspase activation also comes from a recent report that in the C.elegans cell death model, CED-4 oligomerization is also required for CED-3 activation (Yang *et al.*, 1998b).

Thus, in receptor and non-receptor mediated caspase activation, a common event is the formation of a complex which then permits the autocatalytic activation of caspases. However, the two pathways differ significantly in that Apaf-1 mediated activation requires two additional co-factors, cytochrome c and dATP. Thus, factors which release cytochrome c and/or increase the concentration of dATP in the cell are potentially capable of activating caspases directly, a property which may have important toxicological and therapeutic consequences if caspase modulators are utilized as drugs (see later). The mechanism by which cytochrome c is released from mitochondria remains controversial but there is little doubt that it is a commonly observed feature in apoptosis. Significantly, cytochrome c release during non-receptor mediated apoptosis is not blocked by caspase inhibitors and appears to occur upstream of the caspase activation event (Bossy-Wetzel *et al.*, 1998). However, inhibition of caspase-8 activity during Fas-induced apoptosis prevents cytochrome c release (Vander Heiden *et al.*, 1997, Kuwana *et al.*, 1998). An explanation for this has been put forward with the finding that Bid, a BH3 domain-containing pro-apoptotic Bcl-2 family member, is cleaved during Fas-induced apoptosis by caspase-8 (Li *et al.*, 1998, Luo *et al.*, 1998). The C-terminal product then translocates to the mitochondria and causes cytochrome c release (Figure 2.3) which then activates the Apaf-1/caspase-9 pathway. These findings suggest an amplification role for caspase-3 in receptor-mediated apoptosis and demonstrates that there is cross-talk or overlap between receptor and non-receptor mediated caspase activation.

In addition to Fas-induced apoptosis, several other death receptors have been identified which on ligation and trimerization of the receptor result in caspase activation. These receptors are related to the TNF receptor and have an additional common adaptor protein TRADD (TNFR-associated death domain) which recruits RIP (receptor-interacting protein), TRAF 2 (TNF receptors associated factor 2) and FADD (Figure 2.4), thereby conveying a dual functionality for this type of receptor. Thus, FADD recruits caspase-8 and activates the caspase cascade while RIP/TRAF2 activate the NF-κB-inducing kinase (NIK), which in turn activates the inhibitor (I-κB) of the κB kinase complex (IKK). I-κB is then

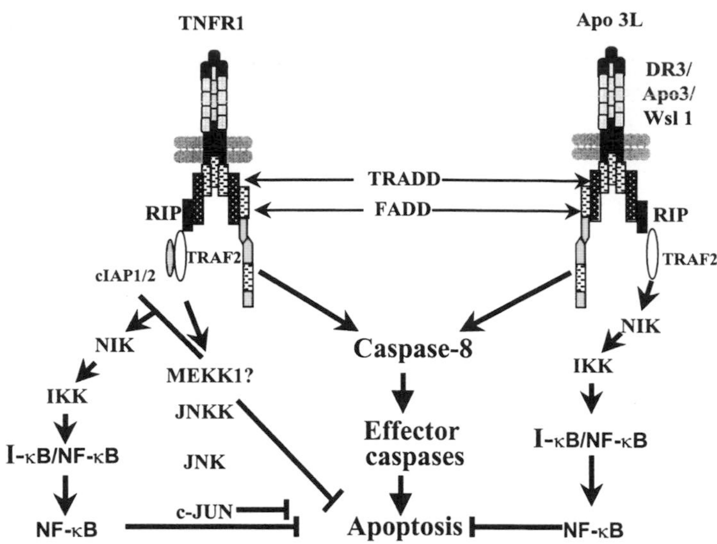


**Figure 2.4** TNFα and DR3 receptor-mediated caspase activation.

phosporylated by IKK allowing the degradation of I-κB and the release of NF-κB which translocates to the nucleus where it initiates transcription and inhibits apoptosis. Thus, for TNFα to induce apoptosis it is usually necessary to inhibit protein transcription, thereby inactivating the NF-κB pathway and allowing the FADD/caspase-8 activating pathway to predominate. TRAF2 can also activate c-JUN by a cascade that includes mitogen-activated protein (MAP) kinases such as MEKK1 (MAP/Erk kinase kinase-1) and JNKK (JNK kinase) to inhibit apoptosis. FADD can also engage an adaptor called RAID or CRADD which binds via a death domain to RIP and via a CARD-CARD interaction to caspase-2, thereby inducing apoptosis (Ahmad *et al.*, 1997, Duan and Dixit, 1997).

Another TNF related ligand TRAIL (Apo 2L) has been identified (Pitti *et al.*, 1996) which binds to two receptors DR4 (Pan *et al.*, 1996) and DR5 (MacFarlane *et al.*, 1997). TRAIL is similar to Fas but its action, surprisingly, is not blocked by overexpression of a FADD dominant negative mutant (Marsters *et al.*, 1996).

## 2.3 MODULATORS OF CASPASE ACTIVITY

There are many natural and synthetic modulators that inhibit caspases either by directly inhibiting the enzyme or alternatively preventing the activation of the caspases. These will be discussed in this section. The major controlling factors for apoptosis appear to be upstream of caspase activation (see Chapter 1 for a discussion of commitment and execution). However, the existence of viral inhibitors suggest that the caspase-inhibited cell remains sufficiently viable for the virus to replicate itself even though the cell death commitment status of the cell has not changed. It is not clear whether cell death has been temporarily delayed or abolished although there is some evidence that caspase inhibited cells still have clonogenic potential (Longthorne and Williams, 1997).

### 2.3.1 Viral inhibitors

Three distinct classes of viral inhibitors have been identified: Crm A, p35 and IAPs (Table 2.2). Cowpox virus Crm A is a potent inhibitor of some activator caspases, whilst the baculovirus p35 protein has no known homologues. Both Crm A and p35 need to be cleaved in order to inhibit the caspases. The IAPs present a more complex story, as they do have mammalian homologues. One of these, x-like IAP is a potent inhibitor of caspase-3 and -7 (Deveraux *et al.*, 1997), suggesting a direct inhibitory action for these proteins in non-receptor mediated apoptosis. However, in the case of receptor mediated apoptosis a more complicated picture emerges in which the human homologues c-IAP1/c-IAP2 bind to TRAF2 and regulate NF-κB activation (Chu *et al.*, 1997, Clem and Duckett, 1997) resulting in the inhibition of the apoptotic pathway. Whilst the mechanisms of action of the IAPs are still largely unknown it is clear that they have an evolutionary conserved regulatory role in apoptosis.



### 2.3.2 Ced-9 like proteins

The growing Bcl-2 family of proteins has an important role in apoptosis (reviewed in Adams and Cory, 1998, Green, 1998, Green and Reed, 1998), but it is only recently that some of the potential mechanisms have become apparent (Table 2.2). The pro-survival members, Bcl-2/Bcl-XL seem to act by either by preventing the release of cytochrome c from the mitochondria (Figure 2.3), abrogating activation of the apoptosome and/or by binding to constituent members of the apoptosome such as Apaf-1 to block caspase activation by preventing oligomerization. The pro-apoptotic members fall into two families (Adams and Cory, 1998) the Bax-like (includes Bak and Bok) family and the BH3 family (Bik, Blk, Hrk, BNIP3, $Bim_L$ Bad, Bid, EGL-1). The latter family is distinguished by the fact that they contain the BH3 domain, which seems to be essential for promoting cell death by stimulating the release of cytochrome c and activating the caspases (Luo *et al.*, 1998, Figure 2.3). The Bax-like family members contain other Bcl-2 domains (BH1 and BH2) which are important in allowing homo- and hetero-dimerization with other Bcl-2 family members. These interactions probably play a direct role in the action of the pro-apoptotic Bax-like proteins by binding/neutralizing the anti-apoptotic effects of Bcl-2 or Bcl-XL. Irrespective of the precise mechanisms of action it is clear that the Bcl-2 family of proteins play a major role in regulating caspase activity.

**TABLE 2.2**
Modulators of caspase activity

| Inhibitor | Specificity | Comments |
|---|---|---|
| **(A) Natural inhibitors of caspase activity** | | |
| Crm A (serpin like inhibitor) | Caspase-1, -4, -8 and -6 | Reversible inhibitor, mode of action in receptor mediated apoptosis probably involves cleavage of Crm A and binding of cleaved product to active site. |
| P35 (baculovirus gene product) | Caspase-1, -2, -3 and -4 | Cleavage of p35 at DQMD↓G site produces inhibitory peptide and blocks receptor mediated apoptosis. |
| IAPs (XIAP, human X-linked) | Caspase-3, -7 | Large family of viral inhibitors, only one known to have mammalian members; reversible inhibitor of caspases binds to active site. |
| IAPs (c-IAP1, c-IAP-2) | Bind to TRAF2 | May modulate NF-κB activation during TNFα signalling and prevent activation of caspases. |
| Bcl-2/Bcl-XL | Non-receptor mediated apoptosis | Prevents release of cytochrome c during apoptosis but may also have other downstream effects by binding to Apaf-1 and caspase-9. |
| **(B) Active site peptide inhibitors** | | |
| Ac-YVAD.CHO* | Caspase-1, -4 and -5 | Reversible inhibitor, relatively specific and potent inhibitor ($K_i$ = < 1–360 nM) but requires much higher ($K_i$ > 10 μM) concentrations to inhibit effector caspases. |
| Ac-DEVD.CHO | Caspase-3, -7 | Reversible inhibitor, with specificity for caspase-3 and -7 ($K_i$ = 0.5 and 35 nM) but inhibits caspase-1 and -4 at only moderately higher concentrations ($K_i$ = 15–135 nM). |
| -CMK, -FMK and AMK | All caspases | Covalent inhibitors, specificity determined by peptide sequences. |
| Z-VAD.FMK | All caspases | Irreversible inhibitor, tripeptide sequence means that it is less specific and will inhibit all caspases. Also methyl group on aspartic acid renders the inhibitor cell permeable. |
| **(C) Decoy proteins** | | |
| vFLIPs and cFLIP (Casper/I-Flice/ MRIT Flame/ CASH/USURPIN) | FADD/caspase-8 | Contain death effector domains similar to FADD and caspase-8. Role and mechanism of FLIPs controversial as it has been reported to both inhibit and activate apoptosis. But recently shown to heterodimerize with caspase-8 preventing recruitment by FADD. |
| Decoy receptor 1 (DcR1/TRID/ TRAIL-R3) | DR4 and DR5 receptors | Prevent activation of caspases by binding TRAIL ligand and preventing DR4 and DR5 receptors being ligated. |
| Decoy receptor 2 (DcR2/TRAIL-R4/TRUND) | DR4 and DR5 receptors | Contain truncated death mains and compete for TRAIL, thereby preventing DR4 and DR5 being ligated. |

## 2.3.3 Synthetic active site-directed peptide based inhibitors

An important development in our understanding of caspases has been the use of peptide site-directed inhibitors that target a particular caspase depending on their peptide sequence. The inhibitors can be divided into either reversible or irreversible inhibitors depending on the chemical identity of the leaving group (X) on the C-terminal side of the aspartate cleavage site (Thornberry and Molineux, 1995). Thus, if X is an aldehyde, nitrile or ketone the inhibitor will undergo nucleophilic addition of the catalytic cysteine to form thiohemiacetals, thioimidates and thiohemiketals, respectively. These intermediates are believed to mimic the acyl-enzyme intermediates formed during enzyme hydrolysis, but unlike the substrate-enzyme ternary complexes, the dissociation of the inhibitor-enzyme complexes is very slow. The irreversible inhibitors usually employed are α-substituted ketones with the general structure R-CO-$CH_2$-X where X is a carboxylate (acyloxymethylketone), diazonium (diazomethylketone) or halogen (halomethylketone) leaving group (Thornberry and Molineux, 1995). The inhibitors inactivate the catalytic cysteine of the caspase to form a thiomethylketone. The irreversible inhibitors are generally described as being very specific, and in the case of caspase-1 like inhibitors (Ac-YVAD.CHO and Ac-WEHD.CHO) (Garcia-Calvo *et al.*, 1998) these are active against the caspases-1, -4 and -5 in the low nM range, but much higher concentrations (i.e > 10 μM) are required to inhibit effector caspases -2, -3, -6 and -7. In contrast Ac-DEVD.CHO is much less specific even though it will inhibit caspase -3, -7 and -8 at low concentrations ($K_I$ between 0.23–1.6 nM) it will inhibit caspase-1 and caspase-4 at only moderately higher concentrations (15–135 nM). It must also be stressed that these differences in specificity are observed *in vitro* assays where the time of incubation and the concentrations are very carefully controlled. Increasing the inhibitor concentration or incubation time will diminish the apparent specificity of the inhibitor.

Many of the peptide inhibitors mentioned above are not particularly cell-permeable. However, Z-VAD.FMK (benzyloxycarbonyl-Val-Ala-Asp (OMe) fluoromethylketone by virtue of its benzyloxycarbonyl and OMe groups is cell permeable and therefore is very good at blocking apoptosis. Also, as it is irreversible and less specific (by virtue of its tripeptide motif) it can be used to inhibit all caspases in a wide variety of systems (Cohen, 1997, Villa *et al.*, 1997).

## 2.3.4 Natural decoy proteins

A final class of caspase inhibitors works in an indirect manner by preventing activating of the caspases during receptor-mediated apoptosis. These inhibitors function either by preventing recruitment of FADD such as cFLIP (Rasper *et al.*, 1998) or by binding the death effector ligand to act as decoy receptors (Pan *et al.*, 1997). The distribution of the decoy and functioning receptor can vary from tissue to tissue and also between normal and cancer cells (Ashkenazi and Dixit, 1998). Thus, this balance between the decoy and functioning receptor may determine whether or not caspase activation and cell death are triggered.

## 2.4 TOXICOLOGICAL CONSEQUENCES OF CASPASE INHIBITION AND ACTIVATION

To date, there are no data on the use of direct inhibitors or inducers of caspase activity. Thus, an assessment of the consequences of the use of such agents must be speculative. The putative use of caspase inhibitors relies on the development of compounds that can selectively target caspases to inhibit apoptosis *in vivo*. In this respect, any experimental drug design will almost certainly be based on peptide inhibitors; it is unlikely that a compound could be synthesized that would be specific for only one caspase. Thus, any use of such a compound raises the possibility of unwanted caspase inhibition with specificity of action and toxicity determined by the pharmacokinetic profile of the drug.

There is good experimental evidence from animal studies to show that caspase inhibitor compounds can protect against excessive apoptosis in the whole animal. When anti-Fas antibody is injected into mice, it produces a rapid and fatal liver failure (Ogasawara *et al.*, 1993) and death of the animal. This can be abrogated by administration of Z.VAD.FMK (Rodriquez *et al.*, 1996, Kuenstle *et al.*, 1997, Chandler *et al.*, 1998) or YVAD.CMK before or after the injection of the anti-Fas antibody (Rouquet *et al.*, 1997). Both Z-VAD.FMK and YVAD.CMK can also block the liver damage induced by TNF$\alpha$ (Rodriguez *et al.*, 1996, Rouquet *et al.*, 1997). These experiments showed that there was a critical time period, during which the Z-VAD.FMK could be administered, in order to abrogate the liver injury. Furthermore, the protective effect of Z-VAD.FMK treatment seemed to be long-lived as the plasma levels of liver enzymes remained normal 4 days after the Z-VAD.FMK injection. These experiments clearly demonstrate that caspase inhibitors can be remarkably effective at attenuating the toxicological consequences of caspase activation and caspase inhibition can apparently rescue cells that are committed to die. However, this, raises a contentious point; several studies (Brown *et al.*, 1997, Amarante-Mendes *et al.*, 1998, Friedlander and Yuan, 1998) have reported that Z-VAD.FMK can block non-receptor mediated apoptosis but does not confer clonogenic viability. Thus, in these experiments, Z-VAD.FMK may act downstream of the commitment point so that cells can still die but by a slower non-apoptotic form of cell death (Ashkenazi and Dixit, 1998). However, Z-VAD.FMK has been reported to maintain clonogenic potential of Jurkat cells (Longthorne and Williams, 1997). A possible explanation for these differences is that in chemical, non-receptor mediated cell death the mitochondrial effects upstream of caspase activation are sufficient in themselves to kill the cell (Ashkenazi and Dixit, 1998). Such a mechanism would not operate in receptor mediated apoptosis since caspase-8 activation is needed for the secondary downstream mitochondrial effects. Additionally, there is good evidence that, *in vivo*, caspase-inhibited cells can recover and switch from a 'committed to die' status to a 'committed to live' status. For example, caspase inhibition has been shown to ameliorate a number of neurodegenerative diseases (reviewed in Friedlander and Yuan, 1998).

There is thus clear experimental evidence that inhibiting caspases either by drugs or genetic manipulation can help to limit or abrogate pathological lesions that involve apoptosis. In these *in vivo* models the rescued cells survive and may proliferate to be replaced by new cells in a controlled and non-damaging manner. This raises the possibility that anti-caspase treatment can in itself lead to toxicity. In most studies where caspase inhibitors have been injected there have been no reported incidences of acute toxicity but data on chronic toxicity are not yet available. Clearly, there need to be long-term studies on caspase inhibitors to assess whether or not these anti-caspase compounds agents can act as non-genotoxic carcinogens (see Chapter 8).

Chronic dosing with caspase inhibitors could result in dysfunction in those tissues where apoptosis is vital to function. For example, the deletion of self-reacting cells during development of the immune repertoire could be inhibited. Evidence from knockout mice studies provide evidence of possible toxicity. As shown in Table 2.3, knockout mice have been generated for caspase-1, -2, -3, -8 , -9 and Apaf-1. In the case of the group I caspases, 1 and 2, the animals develop normally to adulthood. The main defect in caspase-1 knockouts is related to inflammatory reactions, demonstrating a minimal role for caspase-1 in apoptosis. The caspase-2 knockout mouse provides evidence for a role in germ cell deletion and also a sensitization of neuronal cells to apoptosis. The effect of knocking out effector and activator caspases (-3, -8 and -9) and Apaf-1 is dramatic, producing marked embryonic malformations and lethality. Apaf-1 and caspase-9 show marked brain abnormalities and non-receptor mediated (mitochondrial) apoptosis is blocked without affecting cytochrome c release. Thymocytes from these mice still respond to Fas, demonstrating the different pathways involved in receptor mediated caspase activation. The caspase-8 embryos exhibit *in utero* lethality with heart and haemopoietic defects and cells derived from these animals exhibit normal non-receptor mediated apoptosis. These knockout studies are valuable in that they clearly point to different caspase requirement in different tissues at different stages of development. Thus, non-receptor mediated (mitochondrial dependent) apoptosis which requires Apaf-1 and caspase-9 is clearly essential for normal brain development, whilst caspase-8 and receptor mediated apoptosis would appear to be essential for the cardiovascular system. These findings also show that blocking caspase activation *in vivo* leads to excess numbers of cells being retained; cells live even though the death commitment point has been passed.

Another useful example of caspase inhibitor-mediated toxicity comes from Chinnaiyan *et al.*, 1997 and Yoshida *et al.*, 1998 who studied the possible use of caspase inhibitor in AIDS therapy. In this disease, accelerated apoptosis contributes to the $CD4^+$ T-cell deletion and it had been proposed that inhibiting cell death might provide a

**TABLE 2.3**
Phenotype of caspase and Apaf-1 knockout mice

| Knockout genotype | Phenotype of mice | Reference |
|---|---|---|
| Caspase-1$^{-/-}$ | • cannot generate IL-1β, but develop normally and are fertile with no gross abnormalities in apoptosis<br>• resistant to LPS-induced endotoxic shock defects mainly in inflammatory system<br>• protects against neuronal apoptosis in double transgenic model | Brown *et al.*, 1997<br>Kuida *et al.*, 1995 |
| Caspase-2$^{-/-}$ | • develop normally to adulthood and no obvious abnormalities<br>• excess numbers of germ cells and oocytes resistant to chemotherapeutic drugs<br>• B lymphoblasts resistant to granzyme B and perforin<br>• sympathetic neurones more sensitive to NGE withdrawal | Bergeron *et al.*, 1998 |
| Caspase-3$^{-/-}$ | • die at 1–3 weeks of age, brain mass higher, hyperplasia and failed neuronal apoptosis<br>• no noticeable defects in other tissues<br>• caspase-3 plays a dominant non-redundant role in neuronal development | Kuida *et al.*, 1998 |
| Caspase-8$^{-/-}$ | • lethal *in utero* as embryos had impaired heart development and congested accumulation of erthyrocytes<br>• fibroblasts derived from embryos non-responsive to TNFα, Fas and DR3 induced cell death<br>• non-receptor mediated cell death not affected<br>• caspase-8 plays a non-redundant role in embryonal development | Varfolomeev *et al.*, 1998 |
| Caspase-9$^{-/-}$ | • embryonic lethality, defective brain development<br>• embryonic stem cells and fibroblasts resistant to several apoptotic stimuli<br>• thymocytes insensitive to non-receptor mediated apoptosis but cytochrome c release not affected<br>• thymocytes sensitive to UV and CD95 induced apoptosis | Hakem *et al.*, 1998<br>Kuida *et al.*, 1998 |
| Apaf-1$^{-/-}$ | • embryonic lethality with striking craniofacial lesions and brain abnormalities<br>• embryonic stem cells resistant to non-receptor mediated apoptosis even though cytochrome c release unimpaired<br>• thymocytes still respond to Fas-mediated killing<br>• Apaf-1 is essential for mitochondrial related cell death | Cecconi *et al.*, 1998<br>Yoshida *et al.*, 1998 |

therapeutic treatment for HIV infection. However, this study showed that treating peripheral blood mononuclear cells (PBMCs) with Z-VAD.FMK enhanced HIV replication. Furthermore, Z-VAD.FMK treatment of PBMCs isolated from asymptomatic patients stimulated endogenous virus production. These findings vividly illustrate the possibly inadvertent harmful effects of using caspase inhibitors as therapeutic agents.

Finally, there are many anti-cancer agents that induce apoptosis and caspase activation. To date, there has been little evidence of chemicals inducing apoptosis by directly activating the caspases. However, the discovery of Apaf-1 and the role of dATP and cytochrome c suggest that direct activation of caspases could occur and there is a considerable amount of evidence that adenosine and ATP can induce apoptosis possibly via interaction with purinoceptors (reviewed in Chinnaiyan *et al.*, 1997). However, a recent study with fludarabine (2-chloro-2′-deoxyadenosine, 2CdA) has shown that adenosine analogues can kill cells in a much more direct manner (Francheschi *et al.*, 1997). Fludarabine is an anti-metabolite which induces apoptosis in quiescent lymphocytes and is used in treating lymphoproliferative diseases. The fact that an anti-metabolite can induce apoptosis in non-cycling cells is surprising but recent studies have shown that fludarabine is phosphorylated to 2CdATP which then activates caspases by interacting with Apaf-1 and cytochrome c. ATP depletion can also prevent caspase activation stimulated by anti-cancer agents, implicating the Apaf-1/cytochrome c pathway for caspase activation (Leoni *et al.*, 1998). These results demonstrate that caspase activation and cell death can be induced directly and perhaps explains why these compounds are so toxic.

## 2.5 SUMMARY

Caspases are the executioners of cell death and their activation and subsequent actions are tightly controlled. Precise metabolic pathways exist to activate the caspases, which then dismantle the cell. Ultimately, most death-inducing signals result in caspase activation and, as such, the caspases represent a common checkpoint for controlling cell death. *In vivo* studies strongly suggest that blocking caspase activity allows a cell to survive and the consequences of unwanted cells surviving can be dramatic and occasionally lethal to the organism. Equally damaging, however, is the inappropriate activation of caspases. Thus, there will be continued interest in developing drugs which can modulate caspase activity. However, the toxicological consequences of long-term anti-caspase treatments are completely unknown and need further study.

## ACKNOWLEDGEMENTS

The author would like to thank his colleagues Professor G.M. Cohen and Dr M.M. MacFarlane for their support and many helpful discussions.

The editor would like to thank Dr Sabina Cosulich for critical appraisal of this chapter and Zana Dye for invaluable editorial assistance.

## REFERENCES

Adams, J.M. and Cory, S., 1998, The Bcl-2 protein family: Arbiters of cell survival, *Science*, **281**, 1322–1326





LIVERPOOL JOHN MOORES UNIVERSITY
LEARNING & INFORMATION SERVICES

AHMAD, M., SRINIVASULA, S., WANG, L., TALANIAN, R.V., LITWACK, G., FERNANDES-ALNEMRI, T. and ALNEMRI, E.S., 1997, CRADD, a novel human apoptotic adaptor molecule for caspase-2 and FasL/Tumor necrosis factor receptor-interacting protein RIP, *Cancer Research*, **57**, 615–620

ALNEMRI, E.S., LIVINGSTON, D.J., NICHOLSON, D.W., SALVESEN, G., THORNBERRY, N.A., WONG, W.W. and JUAN, Y., 1996, Human ICE/CED-3 protease nomenclature, *Cell*, **87**, 171

AMARANTE-MENDES, G.P., FINUCANE, D.M., MARTIN, S.J., COTTER, T.G., SALVESEN, G.S. and GREEN, D.R., 1998, Anti-apoptotic oncogenes prevent caspase dependent and independent commitment for cell death, *Cell Death and Differentiation*, **5**, 298–306

ASHKENAZI, A. and DIXIT, V.M., 1998, Death receptors: Signalling and modulation, *Science*, **281**, 1305–1308

BERGERON, L., PEREZ, G.L., MACDONALD, G., SHI, L., SUN, Y., JURISICOVA, A., *et al.*, 1998, Defects in regulation of apoptosis in caspase-2-deficient mice, *Genes and Development*, **12**, 1304–1314

BOSSY-WETZEL, E., NEWMEYER, D.D. and GREEN, D.R., 1998, Mitochondrial cytochrome c release in apoptosis occurs upstream of DEVD-specific caspase activation and independently of mitochondrial transmembrane depolarisation. *EMBO Journal*, **17**, 37–49

BROWN, R.H., GALIARDINI, V., WANG, J. and YUAN, J., 1997, Inhibition of ICE slows ALS in mice, *Nature*, **388**, 31

CECCONI, F., ALVAREZ-BOLADO, G., MEYER, B.I., ROTH, K.A. and GRUSS, P., 1998, Apaf1 (Ced-4 homolog) regulates programmed cell death in mammalian development, *Cell*, **94**, 727–737

CHANDLER, J.M., COHEN, G.M. and MACFARLANE, M.M., 1998, Different subcellular distribution of caspase-3 and caspase-7 following Fas-induced apoptosis in mouse liver, *Journal of Biological Chemistry*, **273**, 10815–10818

CHINNAIYAN, A.M., WOFFENDIN, C., DIXIT, V. and NABEL, G.J., 1997, The inhibition of pro-apoptotic ICE-like protease enhances HIV replication, *Nature Medicine*, **3**, 333–337

CHU, Z.-L., MCKINSEY, T.A., LIU, L., GENTRY, J.J., MALIM., M.H. and BALLARD, D.W., 1997, Suppression of tumor necrosis factor-induced cell death by inhibitor of apoptosis c-IAP-2 is under NF-κB control, *Proceedings of the National Academy of Science (USA)*, **94**, 10057–10062

CLEM, R.J. and DUCKETT, C.S., 1997, The IAP genes: unique arbitrators of cell death, *Trends in Cell Biology*, **7**, 337–339

COHEN, G.M., 1997, Caspases: the executioners of apoptosis, *Biochemical Journal*, **326**, 1–16

DEVERAUX, Q.L., TAKAHASHI, R., SALVESEN, G.S. and REED, J.C., 1997, X-linked IAP is a direct inhibitor of cell-death proteases, *Nature*, **388**, 300–303

DUAN, H. and DIXIT, V.M., 1997, RAIDD is a new death adaptor molecule, *Nature*, **385**, 86–89

DUAN, H., ORTH, K., CHINNAIYAN, A.M., POIRIER, G.G., FROELICH, C.J., WE, W.W. and DIXIT, V.M., 1996, Ice-lap6 a novel member of the ICE/Ced-3 gene family, is activated by the cytotoxic T cell protease granzyme B, *Journal of Biological Chemistry*, **271**, 16720–16724

EVAN, G. and LITTLEWOOD, T., 1998, A matter of life and cell death, Science, **281**,1317–1321

FRANCHESCHI, C., ABBRACCHIO, M.P., BARBIERI, D., CERUTI, S., FERRARI, D., ILIOU, J.P., *et al.*, 1997, Purines and cell death, *Drug Development Research*, **39**, 442–449

FRIEDLANDER, R.M. and YUAN, J., 1998, ICE, neuronal apoptosis and neurodegeneration, *Cell Death and Differentiation*, **5**, 823–831

GARCIA-CALVO, M., PETERSON, E.P., LEITING, B., RUEL, R., NICHOLSON, D.W. and THORNBERRY, N.A., 1998, Inhibition of human caspases by peptide-based and macromolecular inhibitors, *Journal of Biological Chemistry*, **273**, 32608–32613

GREEN, D.R., 1998, Apoptotic pathways: the roads to ruin, *Cell*, **94**, 695–698

GREEN, D.R. and REED, J.C., 1998, Mitochondria and apoptosis, *Science*, **281**, 1309–1312

HAKEM, R., HAKEM, A., DUNCAN, G.S., HENDERSON, J.T., WOO, M., SOENGAS, M.S., ELIA, A., DE LA POMPA, J.L., KAGI, D., KHOO, W., POTTER, J., YOSHIDA, R., KAUFMAN, S.A., LOW, S.A., PENNINGER, J.M. and MAK, T.W., 1998, Differential requirement for caspase 9 in apoptotic pathways *in vivo*, *Cell*, **94**, 339–352

HALE, A.J., SMITH, C.A., SUTHERLAND, L.C., STONEMAN, V.E.A., LONGTHORNE, V.L., CULTHAN, A.C. and WILLIAMS, G.T., 1996, Apoptosis: molecular regulation of cell death, *European Journal of Biochemistry*, **236**, 1–26

HENGARTNER, M.O., 1997, CED-4 is a stranger no more, *Nature*, **388**, 714–715

KERR, J.F., WYLLIE, A.H. and CURRIE, A.R., 1972, Apoptosis: a basic biological phenomenon with wide-ranging implications in tissue kinetics, *British Journal of Cancer*, **26**, 239–257

KUENSTLE, G., LEIST, M., UHLIG, S., REVESZ, L., FEIFEL, R., MACKENZIE, A. and WENDEL, A., 1997, ICE-protease inhibitors block murine liver injury and apoptosis caused by CD95 or by TNF-$\alpha$, *Immunology Letters*, **55**, 5–10

KUIDA, K., HAYDAR, T.F., KUAN, C.-Y., GU, Y., TAYA, C., KARASUYAMA, H., *et al.*, 1998, Reduced apoptosis and cytochrome c-mediated caspase activation in mice lacking caspase 9, *Cell*, **94**, 325–337

KUIDA, K., ZHENG, T.S., NA, S., KUAN, C.-Y., KARASUYAMA, H., RAKIC, P. and FLAVELL, R.A., 1995, Altered cytokine export and apoptosis in mice deficient in interleukin-1$\beta$ converting enzyme. *Science*, **267**, 2000–2002

KUWANA, T., SMITH, J.J., MUZIO, M., DIXIT, V., NEWMEYER, D.D. and KORNBLUTH, S., 1998, Apoptosis induction by caspase-8 is amplified through the mitochondrial release of cytochrome c, *Journal of Biological Chemistry*, **273**, 16589–16594

LEONI, L.M., CHAO, Q., COTTAM, H.B., GENINI, D., ROSENBACH, M., CARRERA, C., *et al.*, 1998, Induction of an apoptotic program in cell-free extracts by 2-chloro-2′-deoxyadenosine 5′-triphosphate and cytochrome c, *Proceedings of the National Academy of Science (USA)*, **95**, 9567–9571

LI, H., ZHU, H., XU, C.-J. and YUAN, J., 1998, Cleavage of BID by caspase 8 mediates the mitochondrial damage in the Fas pathway of apoptosis, *Cell*, **94**, 491–501

LI, P., NIJHAWAN, D., BUDIHARDJO, I., SRINIVASULA, S.M., AHMAD, M., ALNEMRI, E.S. and WANG, X., 1997, Cytochrome c and dATP-dependent formation of Apaf-1/caspase-9 complex initiates and apoptotic protease cascade, *Cell*, **91**, 479–489

LIU, X., KIM, C.N., YANG, J., JEMMERSON, R. and WANG, X., 1996, Induction of apoptotic program in cell-free extracts: requirement for dATP and cytochrome c, *Cell*, **86**, 147–157

LONGTHORNE, V.L. and WILLIAMS, G.T., 1997, Caspase activity is required for commitment to Fas-mediated apoptosis, *EMBO Journal*, **16**, 3805–3812

LUO, X., BUDIHARDJO, I., ZOU, H., SLAUGHTER, C. and WANG, X., 1998, Bid, a Bcl2 interacting protein, mediates cytochrome c release from mitochondria in response to activation of cell surface death receptors, *Cell*, **94**, 481–490

MACFARLANE, M.M., AHMAD, M., SRINIVASULA, S.M., FERNANDEZ-ALNEMRI, T., COHEN, G.M. and ALNEMRI, E.S., 1997, Identification and molecular cloning of two novel receptors for the cytotoxic ligand, TRAIL, *Journal of Biological Chemistry*, **272**, 25417–25420

MARSTERS, S.A., PITTI, R.M., DONAHU, C.J., RUPPERT, S., BAUER, K.D. and ASHKENAZI, A., 1996, Activation of apoptosis by Apo-2 ligand is independent of FADD but blocked by Crm A, *Current Biology*, **6**, 750–752

MITTL, P.R.E., DI MARCO, S., KREBS, J.F., BAI, X., KARANEWSKY, D.S., PRIESTLE, J.P., *et al.*, 1996, Structure of recombinant human CPP32 in complex with tetrapeptide acetyl-asp-val-asp fluoromethyl ketone, *Journal of Biological Chemistry*, **272**, 6539–6547

MIURA, M., ZHU, H., ROTELLO, R., HARTWEIG, E.A. and YUAN, J., 1993, Induction of apoptosis in fibroblasts by IL-1β-converting enzyme, a mammalian homolog of the *C.elegans* cell death gene ced-3, *Cell*, **75**, 653–660

MUZIO, M., STOCKWELL, B.R., STENNICKE, H.R., SALVESEN, G.S. and DIXIT, V.M., 1998, An induced proximity model for caspase-8 activation, *Journal of Biological Chemistry*, **273**, 2926–2930

OGASAWARA, J., WATANABE-FUKUNAGA, R., ADACHI, M., MATSUZAWA, A., KASUGAI, T., KIATAMURA, Y., *et al.*, 1993, Lethal effects of the anti-Fas antibody in mice, *Nature*, **364**, 806–809

PAN, G., NI, J., WEI, Y.F., YU, G.-L., GENTZ, R. and DIXIT, V.M., 1997, An antagonist decoy receptor and a death domain-containing receptor for TRAIL, *Science*, **277**, 815–818

PAN, G., O'ROURKE, K., CHINNAIYAN, A.M., GENTZ, R., EBNER, R., NI, J. and DIXIT, V.M., 1996, The receptor for the cytotoxic ligand, TRAIL, *Science*, **276**, 111–113

PITTI, R.M., MARSTERS, S.A., RUPPERT, S., DONAHUE, C.J., MOORE, A. and ASHKENAZI, A., 1996, Induction of apoptosis by Apo-2 ligand, a new member of the tumour necrosis factor cytokine family, *Journal of Biological Chemistry*, **271**, 12687–12690

RASPER, D.M., VAILLANCOURT, J.P., HADANO, S., HOUTZGER, V.M., SEIDEN, I., KEEN, S.L.C., *et al.*, 1998, Cell death attenuation by Usurpin, a mammalian DED-caspase homologue that precludes caspase-8 recruitment and activation by the CD-95 (Fas, Apo-1) receptor complex, *Cell Death and Differentiation*, **5**, 271–288

RODRIGUEZ, I., MATSUURA, K., ODY, C., NAGATA, S. and VASSALLI, P., 1996, Systemic injection of a tripeptide inhibits the intracellular activation of CPP32-like proteases *in vivo* and fully protects mice against Fas-mediated fulminant, *Journal of Experimental Medicine*, **184**, 2067–2072

ROUQUET, N., PAGES, J.-C., MOLINA, T., BRIAND, P. and JOULIN, V., 1997, ICE inhibitor YVAD.CMK is a potent therapeutic agent against *in vivo* liver apoptosis, *Current Biology*, **6**, 1192–1195

SRINIVASULA, S.M., AHMAD, M., FERNANDES-ALNEMRI, T. and ALNEMRI, E.S., 1998, Autoactivation of procaspase-9 by Apaf-1-mediated oligomerisation, *Molecular Cell*, **1**, 949–957

SRINIVASULA, S.M., FERNANDES-ALNEMRI, T., ZANGRILLI, J., ROBERTSON, N., ARMSTRONG, R.C., WANG, L., *et al.*, 1996, The Ced-3/interleukin 1β converting enzyme-like homolog Mch6 and the lamin-cleaving enzyme Mch2α are substrates for the apoptotic mediator CPP32, *Journal of Biological Chemistry*, **271**, 27099–27106

TAKAHASHI, A., ALNEMRI, E.S., LAZEBNIK, Y.A., FERNADES-ALNEMRI, T., LITWACK, G., MOIR, R.D., *et al.*, 1996, Cleavage of lamin A by Mch2α but not CPP32: multiple interleukin 1β-converting enzyme-related proteases with distinct substrate recognition properties are active in apoptosis, *Proceedings of the National Academy of Science* (USA), **93**, 8395–8400

THORNBERRY, N.A. and LAZEBNIK, Y., 1998, Caspases: enemy within, *Science*, **281**, 1312–1316

THORNBERRY, N.A. and MOLINEUX, S.M., 1995, Interleukin-1β converting enzyme: a novel cysteine protease required for IL-1β production and implicated in programmed cell death, *Protein Science*, **4**, 3–12

THORNBERRY, N.A., RANO, T.A., PETERSON, E.P., RASPER, D.M., TIMKEY, T., GARCIA CALVO, M., *et al.*, 1997, A combinatorial approach defines specificities of members of the caspase family and granzyme B, *Journal of Biological Chemistry*, **272**, 17907–17911

VANDER HEIDEN, M.G., CHANDEL, N.S., WILLIAMSON, E.K., SCHUMACKER, P.T. and THOMPSON, C.B., 1997, Bcl-XL regulates the membrane potential and volume homeostasis of mitochondria, *Cell*, **91**, 627–637

VARFOLOMEEV, E.E., SCHUCHMANN, M., LURIA, V., CHIANNILKULCHAI, N., BECKMANN, J.S., METT, I.L., *et al.*, 1998, Targeted disruption of mouse caspase-8 gene ablates cell death induction by the TNF receptors, Fas/Apo1 and DR3 and is lethal prenatally, *Immunity*, **9**, 267–276

VILLA, P., KAUFMANN, S.H. and EARNSHAW, W.C., 1997, Caspases and caspase inhibitors, *Trends in Biochemical Sciences*, **22**, 388–393

WILSON, K.P., BLACK, J.-A.F., THOMSON, J.A., KIM, E.E., GRIFFITH, J.P., NAVIA, M.A., *et al.*, 1994, Structure and mechanism of interleukin-1β converting enzyme, *Nature*, **370**, 270–275

YANG, X., CHANG, H.Y. and BALTIMORE, D., 1998a, Autoproteolytic activation of pro-caspases by oligomerisation, *Molecular Cell*, **1**, 319–325

YANG, X., CHANG, H.Y. and BALTIMORE, D., 1998b, Essential role of CED-4 oligomerisation in CED-3 activation and apoptosis, *Science*, **281**, 1355–1357

YOSHIDA, H., KONG, Y.-Y., YOSHIDA, R., ELIA, A.J., HAKEM, A., HAKEM, R., *et al.*, 1998, Apaf1 is required for mitochondrial pathways of apoptosis and brain development, *Cell*, **94**, 739–750

YUAN, J., SHAHAM, S., LEDOUX, S., ELLIS, H.M. and HORVITZ, H.R., 1993, The C.elegans cell death gene ced-3 encodes a protein similar to mammalian IL-1β-converting enzyme, *Cell*, **75**, 641–652

ZOU, H., HENZEL, W.J., LIU, X., LUTSCHG, A. and WANG, X., 1997, Apaf-1, a human protein homologous to C.elegans CED-4, participates in cytochrome c-dependent activation of caspase-3, *Cell*, **90**, 405–413

CHAPTER

3

# *Modulation of Apoptosis in Immunotoxicology*

**JENS SCHÜMANN AND GISA TIEGS**

Friedrich Alexander Universtät Erlangen-Nürnberg, Fahrstrasse 17, Erlangen D-91054

## *Contents*

## 3.1 INTRODUCTION

Certain microbial or chemical toxicants affect the immune system. They are immunotoxicants in that they are able to cause acute or chronic disease by means of suppression or overstimulation of the immune system. Anergy (the inability to mount an immune response to a particular antigen) and depletion of effector cells such as T lymphocytes or macrophages are characteristic of immunosuppression induced by a foreign agent. As a consequence, the host becomes incapable of producing immune mediators such as the cytokines that play a crucial role in the defense against infection. Thus, the susceptibility of the host towards infection increases. Moreover, the lack of immune mediators that regulate cell proliferation, cell differentiation and cell death may result in malignant growth. For example, impaired host resistance as well as increased incidence of neoplastic disease is seen in patients treated with immunosuppressive drugs. In contrast, overstimulation of the immune system by a bacterial toxicant or a chemical may induce an excess of proinflammatory mediators which can cause death of host cells, autoimmunity due to activation of autoreactive T cells or allergic responses. Interestingly, such an overstimulation of the immune system is often accompanied by a state of hypersensitivity where detrimental synergistic effects are seen in the presence of additional immunostimulators. This can be followed by an immunosuppressed state, that facilitates the survival of invaded pathogens.

## 3.2 ROLE OF TUMOUR NECROSIS FACTOR α AND ITS CELLULAR SOURCES

Tumour necrosis factor α (TNFα), a proinflammatory cytokine which is produced among others by macrophages and T cells, turned out to be one of the most harmful endogenous mediators following overstimulation of the immune system. This cytokine is well known as a modulator of apoptosis in vertebrates and there are examples of both induction and suppression of apoptosis by this cytokine. However, in the context of immunotoxicology, TNFα usually induces cell death. TNFα contributes to rapid multi organ failure due to septic complications (Beutler *et al.*, 1985, Tracey *et al.*, 1987) and it is the key mediator of liver injury and lethality induced by lipopolysaccharide (LPS) (Lehmann *et al.*, 1987, Tiegs *et al.*, 1990, Leist *et al.*, 1995a) or superantigens (Miethke *et al.*, 1992, Miethke *et al.*, 1993a, Nagaki *et al.*, 1994) in mice sensitized with the hepatic transcriptional inhibitor D-galactosamine (GalN) (for further details on the role of sensitization see 3.6). TNFα also plays an important role in promoting liver injury elicited either by the T cell mitogenic plant lectin concanavalin A (ConA) (Mizuhara *et al.*, 1994, Gantner *et al.*, 1995a, Ksontini *et al.*, 1998) or by *Pseudomonas aeruginosa* exotoxin A (PEA), a bacterial toxin primarily known as an inhibitor of protein synthesis and recently described as an inducer of T cell-dependent liver injury (Schümann *et al.*, 1998). Surprisingly, TNFα even contributes to liver failure caused by hepatotoxins such as the deathcup poison α-amanitin (Leist *et al.*, 1997a) or the environmental contaminant 2,3,6,7-tetrachlorodibenzo-p-dioxin (TCDD) (Taylor *et al.*, 1992). In the latter cases, TNFα production is probably a secondary event, caused by inflammation due to TNFα-independent necrotic tissue damage and/or by translocation of LPS from the intestine. At least in the case of TCDD-induced TNFα-dependent lethality, LPS-resistant C3H/HeJ mice were much less sensitive than normal mice (Clark and Taylor, 1994). The role of TNFα in mediating the effects of hepatic nongenotoxic carcinogens is discussed in detail in Chapter 8.

Macrophages are supposed to be the source of TNFα in LPS-treated animals, since transfer of macrophages from LPS-sensitive C3H/HeN mice to LPS-resistant C3H/HeJ mice renders C3H/HeJ mice susceptible to GalN/LPS-induced lethality (Freudenberg *et al.*, 1986). In all other models mentioned above the cellular source of TNFα has not been identified until now. We have shown by means of immunofluorescent staining of liver sections that soon after PEA injection there is a massive staining of TNFα in the liver, co-localized with the resident liver macrophages, the Kupffer cells, but not with T cells (Schümann *et al.*, 1998). This is especially exciting when taking into account that TNFα is not measurable in plasma at this early stage (Schümann *et al.*, 1998). This is clear evidence for the suggestion that circulating cytokine levels only insufficiently reflect local tissue concentrations, well known as the concept of 'the tip of the iceberg' (Cavaillon *et al.*, 1992). The absence of T cells results in complete abolishment of TNFα production by Kupffer cells (Schümann *et al.*, 1998), suggesting that T cells, though not the source of TNFα, activate Kupffer cells to produce TNFα. Intravenous administration of ConA to mice induces TNFα production by Kupffer cells and $CD4^+$ T cells (G. Tiegs, unpublished results) and causes liver injury (Tiegs *et al.*, 1992) that requires the resident liver T cell population, $NK1.1^+$ T cells (Toyabe *et al.*, 1997). This cell type may have wide relevance for Kupffer cell-dependent TNFα production.



## 3.3 HYPERSENSITIVITY AND HYPORESPONSIVENESS

The toxicity of a xenobiotic that works by overstimulating the immune system clearly depends on the immunostatus of the intoxicated mammalian organism. This is equally applicable whether the toxicant is man-made or natural. This status is discussed in the following section.

### 3.3.1 Hypersensitivity

The superantigen staphylococcal enterotoxin B (SEB) that specifically stimulates T cells carrying particular $V_\beta$ regions in their T cell receptors, causes a short-lasting hyperreactive state (1–2 h) of mouse lymph node cells, as assessed by a pronounced increase of proliferation and IL-2 production after *ex vivo* restimulation with SEB (Wahl *et al.*, 1993, Miethke *et al.*, 1993b). Co-administration of SEB together with LPS induces a synergistic and lethal cytokine response syndrome in mice, whereas the same dose of each toxin given alone fails to induce lethality (Blank *et al.*, 1997). The proinflammatory cytokine interferon-γ (IFN-γ) is a key mediator of the synergism since passive immunization with anti-IFN-γ mAb prevented TNFα-dependent lethality of SEB *plus* LPS (Blank *et al.*, 1997). However, the exact role of IFN-γ remains to be elucidated. IFN-γ may have sensitized macrophages to enhanced TNFα production upon LPS stimulation. Moreover, IFN-γ may have also sensitized target cells to TNFα-induced cell death (for further details on IFN-γ-dependent sensitization of target cells see 3.6.5.1).

Another example of hypersensitivity was observed 24 h after injection of a low dose of LPS to mice. Low dose LPS activates macrophages which in turn produce little, *per se*, untoxic amounts of cytokines, such as TNFα and IL-12. As a consequence, IL-12 induces the production of IFN-γ predominantly in $NK1.1^+$ T cells, but not in natural killer (NK) cells (Ogasawara *et al.*, 1998). In turn, IFN-γ primes macrophages and possibly other cell types. Such a priming renders mice highly susceptible to a second injection of LPS that results in massive production of proinflammatory cytokines and lethality, well known as the generalized Shwartzman reaction.

Small amounts of endotoxin also augment liver injury from a variety of hepatotoxicants including carbon tetrachloride, ethanol, halothane, and others (for review see Roth *et al.*, 1997). Vice versa, TCDD- (Rosenthal *et al.*, 1989, Clark *et al.*, 1991) and carbon tetrachloride-dependent hypersensitivity to LPS, accompanied by significantly increased levels of TNFα, have been described. These examples imply that there are fundamental similarities between the immunotoxicologic effects of natural and non-natural xenobiotics.

## 3.3.2 Hyporesponsiveness

As described above, a single dose of SEB causes early short-lasting hyperreactivity of lymph node cells. This state is followed by immunosuppression that is characterized by a $V_\beta$-unrestricted unresponsiveness of T cells, i.e. the inability of spleen cells to release IL-2 upon stimulation with any polyclonal T cell mitogen, also other than SEB (Muraille *et al.*, 1997). This state lasts for about 2–3 days and is followed by selective apoptosis and anergy of superantigen-responsive T cells (Rellahan *et al.*, 1990, Perkins *et al.*, 1993). The general immunosuppressed state is particularly advantageous to the superantigen-producing germ because it facilitates the survival of the invaded bacteria. On the other hand, the immunosuppressed state protects mice from toxicity of other superantigens, such as toxic shock syndrome toxin 1 (TSST-1) (Miethke *et al.*, 1993b) or the T cell-stimulatory toxin PEA (Schümann *et al.*, 1998).

A great deal of effort was made to characterize the hyporesponsive state induced by LPS. Exposure of mice to LPS, even at a very low dose, leads to a down-regulation of the cytokine response to a second high dose LPS challenge (Flohé *et al.*, 1991, Zuckerman and Evans, 1992), and renders mice resistant to GalN/LPS-induced lethality (Freudenberg and Galanos, 1988). Induction of tolerance is mediated by macrophages (Freudenberg and Galanos, 1988). Two possible explanations for LPS-induced hyporesponsiveness may be given: (1) LPS stimulates macrophages to produce anti-inflammatory cytokines, such as IL-10 and TGF-β1, which suppress the production of proinflammatory cytokines, such as TNFα, after a second exposure to LPS. Such a mechanism was suggested by *in vitro* experiments with human peripheral blood mononuclear cells (PBMC), preincubated with anti-IL-10 mAb or anti-TGF-β1 mAb, which both prevented the induction of LPS tolerance (Randow *et al.*, 1995). However, tolerance induction was still possible in IL-10-deficient mice (Berg *et al.*, 1995), pointing to additional factors, other than IL-10, that down-regulate the TNFα response to the second LPS challenge. (2) LPS-induced production of TNFα and IL-1 is responsible for the induction of target cell resistance to TNFα-induced cytotoxicity because pretreatment with LPS, TNFα, or IL-1 protected mice from GalN/TNFα-induced lethality (Libert *et al.*, 1991). Probably, both aspects play a role in LPS-mediated hyporesponsiveness of mice to a second LPS challenge.

Drugs that elevate the second messenger cyclic adenosine monophosphate (cAMP) in leukocytes, e.g. inhibitors of phosphodiesterases (PDEs), protect mice from GalN/LPS- (Fischer *et al.*, 1993, Jilg *et al.*, 1996), GalN/SEB- (Gantner *et al.*, 1997), and ConA- (Gantner *et al.*, 1997) induced liver injury. The underlying mechanism has not been clarified yet. Inhibition of NF-κB activation that participates in the transcription of the TNFα gene (Ollivier *et al.*, 1996, Yao *et al.*, 1997) or up-regulation of TNFα-suppressive cytokines such as IL-10 (Gantner *et al.*, 1997) may be involved. Indeed, plasma levels of IL-10 in mice significantly increased after GalN/LPS (Jilg *et al.*, 1996), GalN/SEB (Gantner *et al.*, 1997) or ConA (Gantner *et al.*, 1997) challenge in the presence of PDE inhibitors. Concomitantly, plasma concentrations of TNFα significantly decreased. Furthermore, endogenously produced (Florquin *et al.*, 1994a) or exogenously administered IL-10 (Bean *et al.*, 1993) has been shown to protect mice from GalN/SEB-induced TNFα-dependent liver injury. Reduced TNFα plasma levels upon PDE inhibition may also be explained by increased shedding of TNF receptors (Jilg *et al.*, 1996). This is supported by the observation that PDE inhibitors also conferred resistance to GalN/TNFα-induced liver injury (Jilg *et al.*, 1996). In this case IL-10-mediated reduction of TNFα production can be excluded as the mechanism of protection. Very recently, dibutyryl cAMP has been shown to protect mice from GalN/TNFα-induced hepatic failure, correlating with an enhanced expression of heat shock protein 70 (Hsp-70) in hepatocytes of GalN/TNFα-treated mice (Takano *et al.*, 1998) (for further details on Hsp-70 as a cytoprotective protein see 3.6.3). This finding provides evidence for an additional mechanism by which cAMP elevation might protect mice from TNFα-mediated hepatocytotoxicity.

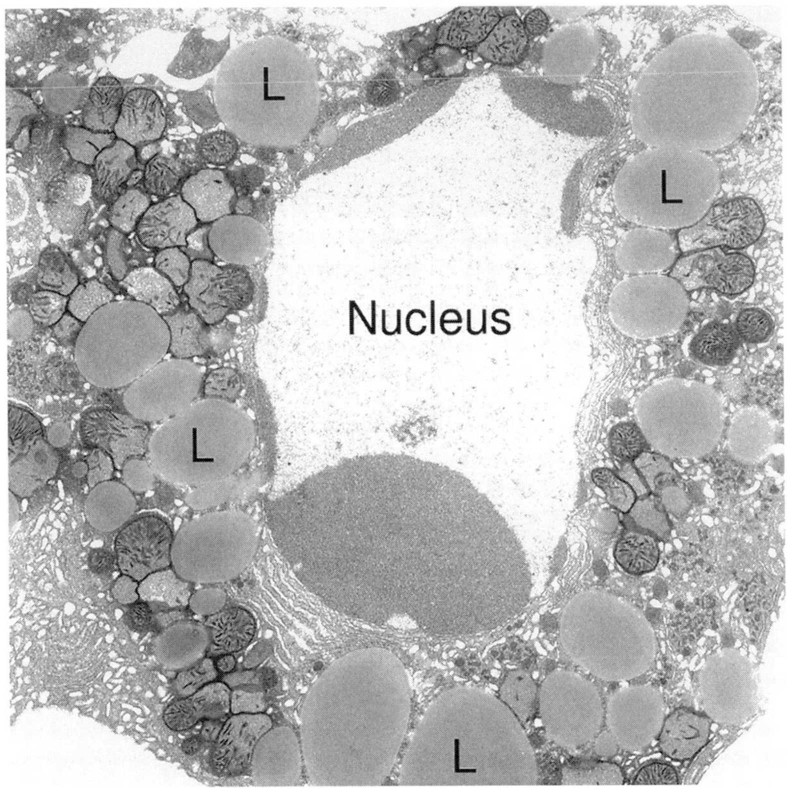


**Figure 3.1** Apoptosis in immunotoxicology. In all TNFα-dependent models of experimental liver injury in mice, hepatocytes die by apoptosis. The figure shows an electron micrograph of a nucleus of an apoptotic hepatocyte in a mouse liver 5 h after treatment with GalN/TNFα, as described elsewhere (Angermüller *et al.*, 1998) (the micrograph was kindly provided by Dr S. Angermüller, Heidelberg; L = lipid droplets). In the nucleus chromatin is condensed at the nuclear wall as it is typical of apoptosis.

## 3.4 MECHANISMS OF CELL DEATH IN MURINE MODELS OF IMMUNE-MEDIATED TOXIC LIVER INJURY

In the TNFα-dependent models of experimental liver injury in mice, liver necrosis-associated transaminase release is preceded by hepatic internucleosomal DNA fragmentation, as quantified in liver homogenates by an ELISA specific for histone-bound DNA (Gantner *et al.*, 1995b, Leist *et al.*, 1995a, 1997a, Schümann *et al.*, 1998). Additionally, the internucleosomal character of DNA cleavage was verified by analysis of liver DNA on agarose gels (Gantner *et al.*, 1995b, Leist *et al.*, 1995a, 1997a, Schümann *et al.*, 1998). In all cases, DNA fragmentation showed the specific ladder pattern, as it is typical of apoptosis. Most importantly, microscopic analysis proved that liver cells die by apoptosis following any treatment mentioned above (see Figure 3.1), i.e. GalN/TNFα (Leist *et al.*,

1995a, Angermüller *et al.*, 1998), GalN/LPS (Leist *et al.*, 1995a), GalN/superantigen (Gantner *et al.*, 1995b), ConA (Leist *et al.*, 1995a), PEA (Schümann *et al.*, 1998), or α-amanitin (Leist *et al.*, 1997a). This was demonstrated by detection of hepatocellular apoptotic bodies in liver tissues of treated mice.

Until now it still remains unclear why apoptotic cell death is accompanied by inflammatory necrosis at later stages. Two explanations of this phenomenon can be discussed: (1) Due to the loss of the inner mitochondrial membrane potential during apoptosis (Zamzami *et al.*, 1995), a process known as mitochondrial permeability transition (PT), excess generation of reactive oxygen intermediates involved in apoptotic signalling (Hockenberry *et al.*, 1993, Kane *et al.*, 1993, Sandström *et al.*, 1994, Zamzami *et al.*, 1995), may cause oxidative stress leading to necrosis. Additionally, loss of ATP due to PT may prevent the energy-dependent process of apoptosis, thereby inevitably changing the apoptotic signal cascade into a necrotic one (Leist *et al.*, 1997b) (2). As evidenced by electron microscopy, it appears that phagocytosis of an increasing amount of apoptotic bodies overtaxes the capacity of the phagocytizing neighbouring hepatocytes, thereby leading to the burst of lysosomes, which in turn causes lytic cell death (S. Angermüller, personal communication).

## 3.5 ROLE OF CASPASES

A very recent work demonstrated a double-edged role of cysteine aspartyl proteases, called caspases (for further details on caspases see Chapter 2), in the scenario of cell death (Vercammen *et al.*, 1998). Primarily, caspases are known as the executioners of apoptosis. Peptide inhibitors of caspases protect cells from apoptosis (for review see Villa *et al.*, 1997). Interestingly, it has been shown that the general tripeptidic caspase inhibitor Z-VAD.fmk protects Jurkat cells not only from apoptosis but also from necrosis triggered by activation of the CD95/Fas receptor molecule (Leist *et al.*, 1997b), a member of the TNF receptor superfamily. However, a novel role of caspases has recently been described, i.e. prevention of necrosis (Vercammen *et al.*, 1998). TNFα-mediated necrosis of murine L929 fibrosarcoma cells was enhanced in the presence of the broad spectrum caspase inhibitors Z-VAD.fmk and Z-D.fmk as well as of the caspase-1-like enzyme inhibitor Ac-YVAD.cmk and the caspase-3-like enzyme inhibitor Z-DEVD.fmk (Vercammen *et al.*, 1998). Caspases seem to downmodulate necrotic pathways in these cells by inhibiting TNFα-induced oxygen radical production. This was concluded from the fact that the oxygen radical scavenger butylated hydroxyanisole inhibited Z-VAD.fmk-dependent sensitization to TNFα-induced cell necrosis (Vercammen *et al.*, 1998). Until now, no *in vivo* data has been available to confirm the hypothesis of the role of caspases in preventing necrosis. In all studies dealing with the influence of caspase inhibitors on TNFα-mediated or agonistic anti-Fas-antibody-triggered liver injury in mice, the inhibitors tested, i.e. Z-VAD.fmk (GalN/TNFα (Künstle *et al.*, 1997), GalN/LPS (Jaeschke *et al.*, 1998) and anti-Fas (Rodriguez *et al.*, 1996, Künstle *et al.*, 1997)), Ac-YVAD.cmk (GalN/TNFα (Schümann and Tiegs, unpublished results) and anti-Fas (Rodriguez *et al.*, 1996, Rouquet *et al.*, 1996)), and Ac-DEVD.CHO (anti-Fas (Suzuki, 1998)), showed protective properties. These protective effects include prevention of apoptosis, necrosis, and death. Furthermore, Z-VAD.fmk also protected mice from PEA-induced apoptotic and necrotic liver damage (Schümann *et al.*, 1998), and caspase-1-deficient mice were resistant to ConA-induced hepatitis (Ksontini *et al.*, 1998). Hence, *in vivo* caspase inhibition failed to enhance necrotic cell damage that usually follows hepatocellular apoptosis in the models of liver injury.

## 3.6 TARGET CELL SENSITIZATION TO TNFα-MEDIATED CYTOTOXICITY AND SELF-PROTECTION

Despite the TNFα-dependence of liver injury observed in the several models mentioned above, sole administration of a reasonable TNFα dose neither induces apoptosis nor necrosis in mice (Tiegs *et al.*, 1989, Leist *et al.*, 1995a). Hence, TNFα is unable to cause cell death in the absence of additional factors that sensitize target cells. A common sensitizing mechanism of liver parenchymal cells to TNFα-induced apoptosis and necrosis is the transcriptional inhibition by the uracil nucleotide depleting aminosugar GalN (Galanos *et al.*, 1979, Tiegs *et al.*, 1989, 1990) or by the cytostatic drug actinomycin D (ActD) (Leist *et al.*, 1994, 1995a). The mechanism by which transcriptional inhibition provides susceptibility of hepatocytes to TNFα is unresolved. One possible explanation may be given in terms of inhibition of the expression of so-called 'protective proteins'. Activation of the transcription factor 'nuclear factor-κB' (NF-κB) by TNFα is a crucial step in the TNFα-induced self-protection from TNFα-induced cell death. The functional role of the 'c-jun N-terminal kinase' (JNK) which is also activated by TNFα is still under debate (Liu *et al.*, 1996, Lee *et al.*, 1997, Yeh *et al.*, 1997). However, if activation of NF-κB is prevented, TNFα-induced cell death occurs without additional sensitization in different cell types (Beg and Baltimore, 1996, Liu *et al.*, 1996, van Antwerp *et al.*, 1996, Wang *et al.*, 1996). The intracellular signalling cascade that leads to the activation of NF-κB is not fully understood. The presence of functional 'TNF receptor associated factor 2' (TRAF-2), that intracellularly associates with TNF receptor 1 as well as TNF receptor 2 (for further details on the TNF receptors see 3.7) upon TNF receptor activation (Rothe *et al.*, 1994, Hsu *et al.*, 1996a,b), has been shown to protect cells from TNFα-induced cell death (Lee *et al.*, 1997, Yeh *et al.*, 1997). It has been assumed that TRAF-2 stimulates the activation of NF-κB, because cells expressing a dominant-negative TRAF-2 molecule were unable to activate NF-κB upon triggering of TNF receptor 1 (Hsu *et al.*, 1996a) or TNF receptor 2 (Rothe *et al.*, 1995a). On the other hand, experiments with cells from TRAF-2-deficient mice (Yeh *et al.*, 1997) and from mice overexpressing a dominant negative form of the TRAF-2 protein in lymphocytes (Lee *et al.*, 1997) have shown that TNFα-induced NF-κB activation is still possible. However, NF-κB activation was delayed in case of TRAF-2-deficiency (Yeh *et al.*, 1997) and possibly compensated by co-existing wild-type TRAF-2 protein in the transgenic mice (Lee *et al.*, 1997). Hence, recruitment of TRAF-2 and activation of NF-κB are possibly partially independent events that both provide the TNFα-induced protection from TNFα-induced cell death.

A very recent publication showed a participation of reactive oxygen intermediates in the activation of NF-κB and the apoptotic protease caspase-3 by means of overexpression of the TNFα-inducible manganese superoxide dismutase (Mn-SOD) in human breast cancer MCF-7 cells. Overexpression of Mn-SOD prevented the activation of both NF-κB and caspase-3 by TNFα (Manna *et al.*, 1998). *In vivo*, TNFα-dependent activation of NF-κB has been observed in livers of ConA-injected mice (Trautwein *et al.*, 1998a). Drug-induced activation of NF-κB correlated with protection from ConA-induced liver injury (Tiegs *et al.*, 1998). The whole set of cytoprotective proteins induced by NF-κB is unknown, but there are some 'hot' candidates, among which are the cellular inhibitor of apoptosis protein 2 (c-IAP-2), the zinc finger protein A20, the heat shock protein 70 (Hsp-70), and the above mentioned manganese superoxide dismutase (Mn-SOD) whose inhibition by antisense-MnSOD-RNA rendered 293 human embryonic kidney cells sensitive to TNFα (Wong *et al.*, 1989). Furthermore, pretreatment of rats with TNFα prevented liver damage caused by carbon tetrachloride. This was associated with an increased activity of Mn-SOD in the liver (Sato *et al.*, 1995), thus pointing to a very central role of cytoprotective proteins in the toxicity of both, natural and non-natural xenobiotics.



### 3.6.1 Cellular inhibitor of apoptosis protein 2 (c-IAP-2)

Overexpression of the NF-κB-dependent c-IAP-2, a protein of the IAP family that associates with TRAF-2 (Rothe *et al.*, 1995b), has been shown to protect HeLa cells from TNFα-induced apoptosis and to contribute to the activation of NF-κB (Chu *et al.*, 1997). Cells overexpressing a mutated, not inactivatable form of the NF-κB-inhibitory I-κB protein loose their c-IAP-2-conferred resistance to TNFα-induced apoptosis (Chu *et al.*, 1997). Furthermore, c-IAP-2 seems to directly inhibit caspase-3-like enzymes, and overexpression of c-IAP-2 has been shown to inhibit etoposide- or Fas-triggered activation of caspase-3-like enzymes and apoptosis in human 293 cells (Roy *et al.*, 1997). Taken together, NF-κB-dependent expression of c-IAP-2 may confer resistance to apoptosis triggered by various stimuli. This may be achieved by enhancing the activation of NF-κB *and* by the inhibition of caspase-3-like enzymes which are regarded as central executioners of many apoptotic pathways.

### 3.6.2 The A20 protein

Although capable of inhibiting TNFα-induced apoptosis (Opipari *et al.*, 1992, Jäättelä *et al.*, 1996, Ferran *et al.*, 1998), the NF-κB-dependent (Krikos *et al.*, 1992), TNFα-inducible

(Dixit *et al.*, 1990, Opipari *et al.*, 1992) A20 protein inhibits further activation of NF-κB (Song *et al.*, 1996, Jäättelä *et al.*, 1996, Cooper *et al.*, 1996). Inhibition of NF-κB activation by A20 may therefore be a mechanism of down-regulating (e.g. TNFα-induced) NF-κB activation. The mechanism by which A20 inhibits NF-κB activation and apoptosis remains unknown. Hence, A20 may provide a very interesting example of a molecule that possibly attenuates inflammation. The synthesis of proinflammatory cytokines, such as TNFα and IFN-γ, depends on the activation of NF-κB, while at the same time A20 has a negative effect on TNFα-induced apoptosis.

### 3.6.3 Heat shock protein 70 (Hsp-70)

Induction of Hsp-70 following exposure to TNFα has been observed in cardiac myocytes (Nakano *et al.*, 1996), hepatocytes (G. Tiegs, unpublished results), and other cell types. An increase in Hsp-70 protein also occurs after incubation of rat hepatocytes with IFN-γ *plus* IL-1β (Kim *et al.*, 1997). In the latter case, accumulation of Hsp-70 turned out to depend on the formation of nitric oxide (NO). This was concluded from the fact that the NO synthase inhibitor $N^G$-monomethyl-L-arginine (NMMA) prevented the cytokine-dependent accumulation of Hsp-70 (Kim *et al.*, 1997). Accordingly, treatment of rat hepatocytes with the NO donor *S*-nitroso-*N*-acetylpenicillamine (SNAP) induced the expression of Hsp-70, and furthermore protected the cells from apoptosis triggered by TNFα *plus* an inhibitor of transcription (Kim *et al.*, 1997). The functional role of Hsp-70 as a cytoprotective component has been proven by an antisense oligomer to Hsp-70 that blocked Hsp-70 expression and prevented the NO-dependent protection from TNFα-induced apoptosis of sensitized cells (Kim *et al.*, 1997). The combination of IFN-γ and IL-1β is a well known inducer of the NF-κB-dependent inducible NO synthase (Geller *et al.*, 1993, Xie *et al.*, 1994). Hence, TNFα may also cause an increase of Hsp-70 via NF-κB-dependent induction of the inducible NO synthase, thereby protecting the exposed cell from TNFα and possibly other apoptotic stimuli. This hypothesis is supported by a life-saving role of TNFα pretreatment (Libert *et al.*, 1991) and a hepatoprotective effect of NO donor pretreatment (Bohlinger *et al.*, 1995a,b) in the *in vivo* model of GalN/TNFα-induced liver injury, although a causal relationship between TNFα pretreatment, NO synthesis and induction of Hsp-70 has not yet been proven. A possible protective role of Hsp-70 induction *in vivo* is evidenced by a correlation of dibutyryl cAMP-elicited Hsp-70 expression and protection from GalN/TNFα-induced liver injury (Takano *et al.*, 1998).

### 3.6.4 The role of a proliferation signal induced by TNFα

It has been demonstrated that TNFα not only protects cells from its own deleterious effects, but also induces proliferation of hepatocytes via activation of TNF receptor 1. This has been demonstrated by an impaired mitogenic response of hepatocytes during

$CCl_4$-induced liver injury in TNF receptor 1-deficient mice (Yamada and Fausto, 1998). The impaired mitogenic response correlated with reduced DNA binding of the transcription factors NF-κB and 'signal transducer and activator of transcription 3' (STAT3), as well as with lower plasma levels of IL-6 (Yamada and Fausto, 1998). These components may be important for the TNFα-dependent hepatocyte mitogenic response contributing to liver regeneration that counteracts apoptosis. Mechanistic studies on liver regeneration have been performed in the model of ConA-induced hepatitis in mice. They revealed that the nuclear expression of a protein responsible for cell cycle arrest at the G1/S checkpoint, but presumably necessary for G0/G1 shift, i.e. CCAAT/enhancer-binding protein-β/liver-enriched activating protein (C/EBP-β/LAP), is induced soon after ConA injection as a consequence of TNFα production, since passive immunization with anti-TNFα Ab prevented expression of C/EBP-β/LAP (Trautwein *et al.*, 1998b). The concentration of C/EBP-β/LAP returned to normal levels 24 h after ConA administration, and the hepatocytes entered the S-phase of the cell cycle (Trautwein *et al.*, 1998b). This example shows that cell cycle and apoptosis are contrarily regulated. A mediator possibly responsible for the switch from cell destruction to cell proliferation is IL-6. This cytokine is produced in a TNFα-dependent manner via activation of TNF receptor 1 upon partial hepatectomy in mice, and its lack in TNF receptor 1-deficient mice correlates with an impaired nuclear translocation of STAT3 and impaired DNA synthesis (Yamada *et al.*, 1997). DNA binding of STAT3 is also prevented in IL-6-deficient mice after partial hepatectomy (Cressman *et al.*, 1996). Taken together, one might hypothesize that TNFα induces C/EBP-β/LAP and IL-6, IL-6 in turn induces nuclear translocation of STAT3, and STAT3 contributes to cell proliferation and possibly also provides yet unknown anti-apoptotic signals, shifting the situation from cell destruction to cell survival and proliferation, which results in liver regeneration. This hypothesis is supported by the finding that mice pretreated with IL-6 were resistant to ConA-induced liver injury (Mizuhara *et al.*, 1994). Furthermore, knockout mice lacking the functional Rel A component of NF-κB have a defect in normal liver development (Bellas *et al.*, 1997). Since NF-κB is associated with cell cycle progression in hepatocytes during liver regeneration, NF-κB also seems to contribute to enhanced proliferation of hepatocytes. It may be assumed that the sequence of events for the initiation of liver regeneration is: TNFα → TNF receptor 1 → NF-κB → IL-6 → STAT3 → DNA synthesis (Yamada *et al.*, 1997).



### 3.6.5 Cryptic sensitizations and the role of additional mediators

PEA, in contrast to LPS or superantigen, is able to induce TNFα-mediated hepatotoxicity without further sensitization (Schümann *et al.*, 1998). This could be easily explained by the immanent protein synthesis inhibitory property of PEA (Pavlovskis and Shackelford, 1974, Iglewski and Kabat, 1975) which mono-ADP-ribosylates eukaryotic translation elongation factor 2 (eEF-2) (Iglewski *et al.*, 1977), thus providing a similar sensitization as shown by GalN or ActD. This example shows that a naturally occurring toxin, such as PEA, can possess different properties, i.e. in this case immunostimulation and protein synthesis inhibition, that contribute to its toxicity. TNFα is not the only mediator of PEA-induced liver injury, since perforin-deficient mice showed a significantly reduced susceptibility to PEA (Schümann *et al.*, 1998). Hence, PEA is an excellent example of a toxin that evokes multiple immunotoxic reactions. In particular, TNFα might preactivate antigen presenting cells such as dendritic cells thereby facilitating the activation of cytotoxic T cells (Bennett *et al.*, 1998, Ridge *et al.*, 1998, Schoenberger *et al.*, 1998) that in turn produce cytotoxic mediators like perforin.

### 3.6.5.1 The dual role of interferon-γ

ConA induces liver injury independently of sensitization (Tiegs *et al.*, 1992). ConA neither inhibits transcription nor translation (Leist *et al.*, 1995a), as do GalN, ActD, or PEA. The mechanisms by which ConA sensitizes the mouse liver to TNFα are unknown. In contrast to the GalN models, passive immunization with anti-IFN-γ Ab protected mice from ConA-induced liver failure (Küsters *et al.*, 1996). IFN-γ is known to sensitize p53-negative human colon adenocarcinoma cells (HT-29) and other cell lines to different apoptotic stimuli, including TNFα and agonistic Fas Ab (Fransen *et al.*, 1986, Ossina *et al.*, 1997). IFN-γ also sensitizes mice to TNFα *in vivo* and it enhances the effects of very high doses of LPS or TNFα (Doherty *et al.*, 1992). Following incubation with IFN-γ, a lot of potentially apoptosis-promoting proteins accumulate: TNF receptor 1 (Tsujimoto *et al.*, 1986, Ossina *et al.*, 1997), Fas (Ossina *et al.*, 1997), several caspases (Tamura *et al.*, 1996, Chin *et al.*, 1997, Ossina *et al.*, 1997), and Bak, a pro-apoptotic member of the Bcl-2 protein family (Ossina *et al.*, 1997). p53-positive cells, e.g. primary cultured hepatocytes, react with enhanced expression of the tumoursuppressor and apoptosis-mediator p53 in response to IFN-γ (Kano *et al.*, 1997). IFN-γ also induces the expression of the 2′-5′ oligoadenylate (2-5A) synthetase. Its product, 2-5A, activates the rRNA degrading RNase L (Johnston and Torrence, 1984), thereby indirectly inhibiting protein synthesis. Cells from RNase L-knockout mice are less susceptible to different apoptotic stimuli, including TNFα (Zhou *et al.*, 1997). In addition to the 2-5A synthetase, IFN-γ induces the expression of the 'double-stranded RNA-dependent protein kinase' (PKR), a serine/threonine kinase that phosphorylates eukaryotic translation initiation factor 2 (eIF-2), thereby inhibiting protein biosynthesis. Expression of a dominant-negative, catalytically inactive PKR mutant protected NIH3T3 cells from apoptosis induced by different stimuli, including TNFα (Srivastava *et al.*, 1998), and overexpression of a nonphosphorylatable eIF-2 partially protected cells from TNFα-induced apoptosis (Srivastava *et al.*, 1998). The so-called 'death associated protein kinase' (DAP kinase) has also been implicated in the IFN-γ-increased sensitivity of cells to apoptotic stimuli. The DAP kinase is induced by IFN-γ (Deiss *et al.*, 1995). DAP kinase overexpressing tumour cells displayed higher sensitivity to the apoptotic effects of TNFα (Inbal *et al.*, 1997), and a catalytically inactive DAP kinase mutant, displaying dominant-negative features, protected IFN-γ-sensitive HeLa cells from IFN-γ-induced cell death (Cohen *et al.*, 1997). IFN-γ does not only sensitize cells to apoptosis, but also inhibits cell proliferation via activation of STAT1. STAT1-deficient human U3A cells show IFN-γ-induced growth inhibition only when they express complete STAT1 activity (Bromberg *et al.*, 1996). Taken together, there exist many possible mechanisms by which IFN-γ may contribute to ConA-induced TNFα-mediated hepatocellular death *in vivo*.

IFN-γ, however, is not necessarily an enhancer of TNFα toxicity. Under appropriate circumstances it can also slow down cytotoxic processes, probably by induction of the inducible NO synthase and production of NO. Mice deficient in the IFN-γ receptor α chain are significantly more sensitive to the administration of agonistic anti-CD3 T cell activating antibodies. These mice, in contrast to wild-type mice, die as a consequence of T cell activation (Matthys *et al.*, 1995). The higher susceptibility of the IFN-γ receptor knockout mice to anti-CD3-induced death is associated with an impaired NO production, and wild-type mice can be sensitized to anti-CD3 treatment by the co-administration of an inhibitor of NO synthase (Matthys *et al.*, 1995). We have shown that passive immunization with anti-IFN-γ Ab enhances liver injury after treatment of mice with PEA (Schümann and Tiegs, unpublished results). These findings are in contrast to the results obtained with ConA-treated mice (Küsters *et al.*, 1996). A protective role of IFN-γ-dependent NO synthesis was also observed in mice challenged with SEB. Antibodies to IFN-γ inhibited SEB-induced NO formation (Florquin *et al.*, 1994b). An otherwise non-lethal SEB dose turned out to be lethal when co-injected with an inhibitor of NO synthase (*N*-nitroso-L-arginine methyl ester (L-NAME)) (Florquin *et al.*, 1994b). L-NAME/SEB-induced lethality was accompanied by enhanced synthesis of TNFα and IFN-γ. TNF as well as IFN-γ contributed to death as shown by the protective effect of passive immunization with anti-TNF or anti-IFN-γ Ab (Florquin *et al.*, 1994b).

In summary, it becomes clear that IFN-γ is able to play distinct roles in the process of intoxication. Toxicity enhancing as well as toxicity decreasing effects have been described. The role of IFN-γ in immunotoxicology can therefore not be clearly defined and has to be tested in each individual case.

### 3.6.5.2 The role of Fas

One of the most prominent cell death-mediating receptors is Fas (= CD95). Fas is constitutively expressed in a wide range of tissues, including the liver (for review see Galle and Krammer, 1998). The Fas ligand (FasL) is predominantly expressed in T cells, but can also be induced in other cell types. Administration of an agonistic anti-Fas antibody in mice results in rapid liver failure and death of the animals (Ogasawara *et al.*, 1993). Furthermore, Fas-mediated target cell killing has also been implicated in a murine hepatitis B model: transgenic mice, constitutively expressing hepatitis B virus (HBV) envelope proteins in hepatocytes, develop severe acute necroinflammatory liver disease after injection of an MHC-restricted $CD8^+$ CTL clone recognizing the HBV surface antigen. Liver injury and death were prevented by a soluble Fas-Fc construct (Kondo *et al.*, 1997). Immunohistology of liver tissue from patients with chronic HBV infection showed highly elevated Fas expression on hepatocytes compared to liver tissue from healthy individuals (Galle *et al.*, 1995). Hence, Fas-mediated target cell killing might also be involved in one of the described experimental models of immunotoxicologically mediated liver injury.

Many functional studies have been performed using Fas-mutated *lpr/lpr* mice. These mice differ from wild-type mice in many aspects. It has been demonstrated, for example, that peritoneal macrophages from MRL *lpr/lpr* mice overproduce inflammatory cytokines, including TNFα, upon stimulation with SEB (Edwards III. *et al.*, 1996). This is possibly a result of impaired Fas-mediated apoptosis. Therefore, experimental results concerning the role of Fas in target cell killing, obtained from *lpr/lpr* mice, have to be treated with caution. Better tools for investigation of the role of Fas in target cell killing are Fas or FasL neutralizing molecules. For example, monoclonal antibodies neutralizing the murine Fas ligand, reduced transaminase release after treatment of mice with ConA (Seino *et al.*, 1997).

There is also another aspect to the cytotoxic potential of FasL/Fas interactions in immunotoxicologically mediated organ damage. Blockade of FasL with a soluble Fas immunoadhesin failed to prevent liver injury in mice treated with ConA (in contrast to the above described results of Seino *et al.*, 1997). However, when co-administered with an inhibitor of matrix metalloproteinases, the latter being responsible for the shedding of the transmembrane FasL (Tanaka *et al.*, 1996), soluble Fas immunoadhesin reduced the metalloproteinase inhibitor-enhanced ConA-induced liver injury (Ksontini *et al.*, 1998). It has been also demonstrated by others that only the membrane-bound form, but not the soluble form, of FasL has the capacity to kill target cells (Schneider *et al.*, 1998, Tanaka *et al.*, 1998), and that soluble FasL even prevents cytotoxicity of transmembrane FasL (Tanaka *et al.*, 1998). Hence, the impact of FasL-Fas interactions on immunologically mediated toxicity seems to depend on the degree of FasL shedding.

## 3.7 ROLES OF THE TWO TNFα RECEPTORS

The effects of TNF are mediated by two different receptors. Both are expressed in almost all cells: TNF receptor 1 (TNFR-1), also known as p55, and the p75 TNF receptor 2 (TNFR-2), (for review see Smith *et al.*, 1994). Additionally, two forms of TNF have to be distinguished, i.e. the 26 kDa transmembrane protein (mTNF) and the 17 kDa soluble cytokine (sTNF), the latter originating from the membrane-expressed form through proteolytic cleavage by a matrix metalloproteinase (Gearing *et al.*, 1994, McGeehan *et al.*, 1994). Both forms of TNF are biologically active, and bind to the TNF receptors with different affinities. Under physiological conditions, sTNF preferentially binds to TNFR-1 ($K_d = 1.9 \times 10^{-11}$ M *vs.* $K_d = 4.2 \times 10^{-10}$ M of TNFR-2) (Grell *et al.*, 1998a), while the transmembrane form of TNF preferentially binds to TNFR-2 (Grell *et al.*, 1995). Both receptors are capable of independent signalling, but the signalling pathways also interfere.

Mice deficient in TNFR-1 are protected from GalN/TNF-induced hepatocyte death, and wild-type BALB/c mice, injected with human TNFα, that only activates murine TNFR-1 but not TNFR-2, develop severe liver injury (Leist *et al.*, 1995b). Furthermore, TNFR-1-deficient mice are clearly resistant to GalN/LPS-induced liver failure (Pfeffer *et al.*, 1993) and lethality (Pfeffer *et al.*, 1993, Rothe *et al.*, 1993) as well as to GalN/SEB-induced lethality (Pfeffer *et al.*, 1993). These results demonstrate the central role of TNFR-1 in mediating cell death. TNFR-1-deficient mice are also resistant to ConA- (Küsters *et al.*, 1997) or acute PEA-induced liver injury (Schümann *et al.*, 1998). However, TNFR-1 is not only disadvantageous. For example, it plays an important role in the defense of the facultative intracellular bacterium *Listeria monocytogenes* (Pfeffer *et al.*, 1993, Rothe *et al.*, 1993).

Until now, little is known about the biological role of TNFR-2. Recently, Grell *et al.* have shown that TNFα-dependent thymocyte proliferation and secretion of granulocyte-macrophage colony-stimulating factor (GM-CSF) still occurs in TNFR-1-deficient mice (Grell *et al.*, 1998b). Thus, functional TNFR-2 signalling can occur independently of TNFR-1. Lucas *et al.* have revealed an important role of the mTNF-triggered TNFR-2 in the up-regulation of the intercellular adhesion molecule-1 (ICAM-1) during T cell-dependent experimental cerebral malaria (Lucas *et al.*, 1997). TNFR-2-deficient mice, in contrast to wild-type mice, failed to up-regulate ICAM-1 following injection of *Plasmodium berghei* ANKA-parasitized erythrocytes, leukocytes did not infiltrate into brain tissue, and mice were protected from cerebral malaria (Lucas *et al.*, 1997). However, this is not a general mechanism, since up-regulation of ICAM-1 within the liver after injection of recombinant TNF to mice only depended on the TNFR-1 (G. Tiegs, unpublished results). ICAM-1 has been implicated in the pathogenesis of ConA-induced liver injury (Watanabe *et al.*, 1996). Thus, prevention of ICAM-1 expression, besides suppression of cell death triggering, could be one reason for the resistance of TNFR-1-deficient mice to ConA.



However, TNFR-2-deficient mice exhibit decreased sensitivity to high dose TNFα-induced lethality (Erickson *et al.*, 1994), and they are protected from acute liver injury following intoxication with ConA (Küsters *et al.*, 1997) or PEA (Schümann *et al.*, 1998). The importance of the TNFR-2 for cytotoxicity in the ConA model was supported by the fact that transgenic mice, only expressing mTNF, but not sTNF, were still susceptible to ConA-induced liver failure, even though liver damage was less pronounced than in wild-type mice (Küsters *et al.*, 1997). This result shows (1) that mere expression of the prime TNFR-2 ligand, i.e. mTNF, can mediate hepatotoxicity, and (2) that the coexistence of both TNF forms, i.e. mTNF and sTNF, is necessary for full toxicity. However, in the transgenic mice expressing only mTNF, plasma IFN-γ was significantly elevated after ConA treatment (Küsters *et al.*, 1997). Therefore, it can not be excluded that the apparent susceptibility of these mice to ConA is a result of IFN-γ overload. The role of TNFR-2 in this model was further supported by an enhanced susceptibility of transgenic mice, overexpressing human TNFR-2 in addition to their own TNF receptors, to ConA (Küsters *et al.*, 1997). Since the human TNFR-2 is widely expressed in these mice, including lymphoid and hepatic tissue, and since these transgenic mice again showed elevated plasma IFN-γ (Küsters *et al.*, 1997), the exact functional site(s) of mTNF-TNFR-2 interaction in ConA-hepatitis have not yet been defined. However, with respect to the expression of both TNF receptors by the target cells, i.e. the hepatocytes, a functional role of both receptors for either ConA- or PEA-induced hepatocellular apoptosis *in vivo* should be taken into account.

The participation of both TNF receptors in target cell toxicity can be explained by complex interactions between the signalling pathways of TNFR-1 and -2 (for review on TNF receptor signalling see Darnay and Aggarwal, 1997). TNFR-1-signalling includes the activation of the caspase cascade as well as of the transcription factor NF-κB that protects cells from TNFα-induced cell death. TNFR-1 possesses an intracellular so-called 'death domain', comprising the binding site for the 'TNF receptor associated death domain protein' (TRADD), which in turn binds the caspase-8 recruiting 'Fas associated death domain protein' (FADD) as well as the NF-κB-supporting cytoprotective protein TRAF-2. TNFR-2 binds TRAF-2 directly. A recent publication suggests a mechanism by which TNF receptor 2 could support apoptosis-directed TNFR-1-signalling (Duckett and Thompson, 1997). Human 293 cells were co-transfected with TRAF-2 and a construct encoding a fusion protein consisting of the extracellular and transmembrane parts of CD28 and the intracellular part of TNFR-2. This construct has been shown to form homodimers thereby being activated without additional ligands. Interestingly, in these cells, TRAF-2 was strongly degraded, and activation of NF-κB was prevented (Duckett and Thompson, 1997). Thus, TNFR-2 may contribute to TNFR-1-mediated cell death by sensitizing cells through degradation of TRAF-2, thereby preventing activation of NF-κB.

## 3.8 SUMMARY

Certain microbial or chemical toxicants affect the immune system. They are able to cause acute or chronic disease by means of suppression or overstimulation of the immune system. Immunosuppression may be caused by anergy and apoptosis of immune cells while overstimulation of the immune system may result in apoptosis and secondary necrosis of target cells. One central mediator of such a target cell death is TNFα. Additional mediators or target cell sensitization are necessary in most cases to facilitate TNFα cytotoxicity. Target cell death is modulated on several levels of the intoxication pathway (see Figure 3.2). It depends on (1) the immunostatus, (2) the co-exposure to synergizing or sensitizing agents, and (3) the current equipment of the target cells with cytoprotective proteins. Furthermore, the current cell cycle phase may influence the cell's susceptibility to apoptogenic mediators. The cytoprotective proteins might be of great significance for attenuation of the toxicity of xenobiotics. The investigation of novel drugs that induce cytoprotective proteins could possibly provide new approaches to the treatment of organ damage due to intoxications with immune-active natural or chemical toxicants.

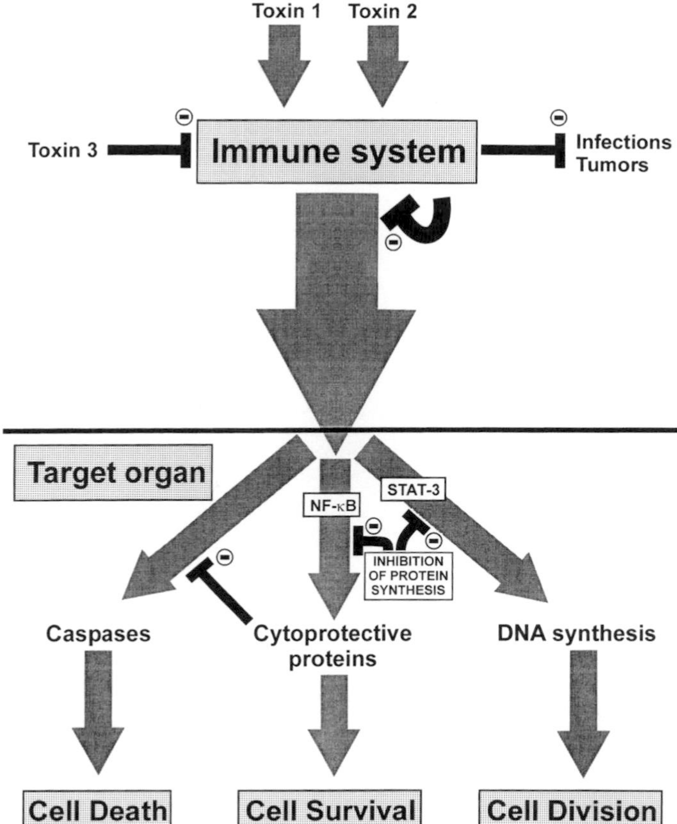




**Figure 3.2** Modulation of apoptosis in immunotoxicology. The immune system can be affected by toxins (toxins 1, 2, and 3). The toxins 1 and 2 symbolize xenobiotics that overstimulate the immune system, and act together in a synergistic manner. Toxin 3, in contrast, is an immunosuppressive toxin or drug that prevents the immune system from controlling infections and tumours. Overstimulation of the immune system is characterized by the production of various proinflammatory immune mediators, most prominently TNFα, IL-1, IL-6, and IFN-γ, that affect certain target cells. The immune reaction is controlled by co-released anti-inflammatory mediators such as IL-10 or TGF-β1. In the target cells, the effector mediators elicit three different pathways that might interfere with each other. (1) The activation of caspases leads to apoptosis, (2) the NF-κB-dependent induction of cytoprotective proteins disrupts the apoptotic pathway thereby facilitating cell survival, and (3) activation of STAT 3 and possibly other transcription factors promotes the cell cycle thereby leading to cell proliferation and possibly inhibition of cell death. Inhibition of protein synthesis represents a mechanism of target cell sensitization, which probably works by inhibiting the induction cytoprotective and cell cycle proteins. The overall balance of the apoptotic, cytoprotective, and proliferative pathways in the target cell should determine the immune-mediated toxicity of a xenobiotic.

## ACKNOWLEDGMENTS

The editor would like to thank Dr Sabina Cosulich for critical appraisal of this chapter and Zana Dye for invaluable editorial assistance.

## REFERENCES

ANGERMÜLLER, S., KÜNSTLE, G. and TIEGS, G., 1998, Pre-apoptotic alterations in hepatocytes of TNFα-treated galactosamine-sensitized mice, *Journal of Histochemistry and Cytochemistry*, **46**, 1–9

BEAN, A.G.D., FREIBERG, R.A., ANDRADE, S., MENON, S. and ZLOTNIK, A., 1993, Interleukin 10 protects mice against staphylococcal enterotoxin B-induced lethal shock, *Infection and Immunity*, **61**, 4937–4939

BEG, A.A. and BALTIMORE, D., 1996, An essential role for NF-κB in preventing TNF-α-induced cell death, *Science*, **274**, 782–784

BELLAS, R.E., FITZGERALD, M.J., FAUSTO, N. and SONENSHEIN, G.E., 1997, Inhibition of NF-κB activity induces apoptosis in murine hepatocytes, *American Journal of Pathology*, **151**, 891–896

BERG, D.J., KUHN, R., RAJEWSKI, K., MÜLLER, W., MENON, S., DAVIDSON, N., *et al.*, 1995, Interleukin-10 is a central regulator of the response to LPS in murine models of endotoxic shock and the Shwartzman reaction but not endotoxin tolerance, *Journal of Clinical Investigation*, **96**, 2339–2347

BENNETT, S.R.M., CARBONE, F.R., KARAMALIS, F., FLAVELL, R.A., MILLER, J.F.A.P. and HEATH, W.R., 1998, Help for cytotoxic-T-cell responses is mediated by CD40 signalling, *Nature*, **393**, 478–480

BEUTLER, B., MILSARK, I.W. and CERAMI, A.C., 1985, Passive immunization against cachectin/tumor necrosis factor protects mice from lethal effect of endotoxin, *Science*, **229**, 869–871

BLANK, C., LUZ, A., BENDIGS, S., ERDMANN, A., WAGNER, H. and HEEG, K., 1997, Superantigen and endotoxin synergize in the induction of lethal shock, *European Journal of Immunology*, **27**, 825–833

BOHLINGER, I., LEIST, M., BARSIG, J., UHLIG, S., TIEGS, G. and WENDEL, A., 1995a, Interleukin-1 and nitric oxide protect against tumor necrosis factor α-induced liver injury through distinct pathways, *Hepatology*, **22**, 1829–1837

BOHLINGER, I., LEIST, M., BARSIG, J., UHLIG, S., TIEGS, G. and WENDEL, A., 1995b, Tolerance against tumor necrosis factor α (TNF)-induced hepatotoxicity in mice: the role of nitric oxide, *Toxicology Letters*, **82/83**, 227–231

BROMBERG, J.F., HORVATH, C.M., WEN, Z., SCHREIBER, R.D. and DARNELL JR, J.E., 1996, Transcriptionally active Stat1 is required for the antiproliferative effects of both interferon α and interferon γ, *Proceedings of the National Academy of Sciences of the United States of America*, **93**, 7673–7678

CAVAILLON, J.M., MUNOZ, C., FITTING, C., MISSET, B. and CARLET, J., 1992, Circulating cytokines: the tip of the iceberg, *Circulatory Shock*, **38**, 145–152

CHIN, Y.E., KITAGAWA, M., KUIDA, K., FLAVELL, R.A. and FU, X.Y., 1997, Activation of the STAT signalling pathway can cause expression of caspase 1 and apoptosis, *Molecular and Cellular Biology*, **17**, 5328–5337

CHU, Z.-L., MCKINSEY, T.A., LIU, L., GENTRY, J.J., MALIM, M.H. and BALLARD, D.W., 1997, Suppression of tumor necrosis factor-induced cell death by inhibitor of apoptosis c-IAP2 is under NF-κB control, *Proceedings of the National Academy of Sciences of the United States of America*, **94**, 10057–10062

CLARK, G.C. and TAYLOR, M.J., 1994, Tumor necrosis factor involvement in the toxicity of TCCD: the role of endotoxin in the response, *Experimental and Clinical Immunogenetics*, **11**, 136–141

CLARK, G.C., TAYLOR, M.J., TRITSCHER, A.M. and LUCIER, G.W., 1991, Tumor necrosis factor involvement in 2,3,7,8-tetrachlorodibenzo-p-dioxin-mediated endotoxin hypersensitivity in C57BL/6J mice congenic at the Ah locus, *Toxicology and Applied Pharmacology*, **111**, 422–431

COHEN, O., FEINSTEIN, E. and KIMCHI, A., 1997, DAP-kinase is a $Ca^{2+}$/calmodulin-dependent, cytoskeletal-associated protein kinase, with cell death-inducing functions that depend on its catalytic activity, *EMBO Journal*, **16**, 998–1008

COOPER, J.T., STROKA, D.M., BROSTJAN, C., PALMETSHOFER, A., BACH, F.H. and FERRAN, C., 1996, A20 blocks endothelial cell activation through a NF-κB-dependent mechanism, *Journal of Biological Chemistry*, **271**, 18068–18073

CRESSMAN, D.E., GREENBAUM, L.E., DEANGELIS, R.A., CILIBERTO, G., FURTH, E.E., POLI, V. and TAUB, R., 1996, Liver failure and defective hepatocyte regeneration in interleukin 6-deficient mice, *Science*, **274**, 1379–1383

DARNAY, B.G. and AGGARWAL, B.B., 1997, Early events in TNF signaling: a story of associations and dissociations, *Journal of Leukocyte Biology*, **61**, 559–566

DEISS, L.P., FEINSTEIN, E., BERISSI, H., COHEN, O. and KIMCHI, A., 1995, Identification of a novel serine/threonine kinase and a novel 15-kD protein as potential mediators of the γ interferon-induced cell death, *Genes and Development*, **9**, 15–30

DIXIT, V.M., GREEN, S., SARMA, V. and PROCHOWNIK, E.V., 1990, Tumor necrosis factor-α induction of novel genes in human endothelial cells including a macrophage specific chemotaxin, *Journal of Biological Chemistry*, **265**, 2973–2978

DOHERTY, G.M., LANGE, J.R., LANGSTEIN, H.N., ALEXANDER, H.R., BURESH, C.M. and NORTON, J.A., 1992, Evidence for IFN-γ as a mediator of the lethality of endotoxin and tumor necrosis factor-α, *Journal of Immunology*, **149**, 1666–1670

DUCKETT, C.S. and THOMPSON, C.B., 1997, CD30-dependent degradation of TRAF2: implications for negative regulation of TRAF signalling and the control of cell survival, *Genes and Development*, **11**, 2810–2821

EDWARDS III, C.K., ZHOU, T., ZHANG, J., BAKER, T.J., DE, M., LONG, R.E., *et al.*, 1996, Inhibition of superantigen-induced proinflammatory cytokine production and inflammatory arthritis in MRL-*lpr/lpr* mice by a transcriptional inhibitor of TNF-α, *Journal of Immunology*, **157**, 1758–1772

ERICKSON, S.L., DE SAUVAGE, F.J., KIKLY, K., CARVER-MOORE, K., PITTS-MEEK, S., GILLETT, N., *et al.*, 1994, Decreased sensitivity to tumour-necrosis factor but normal T-cell development in TNF receptor-2-deficient mice, *Nature*, **372**, 560–563

FERRAN, C., STROKA, D.M., BADRICHANI, A.Z., COOPER, J.T., WRIGHTON, C.J., SOARES, M., *et al.*, 1998, A20 inhibits NF-κB activation in endothelial cells without sensitizing to tumor necrosis factor-mediated apoptosis, *Blood*, **91**, 2249–2258

FISCHER, W., SCHUDT, C. and WENDEL, A., 1993, Protection by phosphodiesterase inhibitors against endotoxin-induced liver injury in galactosamine-sensitized mice, *Biochemical Pharmacology*, **45**, 2399–2404

FLOHÉ, S., HEINRICH, P.C., SCHNEIDER, J., WENDEL, A. and FLOHÉ, L., 1991, Time course of IL-6 and TNFα release during endotoxin-induced endotoxin tolerance in rats, *Biochemical Pharmacology*, **41**, 1607–1614

FLORQUIN, S., AMRAOUI, Z., ABRAMOWICZ, D. and GOLDMAN, M., 1994a, Systemic release and protective role of IL-10 in staphylococcal enterotoxin B-induced shock in mice, *Journal of Immunology*, **153**, 2618–2623

FLORQUIN, S., AMRAOUI, Z., DUBOIS, C., DECUYPER, J. and GOLDMAN, M., 1994b, The protective role of endogenously synthesized nitric oxide in staphylococcal enterotoxin B-induced shock in mice, *Journal of Experimental Medicine*, **180**, 1153–1158

FRANSEN, L., VAN DER HEYDEN, J., RUYSSCHAERT, R. and FIERS, W., 1986, Recombinant tumor necrosis factor: its effect and its synergism with interferon-γ on a variety of normal and transformed human cell lines, *European Journal of Cancer and Clinical Oncology*, **22**, 419–426

FREUDENBERG, M.A. and GALANOS, C., 1988, Induction of tolerance to lipopolysaccharide (LPS)-D-galactosamine lethality by pretreatment with LPS is mediated by macrophages, *Infection and Immunity*, **56**, 1352–1357

FREUDENBERG, M.A., KEPPLER, D. and GALANOS, C., 1986, Requirement for lipopolysaccharide-responsive macrophages in galactosamine-induced sensitization to endotoxin, *Infection and Immunity*, **51**, 891–895

GALANOS, C., FREUDENBERG, M.A. and REUTTER, W., 1979, Galactosamine-induced sensitization to the lethal effects of endotoxin, *Proceedings of the National Academy of Sciences of the United States of America*, **76**, 5939–5943

GALLE, P.R. and KRAMMER, P.H., 1998, CD95-induced apoptosis in human liver disease, *Seminars in Liver Disease*, **18**, 141–151

GALLE, P.R., HOFMANN, W.J., WALCZAK, H., SCHALLER, H., OTTO, G., STREMMEL, W., *et al.*, 1995, Involvement of the CD95 (APO-1/Fas) receptor and ligand in liver damage, *Journal of Experimental Medicine*, **182**, 1223–1230

GANTNER, F., KÜSTERS, S., WENDEL, A., HATZELMANN, A., SCHUDT, C. and TIEGS, G., 1997, Protection from T cell-mediated murine liver failure by phosphodiesterase inhibitors, *Journal of Pharmacology and Experimental Therapeutics*, **280**, 53–60

GANTNER, F., LEIST, M., LOHSE, A.W., GERMANN, P.G. and TIEGS, G., 1995a, Concanavalin A-induced T cell-mediated hepatic injury in mice: the role of tumor necrosis factor, *Hepatology*, **21**, 190–198

GANTNER, F., LEIST, M., JILG, S., GERMANN, P.G., FREUDENBERG, M.A. and TIEGS, G., 1995b, Tumor necrosis factor-induced hepatic DNA fragmentation as an early marker of T cell-dependent liver injury in mice, *Gastroenterology*, **109**, 166–176

GEARING, A.J.H., BECKETT, P., CHRISTODOULOU, M., CHURCHILL, M., CLEMENTS, J., DAVIDSON, A.H., *et al.*, 1994, Processing of tumor necrosis factor-α precursor by metalloproteinases, *Nature*, **370**, 555–557

GELLER, D.A., NUSSLER, A.K., DI SILVIO, M., LOWENSTEIN, C.J., SHAPIRO, R.A., WANG, S.C., *et al.*, 1993, Cytokines, endotoxin, and glucocorticoids regulate the expression of inducible nitric oxide synthase in hepatocytes, *Proceedings of the National Academy of Sciences of the United States of America*, **90**, 522–526

GRELL, M., DOUNI, E., WAJANT, H., LÖHDEN, M., CLAUSS, M., MAXEINER, B., *et al.*, 1995, The transmembrane form of tumor necrosis factor is the prime activating ligand of the 80 kDa tumor necrosis factor receptor, *Cell*, **83**, 793–802

GRELL, M., WAJANT, H., ZIMMERMANN, G. and SCHEURICH, P., 1998a, The type 1 receptor (CD120a) is the high-affinity receptor for soluble tumor necrosis factor, *Proceedings of the National Academy of Sciences of the United States of America*, **95**, 570–575

GRELL, M., BECKE, F.M., WAJANT, H., MÄNNEL, D.N. and SCHEURICH, P., 1998b, TNF receptor 2 mediates thymocyte proliferation independently of TNF receptor type 1, *European Journal of Immunology*, **28**, 257–263

HOCKENBERRY, D.M., OLTVAI, Z.N., YIN, X.-M., MILLIMAN, C.L. and KORSMEYER, S.J., 1993, Bcl-2 functions in an antioxidant pathway to prevent apoptosis, *Cell*, **75**, 241–251

HSU, H.-B., TAKEUCHI, M. and GOEDDEL, D.V., 1996b, The tumor necrosis factor receptor 2 signal transducers TRAF2 and c-IAP1 are components of the tumor necrosis factor receptor 1 signaling complex, *Proceedings of the National Academy of Sciences of the United States of America*, **93**, 13973–13978

Hsu, H., Shu, H.-B., Pan, M.-G. and Goeddel, D.V., 1996a, TRADD-TRAF2 and TRADD-FADD interactions define two distinct TNF receptor 1 signal transduction pathways, *Cell*, **84**, 299–308

Iglewski, B.H. and Kabat, D., 1975, NAD-dependent inhibition of protein synthesis by *Pseudomonas aeruginosa* toxin, *Proceedings of the National Academy of Sciences of the United States of America*, **72**, 2284–2288

Iglewski, B.H., Liu, P.V. and Kabat, D., 1977, Mechanism of action of *Pseudomonas aeruginosa* exotoxin A: adenosine diphosphate-ribosylation of mammalian elongation factor 2 in vitro and in vivo, *Infection and Immunity*, **15**, 138–144

Inbal, B., Cohen, O., Polak-Charcon, S., Kopolovic, J., Vadai, E., Eisenbach, L. and Kimchi, A., 1997, DAP kinase links the control of apoptosis to metastasis, *Nature*, **390**, 180–184

Jäättelä, M., Mouritzen, H., Elling, F. and Bostholm, L., 1996, A20 zinc finger protein inhibits TNF and IL-1 signaling, *Journal of Immunology*, **156**, 1166–1173

Jaeschke, H., Fischer, M.A., Lawson, J.A., Simmons, C.A., Farhood, A. and Jones, D.A., 1998, Activation of caspase 3 (CPP32)-like proteases is essential for TNF-α-induced hepatic parenchymal cell apoptosis and neutrophil-mediated necrosis in a murine endotoxin shock model, *Journal of Immunology*, **160**, 3480–3486

Jilg, S., Barsig, J., Leist, M., Küsters, S., Volk, H.-D. and Wendel, A., 1996, Enhanced release of interleukin-10 and soluble tumor necrosis factor receptors as novel principles of methylxanthine action in murine models of endotoxic shock, *Journal of Pharmacology and Experimental Therapeutics*, **278**, 421–431

Johnston, M.I. and Torrence, P.F., 1984, The role of interferon-induced proteins, double-stranded RNA and 2′,5′-oligoadenylate in the interferon-mediated inhibition of viral translation, in Friedman, R.M. (ed.) *Interferon: mechanisms of production and action*, Amsterdam: Elsevier Science Publishers, pp. 189–298

Kane, D.J., Sarafian, T.A., Anton, R., Hahn, H., Gralla, E.B., Valentine, J.S., *et al.*, 1993, Bcl-2 inhibition of neural death: decreased generation of reactive oxygen species, *Science*, **262**, 1274–1277

Kano, A., Watanabe, Y., Takeda, N., Aizawa, S.-I. and Akaike, T., 1997, Analysis of IFN-γ-induced cell cycle arrest and cell death in hepatocytes, *Journal of Biochemistry*, **121**, 677–683

Kim, Y.-M., de Vera, M.E., Watkins, S.C. and Billiar, T.R., 1997, Nitric oxide protects cultured rat hepatocytes from tumor necrosis factor-α-induced apoptosis by inducing heat shock protein 70 expression, *Journal of Biological Chemistry*, **272**, 1402–1411

KONDO, T., SUDA, T., FUKUYAMA, H., ADACHI, M. and NAGATA, S., 1997, Essential roles of the Fas ligand in the development of hepatitis, *Nature Medicine*, **3**, 409–413

KRIKOS, A., LAHERTY, C.D. and DIXIT, V.M., 1992, Transcriptional activation of the tumor necrosis factor-inducible zinc finger protein, A20, is mediated by κB elements, *Journal of Biological Chemistry*, **267**, 17971–17976

KSONTINI, R., COLAGIOVANNI, D.B., JOSEPHS, M.D., EDWARDS III, C.K., TANNAHILL, C.L., SOLORZANO, C.C., *et al.*, 1998, Disparate roles for TNF-α and Fas ligand in concanavalin A-induced hepatitis, *Journal of Immunology*, **160**, 4082–4089

KÜNSTLE, G., LEIST, M., UHLIG, S., REVESZ, L., FEIFEL, R., MACKENZIE, A. and WENDEL, A., 1997, ICE-protease inhibitors block murine liver injury and apoptosis caused by CD95 or by TNF-α, *Immunology Letters*, **55**, 5–10

KÜSTERS, S., GANTNER, F., KÜNSTLE, G. and TIEGS, G., 1996, Interferon gamma plays a critical role in T cell-dependent liver injury in mice initiated by concanavalin A, *Gastroenterology*, **111**, 462–471

KÜSTERS, S., TIEGS, G., ALEXOPOULOU, L., PASPARAKIS, M., DOUNI, E., KÜNSTLE, G., *et al.*, 1997, *In vivo* evidence for a functional role of both tumor necrosis factor (TNF) receptors and transmembrane TNF in experimental hepatitis, *European Journal of Immunology*, **27**, 2870–2875

LEE, S.Y., REICHLIN, A., SANTANA, A., SOKOL, K.A., NUSSENZWEIG, M.C. and CHOI, Y., 1997, TRAF2 is essential for JNK but not NF-κB activation and regulates lymphocyte proliferation and survival, *Immunity*, **7**, 703–713

LEHMANN, V., FREUDENBERG, M.A. and GALANOS, C., 1987, Lethal toxicity of lipopolysaccharide and tumor necrosis factor in normal and D-galactosamine-treated mice, *Journal of Experimental Medicine*, **165**, 657–663

LEIST, M., GANTNER, F., JILG, S. and WENDEL, A., 1995b, Activation of the 55 kDa TNF receptor is necessary and sufficient for TNF-induced liver failure, hepatocyte apoptosis, and nitrite release, *Journal of Immunology*, **154**, 1307–1316

LEIST, M., SINGLE, B., CASTOLDI, A.F., KÜHNLE, S. and NICOTERA, P., 1997b, Intracellular adenosine triphosphate (ATP concentration: a switch in the decision between apoptosis and necrosis, *Journal of Experimental Medicine*, **185**, 1481–1486

LEIST, M., GANTNER, F., BOHLINGER, I., GERMANN, P.G., TIEGS, G. and WENDEL, A., 1994, Murine hepatocyte apoptosis induced in vitro and in vivo by TNF-α requires transcriptional arrest, *Journal of Immunology*, **153**, 1778–1788

LEIST, M., GANTNER, F., BOHLINGER, I., TIEGS, G., GERMANN, P.G. and WENDEL, A., 1995a, Tumor necrosis factor-induced hepatocyte apoptosis precedes liver failure in experimental murine shock models, *American Journal of Pathology*, **146**, 1220–1234

LEIST, M., GANTNER, F., NAUMANN, H., BLUETHMANN, H., VOGT, K., BRIGELIUS-FLOHÉ, R., *et al.*, 1997a, Tumor necrosis factor-induced apoptosis during the poisoning of mice with hepatotoxins, *Gastroenterology*, **112**, 923–934

LIBERT, C., VAN BLADEL, S., BROUCKAERT, P., SHAW, A. and FIERS, W., 1991, Involvement of the liver, but not IL-6, in IL-1-induced desensitization to the lethal effects of tumor necrosis factor, *Journal of Immunology*, **146**, 2625–2632

LIU, Z.-G., HSU, H., GOEDDEL, D.V. and KARIN, M., 1996, Dissection of TNF receptor 1 effector functions: JNK activation is not linked to apoptosis while NF-κB activation prevents cell death, *Cell*, **87**, 565–576

LUCAS, R., JUILLARD, P., DECOSTER, E., REDARD, M., BURGER, D., DONATI, Y., *et al.*, 1997, Crucial role of tumor necrosis factor (TNF) receptor 2 and membrane-bound TNF in experimental malaria, *European Journal of Immunology*, **27**, 1719–1725

MANNA, S.K., ZHANG, H.J., YAN, T., OBERLEY, L.W., AGGARWAL, B.B., 1998, Overexpression of manganese superoxide dismutase suppresses tumor necrosis factor-induced apoptosis and activation of nuclear transcription factor-κB and activated protein-1, *Journal of Biological Chemistry*, **273**, 13245–13254

MATTHYS, P., FROYEN, G., VERDOT, L., HUANG, S., SOBIS, H., VAN DAMME, J., *et al.*, 1995, IFN-γ receptor-deficient mice are hypersensitive to the anti-CD3-induced cytokine release syndrome and thymocyte apoptosis, *Journal of Immunology*, **155**, 3823–3829

MCGEEHAM, G.M., BECHERER, J.D., BAST JR, R.C., BOYER, C.M., CHAMPION, B., CONNOLLY, K.M., *et al.*, 1994, Regulation of tumor necrosis factor-α processing by a metalloproteinase inhibitor, *Nature*, **370**, 558–561

MIETHKE, T., WAHL, C., HEEG, K. and WAGNER, H., 1993b, Acquired resistance to superantigen-induced T cell shock, *Journal of Immunology*, **150**, 3776–3784

MIETHKE, T., DUSCHEK, K., WAHL, C., HEEG, K. and WAGNER, H., 1993a, Pathogenesis of the toxic shock syndrome: T cell mediated lethal shock caused by the superantigen TSST-1, *European Journal of Immunology*, **23**, 1494–1500

MIETHKE, T., WAHL, C., HEEG, K., ECHTENACHER, B., KRAMMER, P.H. and WAGNER, H., 1992, T cell-mediated lethal shock triggered in mice by the superantigen staphylococcal enterotoxin B: critical role of tumor necrosis factor, *Journal of Experimental Medicine*, **175**, 91–98

MIZUHARA, H., O'NEILL, E., SEKI, N., OGAWA, T., KUSUNOKI, C., OTSUKA, K., *et al.*, 1994, T cell activation-associated hepatic injury: mediation by tumor necrosis factors and protection by interleukin 6, *Journal of Experimental Medicine*, **179**, 1529–1537

MURAILLE, E., DE SMEDT, T., ANDRIS, F., PAJAK, B., ARMANT, M., URBAIN, J., *et al.*, 1997, Staphylococcal enterotoxin B induces an early and transient state of immunosuppression characterized by $V_\beta$-unrestricted T cell unresponsiveness and defective antigen-presenting cell functions, *Journal of Immunology*, **158**, 2638–2647

NAGAKI, M., MUTO, Y., OHNISHI, H., YASUDA, S., SANO, K., NAITO, T., *et al.*, 1994, Hepatic injury and lethal shock in galactosamine-sensitized mice induced by the superantigen staphylococcal enterotoxin B, *Gastroenterology*, **106**, 450–458

NAKANO, M., KNOWLTON, A.A., YOKOYAMA, T., LESSLAUER, W. and MANN, D.L., 1996, Tumor necrosis factor-α-induced expression of heat shock protein 72 in adult feline cardiac myocytes, *American Journal of Physiology*, **270**, H1231–1239

OGASAWARA, K., TAKEDA, K., HASHIMOTO, W., SATOH, M., OKUYAMA, R., YANAI, N., *et al.*, 1998, Involvement of NK1$^+$ T cells and their IFN-γ production in the generalized Shwartzman reaction, *Journal of Immunology*, **160**, 3522–3527

OGASAWARA, J., WATANABE-FUKUNAGA, R., ADASHI, M., MATSUZAWA, A., KASUGAI, T., KITAMURA, Y., *et al.*, 1993, Lethal effect of the anti-Fas antibody in mice, *Nature*, **364**, 806–809

OLLIVIER, V., PARRY, G.C.N., COBB, R.R., DE PROST, D. and MACKMAN, N., 1996, Elevated cyclic AMP inhibits NF-κB-mediated transcription in human monocytic cells and endothelial cells, *Journal of Biological Chemistry*, **271**, 20828–20835

OPIPARI JR, A.W., HU, H.M., YABKOWITZ, R. and DIXIT, V.M., 1992, The A20 zinc finger protein protects cells from tumor necrosis factor cytotoxicity, *Journal of Biological Chemistry*, **267**, 12424–12427

OSSINA, N.K., CANNAS, A., POWERS, V.C., FITZPATRICK, P.A., KNIGHT, J.D., GILBERT, J.R., *et al.*, 1997, Interferon-γ modulates a p53-independent apoptotic pathway and apoptosis-related gene expression, *Journal of Biological Chemistry*, **272**, 16351–16357

PAVLOVSKIS, O.R. and SHACKELFORD, A.H., 1974, *Pseudomonas aeruginosa* exotoxin in mice: localization and effect on protein synthesis, *Infection and Immunity*, **9**, 540–546

PERKINS, D.L., WANG, Y., HO, S.S., WIENS, G.R., SEIDMAN, J.G. and RIMM, I.J., 1993, Superantigen-induced peripheral tolerance inhibits T cell responses to immunogenic peptides in TCR (β-chain) transgenic mice, *Journal of Immunology*, **150**, 4284–4291

PFEFFER, K., MATSUYAMA, T., KÜNDIG, T.M., WAKEHAM, A., KISHIHARA, K., SHAHINIAN, A., *et al.*, 1993, Mice deficient for the 55 kD tumor necrosis factor receptor are resistant to endotoxic shock, yet succumb to L. monocytogenes infection, *Cell*, **73**, 457–467

RANDOW, F., SYRBE, U., MEISEL, C., KRAUSCH, D., ZUCKERMANN, H., PLATZER, C. and VOLK, H.-D., 1995, Mechanism of endotoxin desensitization: involvement of interleukin 10 and transforming growth factor β, *Journal of Experimental Medicine*, **181**, 1887–1892

RELLAHAN, B.L., JONES, L.A., KRUISBEEK, A.M., FRY, A.M. and MATIS, L.A., 1990, In vivo induction of anergy in peripheral Vβ8$^+$ T cells by staphylococcal enterotoxin B, *Journal of Experimental Medicine*, **172**, 1091–1100

RODRIGUEZ, I., MATSUURA, K., ODY, C., NAGATA, S. and VASSALLI, P., 1996, Systemic injection of a tripeptide inhibits the intracellular activation of CPP32-like proteases in vivo and fully protects mice against Fas-mediated fulminant liver destruction and death, *Journal of Experimental Medicine*, **184**, 2067–2072

ROSENTHAL, G.J., LEBETKIN, E., THIGPEN, J.E., WILSON, R., TUCKER, A.N. and LUSTER, M.I., 1989, Characteristics of 2,3,7,8-tetrachlorodibenzo-p-dioxin induced endotoxin hypersensitivity: association with hepatotoxicity, *Toxicology*, **56**, 239–251

ROTH, R.A., HARKEMA, J.R., PESTKA, J.P. and GANEY, P.E., 1997, Is exposure to bacterial endotoxin a determinant of susceptibility to intoxication from xenobiotic agents?, *Toxicology and Applied Pharmacology*, **147**, 300–311

ROTHE, M., SARMA, V., DIXIT, V.M. and GOEDDEL, D.V., 1995a, TRAF2-mediated activation of NF-κB by TNF receptor 2 and CD40, *Science*, **269**, 1424–1427

ROTHE, M., WONG, S.C., HENZEL, W.J. and GOEDDEL, D.V., 1994, A novel family of putative signal transducers associated with the cytoplasmic domain of the 75 kDa tumor necrosis factor receptor, *Cell*, **78**, 681–692

ROTHE, M., PAN, M.-G., HENZEL, W.J., AYRES, T.M. and GOEDDEL, D.V., 1995b, The TNFR2-TRAF signalling complex contains two novel proteins related to baculoviral inhibitor of apoptosis proteins, *Cell*, **83**, 1243–1252

ROTHE, J., LESSLAUER, W., LÖTSCHER, H., LANG, Y., KOEBEL, P., KÖNTGEN, F., *et al.*, 1993, Mice lacking the tumour necrosis factor receptor 1 are resistant to TNF-mediated toxicity but highly susceptible to infection by *Listeria monocytogenes, Nature*, **364**, 798–802

ROUQUET, N., PAGÈS, J.-C., MOLINA, T., BRIAND, P. and JOULIN, V., 1996, ICE inhibitor YVADcmk is a potent therapeutic agent against *in vivo* liver apoptosis, *Current Biology*, **6**, 1192–1195

ROY, N., DEVERAUX, Q.L., TAKAHASHI, R., SALVESEN, G.S. and REED, J.C., 1997, The c-IAP-1 and c-IAP-2 proteins are direct inhibitors of specific caspases, *EMBO Journal*, **16**, 6914–6925

SANDSTRÖM, P.A., MANNIE, M.D. and BUTTKE, T.M., 1994, Inhibition of activation-induced death in a T cell hybridoma by thiol antioxidants: oxidative stress as a mediator of apoptosis, *Journal of Leukocyte Biology*, **55**, 221–226

SATO, M., SASAKI, M. and HOJO, H., 1995, Antioxidative roles of metallothionein and manganese superoxide dismutase induced by tumor necrosis factor-α and interleukin-6, *Archives of Biochemistry and Biophysics*, **316**, 738–744

SCHNEIDER, P., HOLLER, N., BODMER, J.-L., HAHNE, M., FREI, K., FONTANA, A. and TSCHOPP, J., 1998, Conversion of membrane-bound Fas(CD95) ligand to its soluble form is associated with downregulation of its proapoptotic activity and loss of liver toxicity, *Journal of Experimental Medicine*, **187**, 1205–1213

SCHOENBERGER, S.P., TOES, R.E.M., VAN DER VOORT, E.I.H., OFFRINGA, R. and MELIEF, C.J., 1998, T-cell help for cytotoxic T lymphocytes is mediated by CD40-CD40L interactions, *Nature*, **393**, 480–483

SCHÜMANN, J., ANGERMÜLLER, S., BANG, R., LOHOFF, M. and TIEGS, G., 1998, Acute hepatotoxicity of *Pseudomonas aeruginosa* exotoxin A in mice depends on T cells and tumor necrosis factor, *Journal of Immunology*, **161**, 5745–5754

SEINO, K.-I., KAYAGAKI, N., TAKEDA, K., FUKAO, K., OKUMURA, K. and YAGITA, H., 1997, Contributions of Fas ligand to T cell-mediated hepatic injury in mice. *Gastroenterology*, **113**, 1315–1322

SMITH, C.A., FARRAH, T. and GOODWIN, R.G., 1994, The TNF receptor superfamily of cellular and viral proteins: activation, costimulation, and death, *Cell*, **76**, 959–962

SONG, H.Y., ROTHE, M. and GOEDDEL, D.V., 1996, The tumor necrosis factor-inducible zinc finger protein A20 interacts with TRAF1/TRAF2 and inhibits NF-κB activation, *Proceedings of the National Academy of Sciences of the United States of America*, **93**, 6721–6725

SRIVASTAVA, S.P., KUMAR, K.U. and KAUFMAN, R.J., 1998, Phosphorylation of eukaryotic translation initiation factor 2 mediates apoptosis in response to activation of the double-stranded RNA-dependent protein kinase, *Journal of Biological Chemistry*, **273**, 2416–2423

SUZUKI, A., 1998, The dominant role of CPP32 subfamily in fas-mediated hepatitis, *Proceedings of the Society for Experimental Biology and Medicine*, **217**, 450–454

TAKANO, M., ARAI, T., MOKUNO, Y., NISHIMURA, H., NIMURA, Y. and YOSHIKAI, Y., 1998, Dibutyryl cyclic adenosine monophosphate protects mice against tumor nedrosis factor-α-induced hepatocyte apoptosis accompanied by increased heat shock protein 70 expression, *Cell Stress Chaperones*, **3**, 109–117

TAMURA, T., UEDA, S., YOSHIDA, M., MATSUZAKI, M., MOHRI, H. and OKUBO, T., 1996, Interferon-γ induces ice gene experssion and enhances cellular susceptibility to apoptosis in the U937 leukemia cell line, *Biochemical and Biophysical Research Communications*, **229**, 21–26

TANAKA, M., ITAI, T., ADACHI, M. and NAGATA, S., 1998, Downregulation of Fas ligand by shedding, *Nature Medicine*, **4**, 31–36

TANAKA, M., SUDA, T., HAZE, K., NAKAMURA, N., SATO, K., KIMURA, F., *et al.*, 1996, Fas ligand in human serum, *Nature Medicine*, **2**, 317–322

TAYLOR, M.J., LUCIER, G.W., MAHLER, J.F., THOMPSON, M., LOCKHART, A.C. and CLARK, G.C., 1992, Inhibition of acute TCDD toxicity by treatment with anti-tumor necrosis factor antibody or dexamethasone, *Toxicology and Applied Pharmacology*, **117**, 126–132

TIEGS, G., HENTSCHEL, J. and WENDEL, A., 1992, A T cell-dependent experimental liver injury in mice inducible by concanavalin A, *Journal of Clinical Investigation*, **90**, 196–203

TIEGS, G., NIEHÖRSTER, M. and WENDEL, A., 1990, Leukocyte alterations do not account for hepatitis induced by endotoxin or TNFα in galactosamine-sensitized mice, *Biochemical Pharmacology*, **40**, 1317–1322



LIVERPOOL
JOHN MOORES UNIVERSITY
AVRIL ROBARTS LRC
TITHEBARN STREET
LIVERPOOL L2 2ER
TEL. 0151 231 4022

TIEGS, G., WOLTER, M. and WENDEL, A., 1989, Tumor necrosis factor is a terminal mediator in galactosamine/endotoxin-induced hepatitis in mice, *Biochemical Pharmacology*, **38**, 627–631

TIEGS, G., KÜSTERS, S., KÜNSTLE, G., HENTZE, H., KIEMER, A.K. and WENDEL, A., 1998, Ebselen protects mice against T cell-dependent, TNF-mediated apoptotic liver injury, *Journal of Pharmacology and Experimental Therapeutics*, **287**, 1098–1104

TOYABE, S., SEKI, S., IIAI, T., TAKEDA, K., SHIRAI, K., WATANABE, H., *et al.*, 1997, Requirement of IL-4 and liver NK1$^+$ T cells for concanavalin A-induced hepatic injury in mice, *Journal of Immunology*, **159**, 1537–1542

TRACEY, K.J., FONG, Y., HESSE, D.G., MANOGUE, K.R., LEE, A.T., KUO, G.C., *et al.*, 1987, Anti-cachectin/TNF monoclonal antibodies prevent septic shock during lethal bacteremia, *Nature*, **330**, 662–664

TRAUTWEIN, C., RAKEMANN, T., BRENNER, D.A., STREETZ, K., LICATO, L., MANNS, M.P. and TIEGS, G., 1998a, Concanavalin A-induced liver cell damage: activation of intracellular pathways triggered by tumor necrosis factor in mice, *Gastroenterology*, **114**, 1035–1045

TRAUTWEIN, C., RAKEMANN, T., MALEK, N.P., PLÜMPE, J., TIEGS, G. and MANNS, M.P., 1998b, Concanavalin A-induced liver injury triggers hepatocyte proliferation, *Journal of Clinical Investigation*, **101**, 1960–1969

TSUJIMOTO, M., YIP, Y.K. and VILCEK, J., 1986, Interferon-γ enhances expression of cellular receptors for tumor necrosis factor, *Journal of Immunology*, **136**, 2441–2444

VAN ANTWERP, D.J., MARTIN, S.J., KAFRI, T., GREEN, D.R. and VERMA, I.M., 1996, Suppression of TNF-α-induced apoptosis by NF-κB, *Science*, **274**, 787–789

VERCAMMEN, D., BEYAERT, R., DENECKER, G., GOOSSENS, V., VAN LOO, G., DECLERCQ, W., GROOTEN, J., FIERS, W. and VANDENABEELE, P., 1998, Inhibition of caspases increases the sensitivity of L929 cells to necrosis mediated by tumor necrosis factor, *Journal of Experimental Medicine*, **187**, 1477–1485

VILLA, P., KAUFMANN, S.H. and EARNSHAW, W.C., 1997, Caspases and caspase inhibitors, *Trends in Biological Sciences*, **22**, 388–393

WAHL, C., MIETHKE, T., HEEG, K. and WAGNER, K., 1993, Clonal deletion as direct consequence of an *in vivo* T cell response to bacterial superantigen, *European Journal of Immunology*, **23**, 1197–1200

WANG, C.-Y., MAYO, M.W. and BALDWIN JR, A.S., 1996, TNF- and cancer therapy-induced apoptosis: potentiation by inhibition of NF-κB, *Science*, **274**, 784–787

WATANABE, Y., MORITA, M. and AKAIKE, T., 1996, Concanavalin A induces perforin-mediated but not Fas-mediated hepatic injury, *Hepatology*, **24**, 702–710

WONG, G.H.W., ELWELL, J.H., OBERLEY, L.W. and GOEDDEL, D.V., 1989, Manganous Superoxide dismutase is essential for cellular resistance to cytotoxicity of tumor necrosis factor, *Cell*, **58**, 923–932

XIE, Q.W., KASHIWABARA, Y. and NATHAN, C., 1994, Role of transcription factor NF-κB/Rel in induction of nitric oxide synthase, *Journal of Biological Chemistry*, **269**, 4705–4708

YAMADA, Y. and FAUSTO, N., 1998, Deficient liver regeneration after carbon tetrachloride injury in mice lacking type 1, but not type 2 tumor necrosis factor, *American Journal of Pathology*, **152**, 1577–1589

YAMADA, Y., KIRILLOVA, I., PESCHON, P.J. and FAUSTO, N., 1997, Initiation of live growth by tumor necrosis factor: deficient liver regeneration in mice lacking type I tumor necrosis factor receptor, *Proceedings of the National Academy of Sciences of the United States of America*, **94**, 1441–1446

YAO, J., MACKMAN, N., EDGINGTON, T.S. and FAN, S.T., 1997, Lipopolysaccharide induction of the tumor necrosis factor-α promoter in human monocytic cells. Regulation by Egr-1, c-Jun, and NF-κB transcription factors, *Journal of Biological Chemistry*, **272**, 17795–17801

YEH, W.-C., SHAHINIAN, A., SPEISER, D., KRAUNUS, J., BILLIA, F., WAKEHAM, A., *et al.*, 1997, Early lethality, functional NF-κB activation, and increased sensitivity to TNF-induced cell death in TRAF2-deficient mice, *Immunity*, **7**, 715–725

ZAMZAMI, N., MARCHETTI, P., CASTEDO, M., DECAUDIN, D., MACHO, A., HIRSCH, T., *et al.*, 1995, Sequential reduction of mitochondrial transmembrane potential and generation of reactive oxygen species in early programmed cell death, *Journal of Experimental Medicine*, **182**, 367–377

ZHOU, A., PARANJAPE, J., BROWN, T.L., NIE, H., NAIK, S., DONG, B., *et al.*, 1997, Interferon action and apoptosis are defective in mice devoid of 2′,5′-oligoadenylate-dependent RNase L, *EMBO Journal*, **16**, 6355–6363

ZUCKERMAN, S.H. and EVANS, G.F., 1992, Endotoxin tolerance: in vivo regulation of tumor necrosis factor and interleukin-1 synthesis at the transcriptional level, *Cellular Immunology*, **140**, 513–519

# Apoptosis in Male Reproductive Toxicology

**IAN WOOLVERIDGE[1] AND IAN D. MORRIS[2]**

[1] Virology, Roche Discovery, Welwyn, Herts, AL7 3AY, UK
[2] School of Biological Sciences, University of Manchester, Manchester, M13 9PT, UK

## Contents

## 4.1 INTRODUCTION

The control of spermatogenesis is extremely complex requiring testosterone from the Leydig cells situated within the interstitium and the supportive role of the Sertoli cells within the seminiferous tubules (reviewed in Sharpe, 1994). During spermatogenesis, spermatogonia divide by mitosis and meiosis producing haploid spermatids which then transform into spermatozoa (Figure 4.1). The synchronous development of a clone of spermatogonia should give rise to 4096 spermatids. However the actual numbers fall far short of this, since the average clone contains only about 100 cells (Russel *et al.*, 1990); cell division and cell death are both implicated in controlling the final number of spermatozoa. Industrialization and the use of chemicals for environmental and therapeutic reasons has exposed the male reproductive system to a wide variety of compounds that might pose a threat to male fertility . There is much debate as to whether sperm counts in men have declined in the last 50 years (Carlsen *et al.*, 1992, Bromwich *et al.*, 1994, Ashby *et al.*, 1997, Seino *et al.*, 1997). It is clear that there are a number of sites within the testis and a number of targets in spermatogenesis where toxicants might act (Seino *et al.*, 1997). Many environmental and 'therapeutic' toxicants have been shown to disrupt spermatogenesis

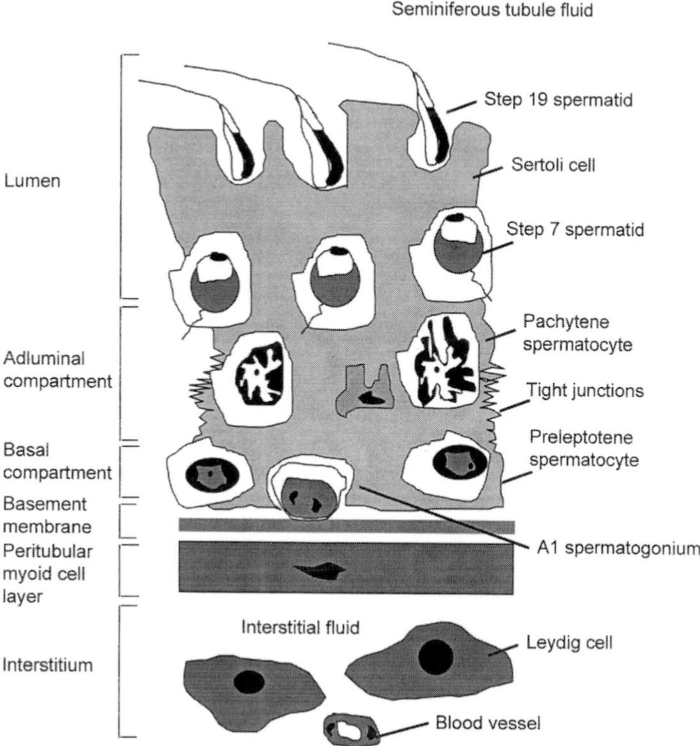


**Figure 4.1** Diagram of a Sertoli cell showing its association with germ cells in the rat testes.

in rodent models. Increasingly, many of them are being shown to induce apoptosis in the testis although in most cases, the molecular mechanisms are not yet defined. This chapter will review current knowledge on the physiological role and genetic control of apoptosis in the testis, then consider toxicants that interfere with this physiology to modulate apoptosis.

## 4.2 ROLE OF SERTOLI AND LEYDIG CELLS

The Sertoli cell is a somatic cell within the seminiferous epithelium of the testis (Figure 4.1). This cell does not divide and, in general, is not susceptible to toxins although some industrial chemicals are known to pose a threat. The Sertoli cell provides the necessary biomolecules for the growth and development of germ cells and structural support for the germ cells so that they can maintain their cellular associations. In addition, tight junctions between the Sertoli cells serve to exclude potentially harmful substances such as toxins and antibodies thus rendering the seminiferous epithelium 'an immune privileged site'. The integrity of this cell is essential for successful spermatogenesis. Reports of Sertoli cell apoptosis are few and far between although Allan *et al.* (1988) claim that 5 Gy irradiation induces death of Sertoli cells by apoptosis in 4-day-old rats. However, when Sertoli cells are cultured, a basement membrane of laminin or matrigel is required for Sertoli cell survival. Sertoli cells cultured on plastic die by apoptosis even in the presence of known regulators of Sertoli cell function (Dirami *et al.*, 1995).

One function of Sertoli cells is to engulf and degrade dead germ cells. This phagocytosis is impaired when liposomes containing acidic but not neutral phospholipids are present *in vitro* (Shiratsuchi *et al.*, 1997). This observation along with the fact that spermatogenic cells expose phosphatidylserine on their surface when maintained in culture would suggest that these acidic phospholipids translocate from the inner to the outer leaflet of the plasma membrane during apoptosis and serve as a signal for phagocytosis by Sertoli cells. Cyclic protein-2 (CP-2) or cathepsin is expressed in the Sertoli cells and is thought to facilitate the degradation of dead germ cells once they are engulfed (Kim and Wright, 1997). The expression of CP-2 increases during testicular maturation and levels are high in all tubules that lose germ cells by apoptosis.

The Leydig cell synthesizes and secretes testosterone thus maintaining the somatic and testicular aspects of male fertility. Like Sertoli cells, somatic Leydig cells do not divide and apoptosis of Leydig cells has not been reported in the normal adult rat testis (Taylor *et al.*, 1998). Toxicants usually exert their effects on Leydig cells by interfering with steroidogenic enzymes and the action of such toxicants is often characterized by their ability to reduce peripheral androgen levels. In addition, toxicants may have indirect effects on Leydig cells perhaps by affecting the neuroendocrine axis at the hypothalamus or pituitary to affect luteinizing hormone (LH) levels. Alternatively, Leydig cells may be killed via an indirect action on Sertoli or germ cells since these three cell types depend on paracrine interactions for survival signals (Morris, 1996).

Some toxicants have been described that act directly on Leydig cells, inducing apoptosis and reducing serum testosterone levels (see section 4.4). Ethane dimethanesulphonate (EDS) has a unique cytotoxic action on the Leydig cells of the rat testis and consequently, has been used to investigate the physiological role of the Leydig cell. Early reports hinted that interstitial cells of the rat testis may undergo apoptosis in response to hypophysectomy or EDS (Morris *et al.*, 1986, Tapanainen *et al.*, 1993, Henriksen *et al.*, 1995) but the findings were far from conclusive. However, work from our laboratory (Morris *et al.*, 1997, Taylor *et al.*, 1998) has shown that Leydig cells will undergo apoptosis in response to EDS treatment both *in vitro* and *in vivo* and the effect is dose and time dependent. It remains to be determined if this apoptotic cell death of Leydig cells in response to EDS is purely toxicological or if it has some physiological significance.

## 4.3 APOPTOSIS IN THE NORMAL TESTES

The numbers of apoptotic cells in rodent testes are low but groups of spermatogonia linked by intercellular bridges undergo apoptosis synchronously (Allan *et al.*, 1992, reviewed in Sharpe, 1994). Indeed, the first clear description was provided by Huckins, 1978, although at the time the phenomenon of apoptosis had not been recognized. Primary and secondary spermatocytes as well as spermatids occasionally undergo apoptosis (Russel and Clermont, 1977, Kerr, 1992, Brinkworth *et al.*, 1995, Blanco-Rodriguez and Martinez-Garcia, 1996a). However, A1, Intermediate and type-B spermatogonia rarely degenerate (Huckins, 1978).

Removal of LH and follicle stimulating hormone (FSH) by hypophysectomy increases the number of degenerating pachytene spermatocytes (Russel and Clermont, 1977) via apoptosis (Tapanainen *et al.*, 1993, reviewed in Sharpe, 1994). This can be attenuated by addition of exogenous LH and FSH alone or in combination (Russel and Clermont, 1977, Tapanainen *et al.*, 1993). The observation that testosterone also prevents hypophysectomy-induced cell death (Tapanainen *et al.*, 1993) suggests a critical role for LH-stimulated testosterone secretion for the maintenance of germ cell viability.

The exact role of FSH in spermatogenesis is still uncertain (reviewed in Sharpe, 1994). FSH can prevent programmed cell death of rat pachytene spermatocytes and spermatids *in vitro* (Henriksen *et al.*, 1996) whilst at the same time stimulating proliferation of spermatogonia and preleptotene spermatocytes. *In vivo*, specific immunoneutralization of FSH induces apoptosis of spermatogonia and pachytene spermatocytes as early as 24 h after administration of the FSH antiserum (Shetty *et al.*, 1996).

In human testes, apoptosis occurs spontaneously as shown in tissues obtained after orchidectomy for prostate cancer (Brinkworth *et al.*, 1997, Woolveridge *et al.*, 1998a) or at autopsy following sudden traumatic death (Sinha Hikim *et al.*, 1998). Apoptosis occurs in a wide variety of germ cell types but not in Sertoli or Leydig cells and there is no significant relationship between the amount of apoptosis and either the age of men or testis weight

(Brinkworth *et al.*, 1997). There are ethnic differences, with a higher incidence of spermatocyte apoptosis in Chinese men compared to Caucasian men, which may explain their greater susceptibility to testosterone-induced spermatogenic suppression for contraceptive purposes (Sinha Hikim *et al.*, 1998).

Men with azoospermia or severe oligozoospermia have an increased frequency of apoptotic germ cells in their testicular biopsies in comparison to men with normal spermatogenesis, suggesting that programmed cell death may play a role in human male infertility (Lin *et al.*, 1997a, b). In addition, there is an increased frequency in the number of spermatozoa with apoptotic DNA in the ejaculates of infertile men in comparison to fertile controls (Gorczyca *et al.*, 1993, Baccetti *et al.*, 1996).

## 4.3.1 Genetic control of apoptosis in the normal testes

### 4.3.1.1 Bcl-2 family members

The role of Bcl-2 family members in the molecular regulation of apoptosis has been described in Chapter 1. In the rat testes, the two *bcl-2* transcripts present are highly expressed (Tilly *et al.*, 1995) but immature rat testis expresses the 7.5 kb species only. Bcl-2 protein is abundant in both the immature and adult rat testis (Woolveridge *et al.*, 1998b). Some groups have shown Bcl-2 immunoreactivity in human and mouse testes (Krajewski *et al.*, 1995, Woolveridge *et al.*, 1998a) whilst others have not (Knudson *et al.*, 1995, Furuchi *et al.*, 1996, Rodriguez *et al.*, 1997). In the former case, immunohistochemistry has localized Bcl-2 to the spermatids and mature sperm in the seminiferous tubules of the testis (Krajewski *et al.*, 1995). Bcl-2 knockout mice do not have any gonadal abnormalities although the grossly disturbed physiology, which includes growth retardation, lymphoid apoptosis and polycystic kidneys usually leading to death, makes it difficult to obtain meaningful data (Nakayama *et al.*, 1993, 1994, Kamada *et al.*, 1995).

*Bax* is expressed as two transcripts in rat tissues, a predominant 1.0 kb form and a less abundant 1.5 kb species. The levels of these transcripts (Tilly *et al.*, 1995) and gene products (Woolveridge *et al.*, 1998b) are highest in reproductive tissues. Bax is present in the mouse and human testis (Knudson *et al.*, 1995, Woolveridge *et al.*, 1998a) although some studies show it is hardly detectable in the mouse (Rodriguez *et al.*, 1997). Immunostaining for Bax is seen only in the germinal cells located near the basement membrane (Krajewski *et al.*, 1994b). Bax-knockout males are infertile as a result of accumulation of atypical premeiotic germ cells, but no mature haploid sperm (Knudson *et al.*, 1995). *In situ* end labelling of apoptotic cells identified a large increase in the number of apoptotic cells in the Bax-deficient mice in comparison with control testes.

Two forms of *Bcl-x* transcript are expressed in the rat testis, a 2.7 kb species and a less abundant 3.5 kb form. A doublet of proteins with an apparent molecular weight of 29–31 kDa is present in the testis (Krajewski *et al.*, 1994a) which represents Bcl-XL and an additional 20 kDa band corresponding to Bcl-XS, is also present. Bcl-X is present in the spermatocytes and spermatids of the human and mouse testis (Krajewski *et al.*, 1994a, Knudson *et al.*, 1995, Woolveridge *et al.*, 1998a) although, again, some studies have shown that it is hardly detectable in the mouse (Rodriguez *et al.*, 1997).

*Bcl-w*, a gene which promotes cell survival (Gibson *et al.*, 1996) is expressed in the elongating spermatids and Sertoli cells of the mouse testis (Ross *et al.*, 1998). Bcl-w knock-out mice are sterile which is a result of progressive testicular degeneration (Ross *et al.*, 1998). The lack of Bcl-w results in a phenotype in which spermatocytes and spermatids are lost by apoptosis resulting in a Sertoli-cell only spermatogenic epithelium. Eventually there is almost complete loss of Sertoli cells and Leydig cell number is reduced although it is not clear if this is an indirect response to the lack of germ cells or a direct effect of the absence of Bcl-w.

Both the immature and adult rat testes express Bak and Bad (Woolveridge *et al.*, 1999). Bad is present in mice testes although Bak is not (Rodriguez *et al.*, 1997). In the human testis, Bak is associated with the Sertoli and Leydig cells (Krajewski *et al.*, 1996) whilst Mcl-1 is found only in the Leydig cells (Krajewski *et al.*, 1995). A new pro-apoptotic member of the Bcl-2 family, Bcl-2-related ovarian killer (Bok) is highly expressed in most reproductive tissues including the testis and could play an important role in spermatogenesis (Hsu *et al.*, 1997).

### 4.3.1.2 p53

The role of p53 in apoptosis has been described in detail in Chapter 1. In the testes, expression is mainly confined to the pachytene spermatocytes in the seminiferous tubules of normal rodents (Almon *et al.*, 1993, Rotter *et al.*, 1993, Schwartz *et al.*, 1994, Sjoblom and Lahdetie, 1996) with some expression in the spermatogonia and spermatids (Stephan *et al.*, 1996). p53 may play a role during meiosis in DNA repair and recombination events because the spermatogenic epithelium in p53 null mice contains multinucleated giant cells which are thought to be the result of the inability of the tetraploid pachytene spermatocytes to complete their meiotic division (Rotter *et al.*, 1993). However, the apoptotic elimination of spermatocytes with synaptic errors has recently been shown to be p53-independent (Odorisio *et al.*, 1998).

### 4.3.1.3 Fas

The testes express Fas ligand (Fas-L) at high levels (Griffith *et al.*, 1995, Suda *et al.*, 1995). Most reports have localized Fas-L to the Sertoli cells (Bellgrau *et al.*, 1995, French *et al.*, 1996, Lee *et al.*, 1997) although some have reported expression in the germ cells (Sugihara *et al.*, 1997). In contrast, some studies have shown that Fas-L expression is virtually absent in both the Sertoli cells and germ cells (Li *et al.*, 1997). We have shown that although Fas-L is expressed in the Sertoli cells, it is predominantly localized in the pachytene spermatocytes and spermatids (Woolveridge *et al.*, 1999). It is thought that the immune-privileged sanctuary within the testes is conferred by Fas-L expression (Bellgrau *et al.*, 1995). It is generally accepted with notable exceptions (Watanabe-Fukunaga *et al.*, 1992, Leithauser *et al.*, 1993) that Fas receptor (Fas-R) is expressed in the seminiferous tubules of the testis (Lee *et al.*, 1997, Li *et al.*, 1997, Sugihara *et al.*, 1997, Ogi *et al.*, 1998), in particular in spermatocytes (Lee *et al.*, 1997) and spermatids (Li *et al.*, 1997) with some reports also showing localization in the Sertoli cells (Sugihara *et al.*, 1997).

## 4.4 TOXICANT-INDUCED TESTICULAR APOPTOSIS

Many diverse toxicants cause apoptosis in the testes. Some may induce germ cell apoptosis directly whereas others act on the Leydig or Sertoli cells to cause androgen ablation and remove trophic support. Examples of such toxicants and a discussion of their site and mode of action is presented below and summarized in Table 4.1.

### 4.4.1 EDS

EDS is a unique testicular toxin with cytotoxic action confined almost exclusively to the Leydig cells (Morris *et al.*, 1986, Morris, 1996) (Table 4.1). EDS selectively eliminates both basal and LH-stimulated testosterone production. As with androgen withdrawal by other pharmacological/physiological manipulation, the predominant germ cell types undergoing apoptosis as a result of androgen withdrawal induced by EDS include pachytene spermatocytes and spermatids in stage VII of spermatogenic cycle (Henriksen *et al.*, 1995). This effect is suppressed by co-administration of testosterone.

During apoptosis of germ cells following androgen withdrawal there are no statistically significant changes in the levels of Bcl-XL, Bak or Bad in the rat testis although Bcl-XL levels do appear to decline (Woolveridge *et al.*, 1999). The decline in Bcl-XL levels might be explained on the basis that it is a survival gene product which might be expected to fall during apoptosis. However, the decline could be a reflection of the loss of Bcl-XL expressing spermatocytes and spermatids (Krajewski *et al.*, 1994a). The expression of Bax and Bcl-2 in the rat testes is up-regulated following withdrawal of androgen (Woolveridge *et al.*, 1999). A rise in Bcl-2 protein levels may reflect an increase in Bcl-2 production by the remaining mature germ cells perhaps as part of a survival mechanism to ensure that the atrophied tissue can respond to restimulation by androgen. At the same time, the pro-apoptotic gene Bax is increased to induce apoptosis. Selective enrichment of Bax-expressing cell types is an alternative explanation.

Despite the marked fall in androgen support to the spermatogenic epithelium following short-term anti-androgen treatment of human subjects, there is no significant change in the expression of the Bcl-2 family members. Similar to the rat, in the long-term androgen depleted human testes, Bcl-XL protein levels decline and Bcl-2 levels are elevated (Woolveridge *et al.*, 1998a). Withdrawal of androgen support to both the rat (Woolveridge *et al.*, 1999) and human (Woolveridge *et al.*, 1998a) testes has no effect on the expression of p53.

The expression of Fas-L declines dramatically 2 days following androgen withdrawal induced by EDS treatment and is barely detectable by 8 days (Woolveridge *et al.*, 1999). The initial drop in Fas-L expression is probably due to loss of Leydig cells within the testis because it is known that these cells also abundantly express Fas-L (Li *et al.*, 1997, Sugihara *et al.*, 1997, Taylor *et al.*, 1999). However, the significant further fall in Fas-L expression between 2 days and 8 days of EDS treatment must be due to changes in the seminiferous epithelium because Leydig cells are absent until at least 14 days after EDS treatment (Jackson *et al.*, 1986). This is confirmed by immunohistochemical data which shows that the number of germ cells expressing Fas-L and the intensity of Fas-L staining in the remaining germ cells is reduced at the later stages of androgen withdrawal (Woolveridge *et al.*, 1999). The expression of Fas-R also declines at 8 days following EDS administration. As with Bcl-XL, the decline in Fas-L and Fas-R expression following androgen ablation might be the result of the loss of the cell types expressing these genes. However, as the co-localization of Fas-L and Fas-R correlates with the germ cell types that die following androgen withdrawal, the potential does exist for apoptosis in the rat spermatogenic epithelium to be regulated by the Fas pathway. Surprisingly, the up-regulation in these gene products at a time before the majority of apoptosis occurs, which has been observed in response to other testicular toxicants (Lee *et al.*, 1997), was not detected following androgen ablation.

Clusterin levels are not directly influenced by androgen levels (Frasoldati *et al.*, 1995, Woolveridge *et al.*, 1999). Clusterin mRNA levels are transiently elevated in the testis 6 h after EDS administration (Frasoldati *et al.*, 1999) perhaps as a secondary event to a direct effect of the alkylating agent on Sertoli cells. However, clusterin expression returns to basal levels and is unchanged until 21 days after EDS administration.

After hypophysectomy, the levels of the caspase substrate, poly (ADP ribose) polymerase (PARP) in the rat testis do not change despite loss of spermatocytes (Alcivar-Warren *et al.*, 1996). Also, short-term anti-androgen treatment in humans does not effect PARP protein levels in the testis (Woolveridge *et al.*, 1998a). In the chronically androgen depleted human testes PARP protein levels decline. This could be explained by the role of PARP in DNA repair so that levels are expected to be suppressed during apoptosis. However, as the cell types expressing PARP are lost in this essentially germ cell depleted testis, the levels in each individual cell may have remained unchanged.

Although the initial events by which EDS triggers the genetic programme of Leydig cell death are unknown at present, it could be suggested that the covalent binding of EDS to glutathione and its cellular depletion is important as this triggers apoptosis in response to cytotoxic treatments in other cells (Beaver and Waring, 1995, Sugimoto *et al.*, 1996). However, *in vitro* studies with Leydig cells suggest that the depletion of cellular glutathione actually protects the cells from EDS-induced death (Kelce and Zirkin, 1993, Kelce, 1994), which is consistent with the findings of one study showing that glutathione levels might not be important for Fas-mediated cell death (Marchetti *et al.*, 1997). Possibly EDS acts through inhibition of protein synthesis (Morris, 1996) as a result of the formation of an active complex with glutathione. Evidence for this comes from observations that cycloheximide, another protein synthesis inhibitor, can also induce an extremely rapid elevation of the protein components of the Fas system (Kimura and Yamamoto, 1997).

### 4.4.2 Estradiol

Estradiol is a natural hormone that, in excess, can act as a toxicant by causing endocrine disruption. In the testes, estradiol mimics some of the effects of hypophysectomy through the suppression of gonadotrophin production. As expected, the predominant cell types which undergo apoptosis in response to estradiol treatment include elongating and round spermatids and to a lesser extent, pachytene spermatocytes (Blanco-Rodriguez and Martinez-Garcia, 1996b, 1997). Cell death is suppressed to varying degrees by testosterone or a combination of FSH and LH (Blanco-Rodriguez and Martinez-Garcia, 1997). This apoptosis occurs before the germ cells detach themselves from the seminiferous epithelium which would suggest that germ cell detachment is not necessary for apoptosis to occur (Blanco-Rodriguez and Martinez-Garcia, 1998). Diethylstilbesterol (DES), an estrogen receptor agonist, induces regression of the testis in male Syrian hamsters. This is associated with an increase in apoptosis of spermatocytes and to a much lesser extent, spermatogonia (Nonclercq *et al.*, 1996).



### 4.4.3 PCBs

Polychlorinated biphenyls (PCBs) are a group of industrial chemicals with a myriad of uses. They are environmentally stable and bioaccumulate in marine, avian and mammalian species particularly in steroid producing organs such as the testes (Brandt, 1977) where they have detrimental effects upon male fertility (Bush *et al.*, 1986) (Table 4.1). It has been proposed that one of the anti-fertility mechanisms is to decrease testicular interstitial cell viability (Johansson, 1989, Kovacevic *et al.*, 1995). PCB169 acts at the aryl hydrocarbon receptor (Poland and Glover, 1977) and can decrease serum testosterone levels by more than 98% (Yeowell *et al.*, 1989). Preliminary data from our laboratory would suggest that *in vitro*, PCB169 can reduce hCG-stimulated testosterone production in Leydig cells at high concentrations. In addition, PCB169 will induce apoptosis of Leydig cells at higher doses. As the dynamics of cell death are different to EDS, the data would suggest that the mechanisms by which PCBs induce apoptosis are different from the alkylating action of EDS and may be related to their activity as aryl hydrocarbon receptor agonists (M. Taylor *et al.*, unpublished observations).

### 4.4.4 Methoxy acetic acid

One member of a family of ethylene glycol ethers whose use as organic solvents is widespread is 2-methoxy ethanol (2-ME) (Table 4.1). This includes incorporation into both domestic and industrial products such as textile dyes, printing inks and paints. There is concern for workers exposed to such chemicals, the majority of whom are male, because these compounds have the potential to produce toxicity in animals in organs that have active cell proliferation such as the bone marrow, thymus and testis (Balasubramanian *et al.*, 1994, Simialowicz *et al.*, 1994, Ku *et al.*, 1995). The active metabolite of 2-ME, 2-methoxy acetic acid (2-MAA), induces lesions in the testes of rats and guinea pigs which are a result of death of pachytene spermatocytes (Anderson *et al.*, 1987) by apoptosis (Brinkworth *et al.*, 1995, Ku *et al.*, 1995), although pachytene spermatocyte degeneration in rats and guinea pigs differs in onset, severity and morphological characteristics (Ku *et al.*, 1995). Apoptosis of rat spermatocytes is very extensive following 2-MAA treatment *in vivo* and dying cells often form rings around the edges of the seminiferous tubule (Clark *et al.*, 1997). The expression of clusterin in the rat testis is elevated in response to 2-MAA induced apoptosis and accumulates in the cytoplasm of dying pachytene spermatocytes as early as 6 h post-2-MAA administration (Clark *et al.*, 1997).

A divalent metal cation-dependent endonuclease activity is present in 2-MAA treated rat testes and increased intracellular calcium might be a stimulus for 2-MAA-induced spermatocyte apoptosis (Ku *et al.*, 1995). An 18 kDa endonuclease was isolated from the pachytene spermatocytes of 2-MAA treated rats (Wine *et al.*, 1997) which is able to cleave DNA into multiples of 200 base pairs. Furthermore, dihydropyridines such as nifedipine or verapamil are able to attenuate 2-MAA-induced apoptosis of pachytene spermatocytes *in vitro* (Li *et al.*, 1996, 1997). All of these compounds inhibit calcium movement through plasma membranes. Drugs which inhibit calcium movement from intracellular stores or block the effect of inositol tri-phosphate have no effect on 2-MAA-induced apoptosis (Li *et al.*, 1997). However, immunohistochemical staining for the dihydropyridine receptor in rat testicular sections proved to be negative and there is no evidence for an increase in intracellular free calcium concentration in germ cells following 2-MAA treatment (Li *et al.*, 1997). The protective effect of calcium channel blockers against the 2-MAA-induced spermatocyte apoptosis is therefore probably not due to blockade of DHP-sensitive calcium channels.

### 4.4.5 2,5-Hexanedione

A metabolite of the commonly used solvents n-hexane and 2-hexanone, 2,5-hexanedione (2,5-HD) is a Sertoli cell toxicant (Table 4.1) which indirectly causes irreversible and widespread germ cell loss and testicular atrophy in rats. The irreversible nature of 2,5-HD toxicity is probably due to direct action that reduced the number of stem cells; those that are remaining appear apoptotic (Allard and Boekelheide, 1996). After administration of 1% 2,5-HD in drinking water over 5 weeks, there is a marked increase in germ cell apoptosis (Blanchard *et al.*, 1996). There is also a differential sensitivity of germ cells to toxicant exposure with apoptotic spermatids appearing first followed by apoptotic spermatogonia and spermatocytes.

### 4.4.6 1,3-Dinitrobenzene/nitrobenzene

The volatile nitroaromatics have a variety of industrial uses and are therefore an occupational hazard to humans. Another Sertoli cell toxicant (Table 4.1), 1,3-dinitrobenzene (DNB), induces apoptosis in the rat testis with a very fast onset following administration *in vivo* (Strandgaard and Miller, 1998) reaching a maximum at 24 to 48 h. The predominant cell types affected are the pachytene spermatocytes. Nitrobenzene affects the same cell types over a similar time frame (Shinoda *et al.*, 1998).

### 4.4.7 Mono-(2-ethylhexyl) phthalate

Phthalate esters are plasticizers present in food packaging and therefore are ideally placed for ingestion by humans. Mono-(2-ethylhexyl) phthalate (MEHP) is a Sertoli cell toxicant (Table 4.1) which also induces testicular germ cell apoptosis *in vivo* in young rats (Richburg and Boekelheide, 1996). Three hours following MEHP administration, Sertoli cell vimentin filaments collapse although tubulin and actin are unaffected. The major cell types undergoing apoptosis following MEHP treatment are the spermatocytes (Lee *et al.*, 1997). Fas is implicated in mediating the effects of MEHP (Lee *et al.*, 1997). Maximal induction of Fas-L and Fas-R expression is after 6 h or 12 h, respectively and the localization pattern of Sertoli cell Fas-L was changed 12 h after MEHP-exposure to a diffuse cytoplasmic staining pattern, particularly at sites of Sertoli cell-germ attachment (Lee *et al.*, 1997). These observations suggest that under normal physiological conditions the Sertoli cell mediates the death of Fas-R bearing germ cells by expressing Fas-L on its cell membrane (Figure 4.2). The initial disruption of the Sertoli cell-germ cell physical contact may result in the uncoupling of the Fas-mediated signal transduction process between these cells leading to a decrease in germ cell apoptosis but at later stages the expression of Fas-L increases to eliminate damaged germ cells. This may be mediated by the generation of a soluble form of Fas-L from Sertoli cells (Figure 4.2C).

### 4.4.8 Cadmium

Exposure to cadmium in dusts and fumes is an occupational hazard for some people. In rodents, cadmium chloride can induce apoptosis in testes *in vivo* at a dose level of 0.03 mmol/Kg and over a time period of 48 hours (Xu *et al.*, 1996, Yan *et al.*, 1997)

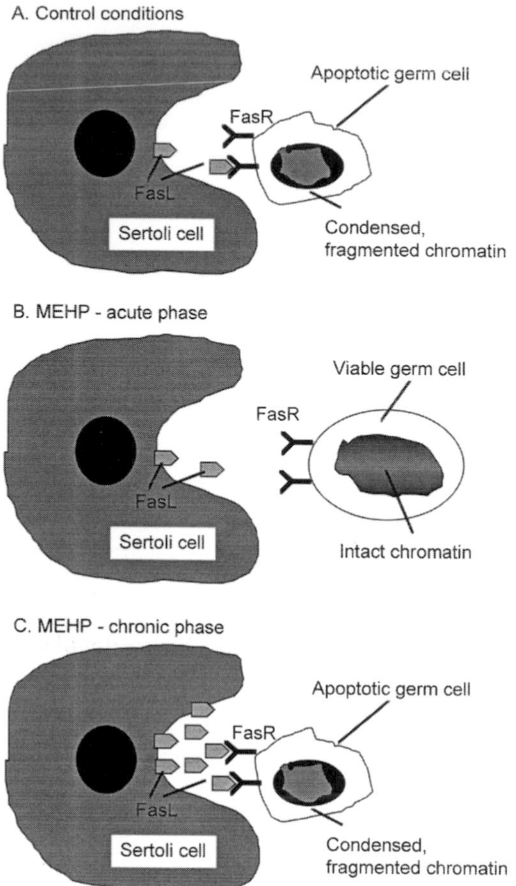


**Figure 4.2** Schematic diagram of the interaction between a Sertoli cell and a germ cell showing induction of apoptosis by MEHP involving release of Fas-L.

(Table 4.1). Cadmium chelating agents can suppress most, if not all, of this apoptosis depending on the time between administration of the cadmium and administration of the chelating agent. Selenium can offset cadmium-induced toxicity in many tissues including the testes although the exact mechanism is unclear but like chelating agents, selenium suppresses cadmium-induced testicular apoptosis if administered prior to or after cadmium (Jones *et al.*, 1997).

## 4.4.9 Phosphamidon

Phosphamidon is one member of the organophosphate family of insecticides which is used widely in agriculture. There is almost a complete absence of differentiating spermatids and spermatozoa in the testes of male rats fed with drinking water containing

**TABLE 4.1**

Toxicants and hormones that induce testicular apoptosis

| Toxicant | Site of action | Mode of action | Reference |
|---|---|---|---|
| Estradiol | • estrogen receptor (ER) | • androgen ablation; gonadotrophin ↓<br>• prevented by testosterone or FSH/LH | Blanco-Rodriguez and Martinez-Garcia, 1997, 1998<br>Nonclercq *et al.*, 1996 |
| Diethyl stibestrol (DES) | • estrogen receptor (ER) | • androgen ablation; gonadotrophin ↓ | Nonclercq *et al.*, 1996 |
| Ethane dimethanesulfonate (EDS) | • Leydig cell toxicant | • Testosterone ↓<br>• loss of trophic support ⇒ germ cell apoptosis | Morris *et al.*, 1986, 1997<br>Morris, 1996<br>Taylor *et al.*, 1998 |
| 2-Methoxy ethanol (2-ME) | • $Ca^{2+}$ regulation? | • extensive apoptosis of spermatocytes<br>• clusterin expression ↑<br>• $Ca^{2+}$↑; ⇒ activation of endonuclease | Ku *et al.*, 1995, Clark *et al.*, 1997, Li *et al.*, 1996, 1997, Wine *et al.*, 1997 |
| 2,5-Hexanedione (2,5-HD) | • Sertoli cell toxicant | • causes widespread germ cell loss | Allard and Boekelheide, 1996, Blanchard *et al.*, 1996 |
| 1,3-Dinitrobenzene (DNB) | • Sertoli cell toxicant | • causes widespread germ cell loss | Strandgaard and Miller, 1998, Shinoda *et al.*, 1998 |
| Mono-(2-ethylhexyl) phthalate (MEHP) | • Sertoli cell toxicant | • causes widespread germ cell loss<br>• FasL implicated in toxicity | Richburg and Boekelheide, 1996, Lee *et al.*, 1997 (see Figure 4.2) |
| Cadmium | | • causes germ cell apoptosis<br>• Offset by cadmium chelating agents or selenium | Xu *et al.*, 1996, Yan *et al.*, 1997, Jones *et al.*, 1977 |
| Phosphamidon | | • ablation of differentiating spermatids | Akbarsha and Sivasamy, 1997 |
| Polychlorinated biphenyls (PCBs) | • aryl hydrocarbon receptor | • testosterone ↓<br>• reduced viability of interstitial cells<br>• loss of trophic support? | Brandt, 1977, Bush *et al.*, 1986, Johansson, 1989, Kovacevic *et al.*, 1995, Poland and Glover, 1977, Yeowell *et al.*, 1989 |
| Gossypol | • inhibits protein kinase C activity | • proliferation ↓<br>• cell death ↑<br>• potential male contraceptive | Lui *et al.*, 1985, Teng, 1995 |

phosphamidon at 35 ppm for 30 days (Akbarsha and Sivasamy, 1997) and numerous uni- and multi-nucleate giant cells with typical apoptotic morphology.

## 4.5 TESTICULAR APOPTOSIS AND DRUGS

Apoptotic death in the testes is caused by two classes of prescription drugs, the male contraceptive and the DNA-damaging chemotherapeutic agents (Table 4.1).

### 4.5.1 Gossypol

Gossypol acetic acid is a drug that has undergone trials as a possible male contraceptive (Lui *et al.*, 1985). It inhibits spermatogenesis in mammalian species by preventing the proliferation of germ cells. Gossypol causes malformations in spermatozoa which include alterations in the chemical composition of chromatin, DNA strand breakage, changes in the shape of the sperm head and a reduction in sperm motility which all result in the death of sperm cells. Gossypol inhibits protein kinase C activity and induces apoptosis in cultured rat spermatocytes which is prevented by activators of protein kinase C (Teng, 1995), which would suggest that maintenance of protein kinase C activity is needed to protect spermatocytes from apoptosis.

### 4.5.2 Chemotherapeutic drugs

The majority of chemotherapeutic agents kill cells in proliferating tissues by damaging DNA. Thus, the testes are very vulnerable to toxicity from this class of agents. Vincristine and vinblastine are microtubule disrupting agents that play an essential role in combination cancer chemotherapy. However, they disrupt spermatogenesis resulting in the formation of multinucleated giant cells. Vincristine affects spermatogenesis in a dose and time-dependent manner at mitotic and meiotic stages resulting in the cells arrested in division dying by apoptosis (Akbarsha *et al.*, 1996).

Mitomycin C (Kobayashi, 1960) alkylates DNA and affects DNA synthesis (Iyer and Szybalski, 1963) to cause chromosome aberrations and dominant-lethal mutations in mouse spermatogonia and spermatocytes. At higher doses, these cells die *in situ* by apoptosis (Nakagawa *et al.*, 1997). Apoptotic elimination of testicular germ cells as a result of anti-cancer therapies may serve to protect subsequent generations from the genotoxic effects of these drugs. Cyclophosphamide, is another cytotoxic DNA and protein alkylating agent that is used in the treatment of cancer but also causes infertility. Cyclophosphamide induces apoptosis in germ cell cells in the seminiferous epithelium of the rat testis in a time-dependent manner reaching a peak at 12 h after a 70 mg/kg dose and declining thereafter, reaching control levels at 48 h (Cai *et al.*, 1997). Lower doses of cyclophosphamide do not induce significant levels of apoptosis but produce transgenerational mutations. The cell types affected include spermatogonia and spermatocytes in all stages of the spermatogenic cycle.

The topoisomerase II inhibitors, etoposide and adriamycin, cause DNA damage by inducing strand breaks. They are well known anti-cancer agents that cause the elimination by apoptosis of spermatogonia, zygotene and early pachytene spermatocytes and meiotically dividing spermatocytes (Sjoblom *et al.*, 1998). Etoposide and adriamycin have no effect on p53 or p21 (WAF1) levels (Sjoblom *et al.*, 1998).

## 4.6 SUMMARY

Clearly a large number of industrial and therapeutic compounds as well as environmental toxicants can disrupt spermatogenesis and induce programmed cell death in the germ and somatic cells of the testis. However, the mechanisms of the induction of germ cell apoptosis are unclear. Further understanding of the toxicology of the testis, future treatments for male infertility and the development of male contraceptives may rely upon a more detailed understanding of the genetic control of toxicant-induced programmed cell death in the testis.



## ACKNOWLEDGMENTS

The editor would like to thank Dr Sabina Cosulich for critical appraisal of this chapter and Zana Dye for invaluable editorial assistance.

## REFERENCES

AKBARSHA, M.A., AVERAL, H.I. and STANLEY, A., 1996, Male reproductive toxicity of vincristine: histopathological changes in the seminiferous tubules in relation to the stages in the spermatogenic cycle, *Biomedical Letters*, **54**, 73–86

AKBARSHA, M.A. and SIVASAMY, P., 1997, Apoptosis in male germinal line cells of rat in vivo: caused by phosphamidon, *Cytobios*, **91**, 33–44

ALCIVAR-WARREN, A., TRASLER, J.M., AWONIYI, C.A., ZIRKIN, B.R. and HECHT, N.B., 1996, Differential expression of ornithine decarboxylase, poly(ADP) ribose polymerase, and mitochondrial mRNAs following testosterone administration to hypophysectomized rats, *Molecular Reproduction and Development*, **43**, 283–289

ALLAN, D.J., GOBE, G.C. and HARMON, B.V., 1988, Sertoli cell death by apoptosis in the immature rat testis following x-irradiation, *Scanning Microscopy*, **2**, 503–512

ALLAN, D.J., HARMON, B.V. and ROBERTS, S.A., 1992, Spermatogonial apoptosis has three morphologically recognizable phases and shows no circadian rhythm during normal spermatogenesis in the rat, *Cell Proliferation*, **25**, 241–250

ALLARD, E.K. and BOEKELHEIDE, K., 1996, Fate of germ-cells in 2,5-hexanedione-induced testicular injury. 2. Atrophy persists due to a reduced stem-cell mass and ongoing apoptosis, *Toxicology and Applied Pharmacology*, **137**, 149–156

ALMON, E., GOLDFINGER, N., KAPON, A., SCHWARTZ, D., LEVINE, A.J. and ROTTER, V., 1993, Testicular tissue-specific expression of the p53 suppressor gene, *Developmental Biology*, **156**, 107–116

ANDERSON, D., BRINKWORTH, M.H., JENKINSON, P.C., CLODE, S.A., CREASY, D.M. and GANGOLLI, S.D., 1987, Effect of ethylene glycol monomethyl ether on spermatogenesis, dominant lethality, and F1 abnormalities in the rat and the mouse after treatment of F0 males, *Teratogenesis Carcinogenesis Mutagenesis*, **7**, 141–158

ASHBY, J., ODUM, J., TINWELL, H. and LEFEVRE, P., 1997, Assessing the risks of adverse endocrine-mediated effects: where to from here, *Regulatory Toxicology and Pharmacology*, **26**, 80–93

BACCETTI, B., COLLODEL, G. and PIOMBONI, P., 1996, Apoptosis in human ejaculated sperm cells (notulae-seminologicae-9), *Journal of Submicroscopic Cytology and Pathology*, **28**, 587–596

BALASUBRAMANIAN, H., CAMPBELL, G.A. and MOSLEN, M.T., 1994, Induction of apoptosis in the thymus by 2-methoxyethanol: rapidity and dose-response in the male rat, *Toxicologist*, **14**, 298

BEAVER, J.P. and WARING, P., 1995, A decrease in intracellular glutathione concentration precedes the onset of apoptosis in murine thymocytes, *European Journal of Cell Biology*, **68**, 47–54

BELLGRAU, D., GOLD, D., SELAWRY, H., MOORE, J., FRANZUSOFF, A. and DUKE, R.C., 1995, A role for CD95 ligand in preventing graft rejection, *Nature*, **377**, 630–632

BLANCHARD, K.T., ALLARD, E.K. and BOEKELHEIDE, K., 1996, Fate of germ cells in 2,5-hexanedione-induced testicular injury. Apoptosis is the mechanism of germ cell death, *Toxicology and Applied Pharmacology*, **137**, 141–148

BLANCO-RODRIGUEZ, J. and MARTINEZ-GARCIA, C., 1996a, Spontaneous germ cell death in the testis of the adult rat takes the form of apoptosis: re-evaluation of cell types that exhibit the ability to die during spermatogenesis, *Cell Proliferation*, **29**, 13–31

BLANCO-RODRIGUEZ, J. and MARTINEZ-GARCIA, C., 1996b, Induction of apoptotic cell-death in the seminiferous tubule of the adult-rat testis – assessment of the germ-cell types that exhibit the ability to enter apoptosis after hormone suppression by estradiol treatment, *International Journal of Andrology*, **19**, 237–247

BLANCO-RODRIGUEZ, J. and MARTINEZ-GARCIA, C., 1997, Apoptosis pattern elicited by estradiol treatment of the seminiferous epithelium of the adult rat, *Journal of Reproduction and Fertility*, **110**, 61–70

BLANCO-RODRIGUEZ, J. and MARTINEZ-GARCIA, C., 1998, Apoptosis precedes detachment of germ cells from the seminiferous epithelium after hormone suppression by short-term estradiol treatment of rats, *International Journal of Andrology*, **21**, 109–115

BRANDT, I., 1977, Tissue localisation of polychlorinated biphenyls. Chemical structure related to pattern of distribution, *Acta Pharmacology Toxicolology*, **40** (Suppl. II), 1–108

BRINKWORTH, M.H., WEINBAUER, G.F., BERGMANN, M. and NIESCHLAG, E., 1997, Apoptosis as a mechanism of germ cell loss in elderly men, *International Journal of Andrology*, **20**, 222–228

BRINKWORTH, M.H., WEINBAUER, G.F., SCHLATT, S. and NIESCHLAG, E., 1995, Identification of male germ cells undergoing apoptosis in adult rats, *Journal of Reproduction and Fertility*, **105**, 25–33

BROMWICH, P., COHEN, J., STEWART, J. and WALKER, A., 1994, Decline in sperm counts: an artefact of changed reference range of 'normal'?, *British Medical Journal*, **199**, 309–310

BUSH, B., BENNETT, A. and SNOW, J., 1986, Polychlorinated biphenyl congeners, p,pí-DDE, and sperm function in humans, *Archives in Environmental Contamination and Toxicology*, **15**, 333–341

CAI, L., HALES, B.F. and ROBAIRE, B., 1997, Induction of apoptosis in the germ cells of adult male rats after exposure to cyclophosphamide, *Biology of Reproduction*, **56**, 1490–1497

CARLSEN, E., GIWERCMAN, A., KEIDING, N. and SKAKKEBAEK, N.E., 1992, Evidence for decreasing quality of semen during the past 50 years, *British Medical Journal*, **305**, 609–612

CLARK, A.M., MAGUIRE, S.M. and GRISWOLD, M.D., 1997, Accumulation of clusterin/sulfated glycoprotein-2 in degenerating pachytene spermatocytes of adult rats treated with methoxyacetic acid, *Biology of Reproduction*, **57**, 837–846

DIRAMI, G., RAVINDRANATH, N., KLEINMAN, H.K. and DYM, M., 1995, Evidence that basement membrane prevents apoptosis of Sertoli cells in vitro in the absence of known regulators of Sertoli cell function, *Endocrinology*, **136**, 4439–4447

FRASOLDATI, A., ZOLI, M., ROMMERTS, F.F.G., BIAGINI, G., FAUSTINI FUSTINI, M., CARANI, C., *et al.*, 1995, Temporal changes in sulphated glycoprotein-2 (clusterin) and ornithine decarboxylase mRNA levels in the rat testis after ethane-dimethane sulphonate-induced degeneration of Leydig cells, *International Journal of Andrology*, **18**, 46–54

FRENCH, L.E., HAHNE, M., VIARD, I., RADLGRUBER, G., ZANONE, R., BECKER, K., *et al.*, 1996, Fas and fas ligand in embryos and adult mice-ligand expression in several immune-privileged tissues and coexpression in adult tissues characterised by apoptotic cell turnover, *Journal of Cell Biology*, **133**, 335–343

FURUCHI, T., MASUKO, K., NISHIMUNE, Y., OBINATA, M. and MATSUI, Y., 1996, Inhibition of testicular germ-cell apoptosis and differentiation in mice misexpressing Bcl-2 in spermatogonia, *Development*, **122**, 1703–1709

GIBSON, L., HOLMGREEN, S.P., HUANG, D.C.S., BERNARD, O., COPELAND, N.G., JENKINS, N.A., *et al.*, 1996, Bcl-w, a novel member of the bcl-2 family, promotes cell survival, *Oncogene*, **13**, 665–675

GORCZYCA, W., TRAGANOS, F., JESIONOWSKA, H. and DARZYNKIEWICZ, Z., 1993, Presence of DNA strand breaks and increased sensitivity of DNA *in situ* to denaturation in abnormal human sperm cells – analogy to apoptosis of somatic-cells, *Experimental Cell Research*, **207**, 202–205

GRIFFITH, T.S., BRUNNER, T., FLETCHER, S.M., GREEN, D.R. and FERGUSON, T.A., 1995, Fas-ligand-induced apoptosis as a mechanism of immune privilege, *Science*, **270**, 1189–1192

HENRIKSEN, K., HAKOVIRTA, H. and PARVINEN, M., 1995, Testosterone inhibits and induces apoptosis in rat seminiferous tubules in a stage-specific manner: in situ quantification in squash preparations after administration of ethane dimethane sulfonate, *Endocrinology*, **136**, 3285–3291

HENRIKSEN, K., KANGASNIEMI, M., PARVINEN, M., KAIPIA, A. and HAKOVIRTA, H., 1996, *In vitro*, follicle-stimulating hormone prevents apoptosis and stimulates deoxyribonucleic acid synthesis in the rat seminiferous epithelium in a stage-specific fashion, *Endocrinology*, **137**, 2141–2149

HSU, S.Y., KAIPIA, A., MCGEE, E., LOMELI, M. and HSUEH, A.J.W., 1997, Bok is a pro-apoptotic Bcl-2 protein with restricted expression in reproductive tissues and heterodimerizes with selective anti-apoptotic Bcl-2 family members, *Proceedings of the National Academy of Sciences of the United States of America*, **94**, 12401–12406

HUCKINS, C., 1978, The morphology and kinetics of spermatogonial degeneration in normal adult rats: an analysis using a simplified classification of the germinal epithelium, *Anatomical Record*, **190**, 905–926

IYER, V.N. and SZYBALSKI, W., 1963, A molecular mechanism of mitomycin action: linking of complementary DNA strands, *Proceedings of the National Academy of Sciences of the United States of America*, **50**, 355–362

JACKSON, N.C., JACKSON, H., SHANKS, J.H., DIXON, J.S. and LENDON, R.G., 1986, Study using in vivo binding of 125I-labeled hCG, light and electron microscopy of the repopulation of rat Leydig cells after destruction due to administration of ethylene-1,2-dimethanesulphonate, *Journal of Reproduction and Fertility*, **76**, 1–10

JOHANSSON, B., 1989, Effects of polychlorinated biphenyls (PCBs) on in vitro biosynthesis of testosterone and cell viability in mouse Leydig cells, *Bulletin of Environmental Contamination and Toxicology*, **42**, 9–14

JONES, M.M., XU, C.Y. and LADD, P.A., 1997, Selenite suppression of cadmium-induced testicular apoptosis, *Toxicology*, **116**, 169–175

KAMADA, S., SHINTO, A.A., TSUJIMURA, Y., TAKAHASHI, T., NODA, T., KITAMURA, Y., *et al.*, 1995, Bcl-2 deficiency in mice leads to pleiotropic abnormalities: accelerated lymphoid cell death in the thymus and spleen, polycystic kidney, hair hypopigmentation, and distorted small intestine, *Cancer Research*, **4**, 354–359

KELCE, W.R., 1994, Buthionine sulfoximine protects the viability of adult rat Leydig cells exposed to ethane dimethanesulphonate, *Toxicology and Applied Pharmacology*, **125**, 237–246

KELCE, W.R. and ZIRKIN, B.R., 1993, Mechanism by which ethane dimethanesulphonate kills adult rat Leydig cells: involvement of intracellular glutathione, *Toxicology and Applied Pharmacology*, **120**, 80–88

KERR, J.B., 1992, Spontaneous degeneration of germ cells in normal rat testis: assessment of cell types and frequency during the spermatogenic cycle, *Journal of Reproduction and Fertility*, **95**, 825–830

KIM, G.H. and WRIGHT, W.W., 1997, A comparison of the effects of testicular maturation and aging on the stage-specific expression of cp-2/cathepsin l messenger ribonucleic acid by sertoli cells of the brown norway rat, *Biology of Reproduction*, **57**, 1467–1477

KIMURA, K. and YAMAMOTO, M., 1997, Rapid induction of fas antigen mRNA expression in vivo by cyclohexamide, *Cell Biochemistry and Function*, **15**, 81–86

KNUDSON, C.M., TUNG, K.S.K., TOURTELLOTE, W.G., BROWN, G.A.J. and KORSMEYER, S.J., 1995, Bax-deficient mice with lymphoid hyperplasia and male germ cell death, *Science*, **270**, 96–99

KOBAYASHI, J., 1960, The cytological effect of chemicals on tumors X on the effect of mitomycin C on tumor cells of two animal tumors, *Cytologia*, **25**, 280–288

KOVACEVIC, R., VOJINOVIC-MILORADOV, M., TEODOROVIC, I. and ANDRIC, S., 1995, Effect of PCBs on androgen production by suspension of adult rat Leydig cells in vitro, *Journal of Steroid Biochemistry and Molecular Biology*, **52**, 595–597

KRAJEWSKI, S., KRAJEWSKA, M. and REED, J.C., 1996, Immunohistochemical analysis of in vivo patterns of Bak expression, a proapoptotic member of the Bcl-2 protein family, *Cancer Research*, **56**, 2849–2855

KRAJEWSKI, S., BODRUG, S., KRAJEWSKA, M., SHABAIK, A., GASCOYNE, R., BEREAN, K. and REED, J.C., 1995, Immunohistochemical analysis of Mcl-1 protein in human tissues, *American Journal of Pathology*, **146**, 1309–1319

KRAJEWSKI, S., KRAJEWSKA, M., SHABAIK, A., MIYASHITA, T., WANG, H.-G. and REED, J.C., 1994b, Immunohistochemical determination of in vivo distribution of Bax, a dominant inhibitor of Bcl-2, *American Journal of Patholology*, **145**, 1323–1336

KRAJEWSKI, S., KRAJEWSKA, M., SHABAIK, A., WANG, H.-G., IRIE, S., FONG, L. and REED, J.C., 1994a, Immunohistochemical analysis of in vivo patterns of Bcl-x expression, *Cancer Research*, **54**, 5501–5507

KU, W.W., WINE, R.N., CHAE, B.Y., GHANAYEM, B.I. and CHAPIN, R.E., 1995, Spermatocyte toxicity of 2-methoxyethanol (ME) in rats and guinea pigs: evidence for the induction of apoptosis, *Toxicology and Applied Pharmacology*, **134**, 100–110

LEE, J., RICHBURG, J.H., YOUNKIN, S.C. and BOEKELHEIDE, K., 1997, The fas system is a key regulator of germ cell apoptosis in the testis, *Endocrinology*, **138**, 2081–2088

LEITHAUSER, F., DHEIN, J., MECHTERSHEIMER, G., KORETZ, K., BRUDERLEIN, S., HENNE, C., *et al.*, 1993, Constitutive and induced expression of APO-1, a new member of the nerve growth factor/tumor necrosis factor receptor superfamily, in normal and neoplastic cells, *Laboratory Investigation*, **69**, 415–429

LI, H., REN, J., DHABUWALA, C.B. and SHICHI, H., 1997, Immunotolerance induced by intratesticular antigen priming: expression of TGF-β, Fas and Fas ligand, *Ocular Immunology and Inflammation*, **5**, 75–84

LI, L.-H., WINE, R.N. and CHAPIN, R.E., 1996, 2-methoxyacetic (MAA)-induced spermatocyte apoptosis in human and rat testes: an in vitro comparison, *Journal of Andrology*, **17**, 538–549

LI, L.-H., WINE, R.N., MILLER, D.S., REECE, J.M., SMITH, M. and CHAPIN, R.E., 1997, Protection against methoxyacetic-acid-induced spermatocyte apoptosis with calcium channel blockers in cultured rat seminiferous tubules: possible mechanisms, *Toxicology and Applied Pharmacology*, **144**, 105–119

LIN, W.W., LAMB, D.J., WHEELER, T.M., LIPSHULTZ, L.I. and KIM, E.D., 1997b, *In situ* end-labeling of human testicular tissue demonstrates increased apoptosis in conditions of abnormal spermatogenesis, *Fertility and Sterility*, **68**, 1065–1069

LIN, W.W., LAMB, D.J., WHEELER, T.M., ABRAMS, J., LIPSHULTZ, L.I. and KIM, E.D., 1997a, Apoptotic frequency is increased in spermatogenic maturation arrest and hypospermatogenic states, *Journal of Urology*, **158**, 1791–1793

LUI, G., LYLE, K.C. and CAO, J., 1985, Trial of gossypol as a male contraceptive, in SEGAL, S.J. (ed.) *Gossypol: a Potential Contraceptive for Men*, New York: Plenum Press, pp. 9–16

MARCHETTI, P., DECAUDIN, D., MACHO, A., ZAMZAMI, N., HIRSCH, T., SUSIN, S.A. and KROEMER, G., 1997, Redox regulation of apoptosis: impact of thiol oxidation status on mitochondrial function, *European Journal of Immunology*, **27**, 289–296

MORRIS, A.J., TAYLOR, M.F. and MORRIS, I.D., 1997, Leydig cell apoptosis in response to ethane dimethanesulphonate after both *in vivo* and *in vitro* treatment, *Journal of Andrology*, **18**, 274–280

MORRIS, I.D., 1996, Leydig cell toxicology, in PAYNE, A.H., HARDY, M.P. and RUSSELL, L.D. (eds) *The Leydig Cell*, Vienna, Ill.: Cache River Press, pp. 573–596

MORRIS, I.D., PHILLIPS, D.M. and BARDIN, C.W., 1986, Ethylene dimethanesulfonate destroys Leydig cells in the rat testis, *Endocrinology*, **118**, 709–719

NAKAGAWA, S., NAKAMURA, N., FUJIOKA, M. and MORI, C., 1997, Spermatogenic cell apoptosis induced by mitomycin c in the mouse testis, *Toxicology and Applied Pharmacology*, **147**, 204–213

NAKAYAMA, K.I., NAKAYAMA, K., NEGISHI, I., KUIDA, K., SAWA, H. and LOH, D.Y., 1994, Targeted disruption of bcl-2 alpha beta in mice: occurrence of grey hair, polycystic kidney disease and lymphocytopenia, *Proceedings of the National Academy of Sciences of the United States of America*, **91**, 3700–3704

NAKAYAMA, K.I., NAKAYAMA, K., NEGISHI, I., KUIDA, K., SHINKAI, Y., LOUIE, M.C., *et al.*, 1993, Disappearance of the lymphoid system in bcl-2 homozygous mutant chimeric mice, *Science*, **261**, 1584–1588

NONCLERCQ, D., REVERSE, D., TOUBEAU, G., BECKERS, J.F., SULON, J., LAURENT, G., *et al.*, 1996, *In situ* demonstration of germinal cell apoptosis during diethylstilbestrol-induced testis regression in adult male syrian-hamsters, *Biology of Reproduction*, **55**, 1368–1376

ODORISIO, T., RODRIGUEZ, T.A., EVANS, E.P., CLARKE, A.R. and BURGOYNE, P.S., 1998, The meiotic checkpoint monitoring synapsis eliminates spermatocytes via p53-independent apoptosis, *Nature Genetics*, **18**, 257–261

OGI, S., TANJI, N., YOKOYAMA, M., TAKEUCHI, M. and TERADA, N., 1998, Involvement of fas in the apoptosis of mouse germ cells induced by experimental cryptorchidism, *Urological Research*, **26**, 17–21

POLAND, A. and GLOVER, E., 1977, Chlorinated biphenyl induction of aryl hydrocarbon hydroxylase activity: a study of the structure-activity relationship, *Molecular Pharmacology*, **13**, 924–938

RICHBURG, J.H. and BOEKELHEIDE, K., 1996, Mono-(2-ethylhexyl) phthalate rapidly alters both sertoli-cell vimentin filaments and germ-cell apoptosis in young-rat testes, *Toxicology and Applied Pharmacology*, **137**, 42–50

RODRIGUEZ, I., ODY, C., ARAKI, K., GARCIA, I. and VASSALLI, P., 1997, An early and massive wave of germinal cell apoptosis is required for the development of functional spermatogenesis, *EMBO Journal*, **16**, 2262–2270

ROSS, A.J., WAYMIRE, K.G., MOSS, J.E., PARLOW, A.F., SKINNER, M.K., RUSSELL, L.D. and MACGREGOR, G.R., 1998, Testicular degeneration in Bcl-w-deficient mice, *Nature Genetics*, **18**, 251–256

ROTTER, V., SCHWARTZ, D., ALMON, E., GOLDFINGER, N., KAPON, A., MESHORER, A., DONEHOWER, L.A. and LEVINE, A.J., 1993, Mice with reduced levels of p53 protein exhibit the testicular giant-cell degenerative syndrome, *Proceedings of the National Academy of Sciences of the United States of America*, **90**, 9075–9079

RUSSEL, L.D. and CLERMONT, Y., 1977, Degeneration of germ cells in normal, hypophysectomized and hormone treated hypophysectomized rats, *Anatomical Record*, **187**, 347–366

Russel, L.D., Weis, T., Goh, J.G. and Curl, J.L., 1990, The effect of Submandibular gland removal on testicular and epididymal parameters, *Tissue Cell*, **22**, 263–268

Schwartz, D., Goldfinger, N. and Rotter, V., 1994, Expression of p53 protein in spermatogenesis is confined to the tetraploid pachytene primary spermatocytes, *Oncogene* **8**, 1487–1494

Seino, K.I., Kayagaki, N., Takeda, K., Fukao, K., Okumura, K. and Yagita, H., 1997, Contribution of Fas ligand to T cell-mediated hepatic injury in mice, *Gastroenterology*, **113**, 1315–1322

Sharpe, R.M., 1994, Regulation of spermatogenesis, in Knobill, E. and Neill, J.D. (eds), *The physiology of reproduction*, New York: Raven Press, 1363–1434

Shetty, J., Marathe, G.K. and Dighe, R.R., 1996, Specific immunoneutralization of FSH leads to apoptotic cell death of the pachytene spermatocytes and spermatogonial cells in the rat, *Endocrinology*, **137**, 2179–2182

Shinoda, K., Mitsumori, K., Yasuhara, K., Uneyama, C., Onodera, H., Takegawa, K., *et al.*, 1998, Involvement of apoptosis in the rat germ cell degeneration induced by nitrobenzene, *Archives of Toxicology*, **72**, 296–302

Shiratsuchi, A., Umeda, M., Ohba, Y. and Nakanishi, Y., 1997, Recognition of phosphatidylserine on the surface of apoptotic spermatogenic cells and subsequent phagocytosis by Sertoli cells of the rat, *Journal of Biological Chemistry*, **272**, 2354–2358

Simialowicz, R.J., Riddle, M.M. and Williams, W.C., 1994, Species and strain comparisons of immunosuppression by 2-methoxyethanol and 2-methoxy acetic acid, *International Journal of Immunopharmacology*, **16**, 695–702

Sinha Hikim, A.P., Wang, C., Lue, Y., Johnson, L., Wang, X.H. and Swerdloff, R.S., 1998, Spontaneous germ cell apoptosis in humans: evidence for ethnic differences in the susceptibility of germ cells to programmed cell death, *Journal of Clinical Endocrinology and Metabolism*, **83**, 152–156

Sjoblom, T., West, A. and Lahdetie, J., 1998, Apoptotic response of spermatogenic cells to the germ cell mutagens etoposide, adriamycin and diepoxybutane, *Environmental and Molecular Mutagenesis*, **31**, 133–148

Stephan, A., Polzar, B., Zanotti, F.R.S. and Mannherz, C.U.H.G., 1996, Distribution of deoxyribonuclease I (DNase I) and p53 in rat testis and their correlation with apoptosis, *Histochemistry and Cell Biology*, **106**, 383–393

Strandgaard, C. and Miller, M.G., 1998, Germ cell apoptosis in rat testis after administration of 1,3-dinitrobenzene, *Reproductive Toxicology*, **12**, 97–103

Suda, T., Okazaki, T., Naito, Y., Yokota, T., Arai, N., Ozaki, S., *et al.*, 1995, Expression of the Fas ligand in cells of T cell lineage, *Journal of Immunology*, **154**, 3806–3813

SUGIHARA, A., SAIKI, S., TSUJI, M., TSUJIMURA, T., NAKATA, Y., KUBOTA, A., *et al.*, 1997, Expression of Fas and Fas ligand in the testes and testicular germ cell tumors: an immunohistochemical study, *Anticancer Research*, **17**, 3861–3865

SUGIMOTO, C., MATSUKAWA, S., FUJIEDA, S., NODA, I., TANAKA, N., TSUZUKI, H. and SAITO, H., 1996, Involvement of intracellular glutathione in induction of apoptosis by cisplatin in a human pharyngeal carcinoma cell line, *Anticancer Research*, **16**, 675–680

TAPANAINEN, J.S., TILLY, J.L., VIHKO, K.K. and HSUEH, A.J.W., 1993, Hormonal control of apoptotic cell death in the testis: gonadotrophins and androgens as testicular cell survival factors, *Molecular Endocrinology*, **7**, 643–650

TAYLOR, M.F., WOOLVERIDGE, I., METCALFE, A., STREULI, C.H., HICKMAN, J.A. and MORRIS, I.D., 1998, Leydig cell apoptosis in the rat testes after the cytotoxin ethane dimethanesulphonate: role of the Bcl-2 family members, *Journal of Endocrinology*, **157**, 317–326

TAYLOR, M.F., WOOLVERIDGE, I., TEERDS, K.J., MORRIS, I.D., 1999, Leydig cell apoptosis after the administration of ethane dimethanesulfonate to the adult male rat is a Fas-mediated process. *Endocrinology*, **140**, 3797–3804

TENG, C.S., 1995, Gossypol-induced apoptotic DNA fragmentation correlates with inhibited protein-kinase-c activity in spermatocytes, *Contraception*, **52**, 389–395

TILLY, J.L., TILLY, K.I., KENTON, M.L. and JOHNSON, A.L., 1995, Expression of members of the Bcl-2 gene family in the immature rat ovary: equine chorionic gonadotrophin-mediated inhibition of granulosa cell apoptosis is associated with decreased Bax and constitutive Bcl-2 and Bcl-xlong messenger ribonucleic acid levels, *Endocrinology*, **136**, 232–241

WATANABE-FUKUNAGA, R., BRANNAN, C.I., ITOH, N., YONEHARA, S., COPELAND, N.G., JENKINS, N.A. and NAGATA, S., 1992, The cDNA structure, expression and chromosomal assignment of the mouse Fas antigen, *Journal of Immunology*, **148**, 1274–1279

WINE, R.N., KU, W.W., LI, L.-H. and CHAPIN, R.E., 1997, Cyclophilin A is present in rat germ cells and is associated with spermatocyte apoptosis, *Biology of Reproduction*, **56**, 439–446

WOOLVERIDGE, I., BRYDEN, A.A.G., TAYLOR, M.F., GEORGE, N.J.R., WU, F.W.C. and MORRIS, I.D., 1998a, Apoptosis and related genes in the human testis following short and long term anti-androgen treatment, *Molecular Human Reproduction*, **4**, 701–707

WOOLVERIDGE, I., TAYLOR, M.F., WU, F.W.C. and MORRIS, I.D., 1998b, Apoptosis and related genes in the rat ventral prostate following androgen ablation in response to ethane dimethanesulfonate, *The Prostate*, **36**, 23–30

WOOLVERIDGE, I., DE BOER-BROUWER, M., TAYLOR, M.F., TEERDS, K.J., WU, F.W.C., MORRIS, I.D., 1999, Apoptosis in the rat spermatogenic epithelium following androgen withdrawal: changes in apoptosis-related genes, *Biology of Reproduction*, **60**, 461–470

XU, C., JOHNSON, J.E., SINGH, P.K., JONES, M.M., YAN, H. and CARTER, C.E., 1996, In vivo studies of cadmium-induced apoptosis in testicular tissue of the rat and its modulation by a chelating agent, *Toxicology*, **107**, 1–8

YAN, H.P., CARTER, C.E., XU, C.Y., SINGH, P.K., JONES, M.M., JOHNSON, J.E. and DIETRICH, M.S., 1997, Cadmium-induced apoptosis in the urogenital organs of the male rat and its suppression by chelation, *Journal of Toxicology and Environmental Health*, **52**, 149–168

YEOWELL, H.N., WAXMAN, D.J., LEBLANC, G.A., LINKO, P. and GOLDSTEIN, J.A., 1989, Suppression of male-specific cytochrome P450 2c and its mRNA by 3,4,5,3í,4í,5í-hexachlorobiphenyl in rat liver is not causally related to changes in serum testosterone, *Archives of Biochemistry and Biophysics*, **271**, 508–514

CHAPTER

5

# *Apoptosis in Female Reproductive Toxicology*

**JONATHAN L. TILLY**

Vincent Center for Reproductive Biology, Massachusetts General Hospital/Harvard Medical School, Boston, MA, USA

## *Contents*

## 5.1 INTRODUCTION

Although the hormonal interplay between components of the human female reproductive system is extremely complex, each organ or tissue that comprises the reproductive apparatus in women contributes towards a single goal: the birth of healthy offspring. A precise and synchronized monthly series of events both produce a mature egg for fertilization and embryogenesis and prepare the uterine endometrial lining for implantation should pregnancy occur. However, this process is unfortunately very sensitive to disruption by a host of environmental insults. Furthermore, due to the active involvement of multiple tissues in achieving successful conception and pregnancy, reproductive toxicology in females covers a vast expanse of research and clinical arenas. Because of this, and the fact that generation of a competent germ cell, an ovulated egg in females, is the first step in species propagation, the present discourse will focus solely on the impact of environmental and clinical toxicants on the ovary. For reviews of toxicant effects on other target tissues associated with female reproductive function, the reader is referred elsewhere (for examples, see Sastry, 1991, Mattison, 1993, Kamrin *et al.*, 1994, Sharara *et al.*, 1998, Spielmann, 1998).

The ovaries of all mammalian species are the powerhouses behind reproductive performance, providing not only the gametes needed for fertilization but also a long list of hormones and bioactive peptides that co-ordinate intraovarian events and those taking place in the hypothalamic-pituitary axis, oviducts and uterus. Together, these are needed for oocyte maturation, ovulation and fertilization, as well as for embryo implantation and gestation. At its most basic level, the female gonad can be roughly divided into four major functional cell types: germ cells (oogonia and oocytes), granulosa cells and theca-interstitial cells of the follicle and luteal cells of the corpus luteum formed following ovulation (Figure 5.1). The ovary is a prime target for disrupted function following toxicant exposure from its

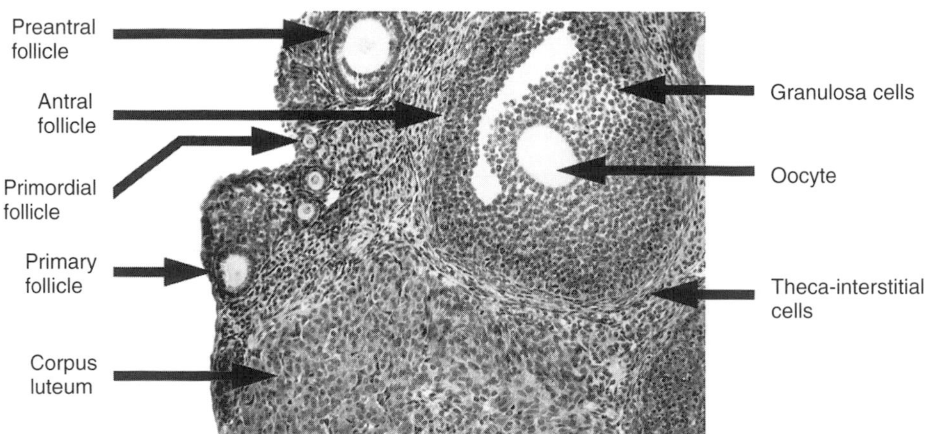


**Figure 5.1** Diagram of the ovary. Follicles in various stages of development are scattered throughout.

earliest stages as a presumptive gonad (genital ridge) housing primordial germ cells to its later stages during the menopausal transition. Furthermore, the impact of any given toxicant on ovarian function and infertility reflects the stage(s) of follicles that are vulnerable to the specific insult. Therefore, this chapter will begin with a brief overview of ovarian development, with particular emphasis on germ cell and follicle dynamics as well as on the role of apoptosis as a normal mechanism by which oocytes, follicles and corpora lutea are deleted. This section will be followed by a more intensive discussion of the literature implicating premature or inappropriate apoptosis as a common denominator in many paradigms of ovarian toxicity.

## 5.2 OVARIAN PHYSIOLOGY

### 5.2.1 Apoptosis and foetal oogenesis

During early embryogenesis in mammals, a relatively small number of primordial germ cells migrate from the yolk sac through the mesentery to the presumptive gonads or genital ridges where a programme of rapid germ cell mitosis is set in motion (Byskov, 1986, Buehr, 1997) (Figure 5.2). In the human female foetus, peak numbers of germ cells



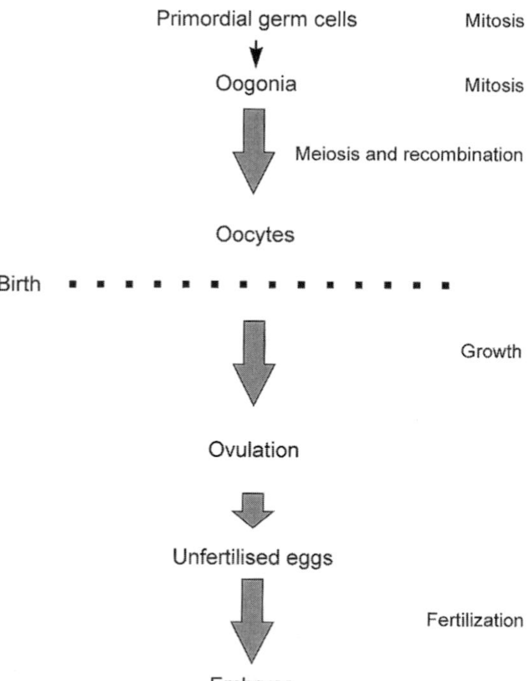


**Figure 5.2** Landmarks of oogenesis. Progression from primordial germ cells during foetal development and from oocytes to fertilized eggs in sexually mature adults.

(approximately $7 \times 10^6$), referred to as oogonia, are produced by the twentieth week of gestation through clonal expansion (Baker, 1963). Many of these oogonia begin the first steps of meiosis, passing through the leptotene, zygotene and pachytene phases, but become arrested in the last step of meiotic prophase (the diplotene stage) during folliculogenesis, at which time the arrested germ cell (now referred to as an oocyte) is surrounded by a single layer of specialized somatic cells termed granulosa cells to form a primordial follicle (Byskov, 1986). In almost all mammalian species, female germ cell mitosis ceases by the time of birth, providing for a finite reserve of oocytes that cannot be replenished in postnatal life. It is this distinguishing feature of a nonrenewable germ cell population that places women, as opposed to men who maintain active spermatogenesis throughout their adult lives, at such a tremendous risk for irreversible reproductive damage upon toxicant exposure.

Female germ cell dynamics is further complicated by the high rate of prenatal oogonium and oocyte attrition known to occur naturally in essentially all mammalian species. For example, in the human female foetus only $1–2 \times 10^6$ of the $7 \times 10^6$ germ cells present at the twentieth week of gestation remain in the ovaries at birth (Baker, 1963, Forabosco *et al.*, 1991), with the other two-thirds of the germ cells being lost via apoptosis (De Pol *et al.*, 1997). Similar events occur in the foetal ovaries of rodents (Borum, 1961, Beaumont and Mandl, 1962, Coucouvanis *et al.*, 1993, Ratts *et al.*, 1995), thus offering relevant and easily-studied model systems to explore the basis of prenatal germ cell death. In fact, the recent development of an organ culture system for maintaining foetal mouse ovaries *in vitro* has already provided intriguing new clues as to the regulation of oogonium and oocyte apoptosis (Morita *et al.*, 1999). This latter study identified the existence of a complex interplay between growth factors and cytokines in determining germ cell fate, and further showed that activation of a key signal transduction enzyme, phosphatidylinositol 3'-kinase (PI3K) (see Chapters 1 and 3), is a critical step for diverse endocrine factors to promote foetal ovarian germ cell survival. These findings of Morita *et al.* (1999) are in agreement with a number of reports from analyses of various somatic cell lineages that have established a central role for PI3K in transducing survival signals to central regulatory steps in the apoptotic cell death pathway governed by Bcl-2 family members (Datta *et al.*, 1997, Kennedy *et al.*, 1997) (Chapter 1).

Aside from this recent investigation (Morita *et al.*, 1999), very little is currently known of the potential involvement of apoptosis-regulatory molecules as determinants of oogonium and oocyte fate during gametogenesis and folliculogenesis. A possible connection between Bcl-2 and ovarian germ cell survival was established several years ago in studies showing that young adult female mice lacking functional expression of the *bcl-2* gene (gene knockout) possess a significantly reduced number of oocyte-containing primordial follicles (Ratts *et al.*, 1995). Although published evidence of *bcl-2* gene expression in the developing foetal ovaries of any species is currently lacking, the conclusions drawn from the analysis of Bcl-2-deficient mice are supported by findings of polyadenylated *bcl-2* mRNA transcripts in postnatal murine oocytes (Jurisicova *et al.*, 1998). Another set of data derived from gene knockout mice, in this case animals lacking expression of functional Bax protein (Knudson *et al.*, 1995), has revealed that Bax may be dispensable for proper execution of apoptosis in the foetal ovarian germ line since the oocyte reserve endowed at birth (which reflects the magnitude of prenatal germ cell attrition) is identical in wild-type and Bax-null females (Perez *et al.*, 1999). However, during postnatal life Bax-deficient female mice exhibit many other pronounced defects in ovarian cell apoptosis (Perez *et al.*, 1999), including germ line resistance to apoptosis induced by toxicant exposure (Perez *et al.*, 1997), that will be expanded on in sections 5.2.2 and 5.3.1 below.

Lastly, a recent set of observations derived from a third gene knockout mouse line has provided the first evidence of a role for caspases (see Chapter 2) as final executioners of germ cell apoptosis during foetal ovarian development. This study reported that female mice lacking expression of functional caspase-2 (previously referred to as Ich-1; Wang *et al.*, 1994) are born with a significantly larger endowment of oocyte-containing primordial follicles when compared with wild-type sisters (Bergeron *et al.*, 1998). These data are consistent with preliminary observations that peptide inhibitors of caspases effectively suppress apoptosis in germ cells of cultured foetal mouse ovaries (Morita *et al.*, 1999a), although the exact caspase family member(s) inhibited by these peptides remains to be clarified. Similar to the comments made regarding Bcl-2, essentially nothing is known of the expression patterns of caspases in the developing foetal ovaries. Nonetheless, it is important to note that polyadenylated *caspase-2* mRNA transcripts are present in postnatal murine oocytes (Jurisicova *et al.*, 1998), consistent with the idea that caspase-2-null oocytes possess a survival advantage *in vivo* (Bergeron *et al.*, 1998). These findings therefore set the stage for the hypothesis that caspases, like Bcl-2 and related proteins, are indeed integral components of a genetic programme of cell death responsible for executing apoptosis in the female germ line (Tilly *et al.*, 1997, Morita and Tilly, 1999).



### 5.2.2 Postnatal ovarian development, follicle atresia and the menopause

For those oocytes that escape death during foetal ovarian development and thus become endowed as the primordial follicle stockpile, there are two possible fates. The first, which happens to a small percentage of the total germ cell pool, is development of the oocyte and the follicle through successive maturational stages (primordial, primary, preantral, antral and, finally, preovulatory). This is followed by release of the oocyte from the preovulatory follicle(s) at ovulation into the reproductive tract for possible fertilization. It is important to point out that along the pathway of maturation, the susceptibility of the oocyte and the supporting somatic cells (granulosa cells and, in latter stages of follicle development, theca-interstitial cells) to damage by a number of toxicants appears to decrease, possibly reflecting the enhanced ability of the increasing number of somatic cells within developing follicles to neutralize potential threats to the oocyte housed within. The second possible fate of oocytes that comprise the primordial follicle reserve is apoptosis, an event that can either trigger or occur as a consequence of a normal degenerative process referred to as follicular atresia.

For as yet unknown reasons, the vast majority of follicles endowed in the ovary at birth will undergo atresia at some point in their developmental programme, thus never producing an ovulated egg for fertilization (Gougeon, 1996, Tilly, 1996). Atresia can occur at any point in the maturational pathway of follicles, although the cellular site of initiation of the event differs quite markedly dependent upon the developmental stage of the follicle. In quiescent (primordial) and slowly-growing immature (primary, early preantral) follicles, fragmentation and death of the oocyte appears prior to any discernible sign of somatic cell demise, suggesting that apoptosis of the germ cell drives atresia. In contrast, follicles progressing through the latter stages of maturation (late preantral, antral) that acquire dependency upon pituitary gonadotropin hormones, such as follicle-stimulating hormone (FSH), for continued survival and development undergo atresia due to apoptosis in the granulosa cell compartment of the follicle (Hughes and Gorospe, 1991, Tilly *et al.*, 1991, Chun *et al.*, 1994, Tilly *et al.*, 1995a, Tilly, 1998). Without direct support from and communication with its surrounding granulosa cells, the oocyte of the dying follicle eventually degenerates, removing itself irreversibly from the ovulatory pathway.

The impact of atresia on the fertile lifespan of the female is enormous when one considers, for example, that of the $1–2 \times 10^6$ primordial follicles endowed in the human ovaries at birth, up to one-half of these may be atretic (Baker, 1963). Furthermore, the number of follicles in the human female drops to less than $4 \times 10^5$ by the time of puberty, arguing that atresia claims over one-half to three-quarters of the oocytes before reproductive life even begins in women. Atresia remains a prominent feature of cyclic ovarian function until nearly complete exhaustion of the follicle reserve is observed at the time of the menopause (Richardson *et al.*, 1987, Tilly and Ratts, 1996).

In this regard, mounting evidence from gene expression and cell culture analyses implicate a number of apoptosis-regulatory molecules as principal signalling factors (Tilly *et al.*, 1997, Tilly, 1998) (Table 5.1). Expression of the pro-apoptotic protein encoded by the *bax* gene has proven to be tightly correlated with, if not absolutely required for, normal ovarian cell death to proceed. For example, *bax* transcripts accumulate in rat granulosa cells of early antral follicles on the verge of atresia, whereas provision of a survival factor such as exogenous gonadotropin *in vivo* reduces *bax* expression concordant with a suppression of granulosa cell apoptosis and follicular atresia (Tilly *et al.*, 1995a). Similar observations of a strong correlation between expression of the Bax protein, as determined by immunohistochemical analysis, and follicular atresia in the human ovary (Kugu *et al.*, 1998), emphasize the evolutionarily-conserved nature of the ovarian cell death programme.

The role of Bax in mediating granulosa cell demise was unequivocally established by morphological assessments of ovaries from young adult Bax-deficient female mice

**TABLE 5.1**
Molecules implicated in the regulation of follicular atresia

| Molecule | Function | Reference |
|---|---|---|
| p53 (see Chapter 1) | Role in sensing DNA damage. Transcription factor. | Tilly *et al.*, 1995b, Hosokawa *et al.*, 1998 |
| Ceramide (see Chapter 1) | Role in signalling execution. | Witty *et al.*, 1996 |
| Reactive oxygen species | Role in signal transduction. | Tilly and Tilly, 1995 |
| Bcl-2 family members (see Chapter 1) | Central interpreters of signals for cell survival. Bax is strongly implicated in ovarian apoptosis. | Tilly *et al.*, 1995a, Knudson *et al.*, 1995, Kugu *et al.*, 1998, Perez *et al.*, 1999 |
| Apaf-1 | Couples Bcl-2 family function to caspase activation. | Robles *et al.*, 1999 |
| Caspases (see Chapter 2) | Executioners. | Flaws *et al.*, 1995, Maravei *et al.*, 1997, Boone and Tsang, 1998, Kugu *et al.*, 1998 |
| Serine proteases | Executioners | Trbovich *et al.*, 1998 |



(Knudson *et al.*, 1995) which revealed the presence of numerous maturing follicles that presumably failed to properly complete the atresia process. A more extensive evaluation of the reproductive features of Bax-null female mice documented a significantly reduced incidence of primordial and primary follicle atresia, leading to a 3-fold surfeit of oocyte-containing primordial follicles, in young adult females (Perez *et al.*, 1999). These findings of a larger ovarian follicle reserve predicted that ovarian senescence would be dramatically delayed, if not absent, as Bax-deficient females reached advanced chronological ages. This postulate was confirmed by evidence that ovaries of aged Bax-null females do in fact retain large numbers of follicles at all stages of development, including large estrogenic antral follicles capable of sustaining uterine hypertrophy (Perez *et al.*, 1999). Aside from the obvious implications of this work, the study by Perez *et al.* (1999) has substantiated the relevance of using the power of mouse genetics to elucidate basic mechanisms responsible for follicle loss leading to the menopause.

## 5.2.3 The corpus luteum and pregnancy maintenance

During the periovulatory period, initiated by the surge of luteinizing hormone from the pituitary gland, granulosa and theca-interstitial cells of the preovulatory follicle(s) undergo a rapid transformation to luteal cells. This event is marked by the shift in steroidogenesis from principally estrogen to progesterone, the latter steroid being absolutely necessary for the maintenance of at least the early gestation period following implantation of the conceptus if fertilization of the ovulated egg occurs (Patton and Stouffer, 1991). The corpus luteum has a pre-programmed lifespan that, unless informed otherwise by as yet incompletely understood mechanisms, ends in collapse of the tissue several days after being formed. The process of corpus luteum regression, or luteolysis, is accomplished in two not necessarily independent phases, the first demarcated by a loss of steroidogenic capacity followed by a second phase of structural involution (Rueda *et al.*, 1997a). Of course, one would predict that impaired progesterone output does not have to precede structural collapse since any agent, toxicant or otherwise, capable of triggering apoptosis in luteal cells would cause an abrupt loss in steroid production by this ovarian structure. As mentioned earlier, such a scenario would place the gestating foetus in grave danger of being aborted since the uterine lining is critically dependent upon ovarian steroid support for maintenance during pregnancy.

Work from a number of laboratories has shown that apoptosis is a prominent feature of structural luteolysis in various mammalian species (Juengel *et al.*, 1993, Dharmarajan *et al.*, 1994, Shikone *et al.*, 1996), as well as postovulatory follicular tissue resorption in the avian ovary (Tilly *et al.*, 1991). The endocrine regulation of this process is probably species-specific, with human chorionic gonadotropin (Dharmarajan *et al.*, 1994), prolactin (Matsuyama *et al.*, 1996), estradiol (Goodman *et al.*, 1998), progesterone (Gaytan *et al.*, 1998) and prostaglandin-$F_{2\alpha}$ (Rueda *et al.*, 1999) identified among others as potential candidates for controlling luteal cell fate. As recently reviewed (Rueda *et al.*, 1997a), the intracellular events set in motion in luteal cells by changes in the extracellular environment are probably the same as those activated in other cell types during apoptosis. For example, an alteration in the intracellular reduction-oxidation state leading to oxidative stress probably accompanies cell death during luteolysis (Rueda *et al.*, 1995, Dharmarajan *et al.*, 1999). Furthermore, altered expression patterns of key apoptosis-regulatory molecules, such as Bcl-2 family members and caspases, have been linked to luteal regression (Rueda *et al.*, 1997b, 1999, Goodman *et al.*, 1998, Dharmarajan *et al.*, 1999).

## 5.3 APOPTOSIS AND OVARIAN TOXICOLOGY

Since the number of chemicals that adversely affect female reproductive performance at the level of ovarian dysfunction is far too great to address in any one chapter (for examples, see Bishop *et al.*, 1997a,b), four classes of toxicants linked to apoptosis induction in the ovary have been selected for discussion. As a preface, it should be stressed that the stage of follicular development affected by exposure to any given toxicant determines its impact on female reproductive function. For example, chemicals that destroy growing and preovulatory follicles cause temporary infertility since these damaged follicles can be replaced by recruitment from the large pool of primordial and primary follicles present in the ovaries. In contrast, toxicants that damage or destroy primordial or primary follicles lead to early menopause and permanent infertility as these follicles are set as a finite number at birth and cannot be replaced (see section 5.2). Of great concern, destruction of primordial follicles may not be detected for weeks, months or even years after exposure to xenobiotics since oestrous or menstrual cyclicity may not be affected until the entire pool of growing follicles has been depleted from the ovary. Therefore, risk assessment for environmental threats to the primordial follicle stockpile is extremely difficult to pursue since the only reliable method to detect such damage is serial section analysis of the ovary for total follicle counts (Bolon *et al.*, 1997, Bucci *et al.*, 1997). As a final comment here, the potential role of apoptosis as a mediator of chemical-induced ovarian damage is a very new concept that has not received sufficient research attention. Therefore, some of what is to follow is hypothetical but supported by published information where possible.

### 5.3.1 Dioxin (2,3,7,8-tetrachlorodibenzo-p-dioxin or TCDD)

Dioxin is one of the most notorious environmental toxicants known (Poland and Knutson, 1982, Silbergeld, 1995). In humans, TCDD has been found at significant levels in ovarian follicular fluid (Tsutsumi *et al.*, 1998) but the consequences of this ovarian TCDD exposure remain to be fully clarified. In rats *in vivo*, TCDD causes abnormal estrous cyclicity and impaired ovulation rates (Li *et al.*, 1995a, Lu *et al.*, 1995) and reduces the numbers of preantral and antral ovarian follicles (Heimler *et al.*, 1998a). Somewhat surprisingly, the reduction in follicle number noted by Heimler *et al.* (1998a) was not associated with an increased incidence of apoptotic cell death in those follicles, raising the possibility that the primordial and/or primary follicle pool (which is the source of maturing preantral and antral follicles) is actually the target of TCDD action. However, previous work has established that primordial oocyte destruction does not occur in female mice following *in vivo* TCDD exposure (Mattison and Nightingale, 1980), leaving open the question as to exactly how TCDD causes a reduction in ovarian follicle numbers.

Species-specific differences in the ovarian cellular response to TCDD could explain the potentially discordant data regarding follicle destruction in mice and rats, and a recent publication supports this possibility. Using cultured human granulosa-luteal cells obtained from follicular aspirates of patients undergoing assisted reproductive technologies, Heimler *et al.* (1998b) reported that dioxin treatment caused a time- and dose-dependent increase in the incidence of apoptosis. The mechanisms underlying the apoptotic response to TCDD were unfortunately not elucidated, although TCDD was noted to also significantly attenuate estrogen production by reducing androgen precursor availability in cultured human granulosa-luteal cells (Heimler *et al.*, 1998b). In light of recent investigations identifying estradiol as a luteal cell survival factor in the rabbit (Goodman *et al.*, 1998), it is possible that the ability of dioxin to suppress estrogen biosynthesis in human granulosa-luteal cells serves as at least one initiating event that leads to apoptosis. Furthermore, in cultured human granulosa-luteal cells TCDD is known to modulate the activity of a number of protein kinases involved in cell signalling (Enan *et al.*, 1996a) and to interfere with glucose transport (Enan *et al.*, 1996b), offering other possible sites of normal cellular function that may be disrupted by TCDD leading to apoptosis. Whatever the case, it is clear that additional studies are needed to fully understand the consequences of dioxin exposure on ovarian function in general and on ovarian cell death in particular.

## 5.3.2 Polycyclic aromatic hydrocarbons (PAH)

Polycyclic aromatic hydrocarbons are formed as a result of the incomplete combustion of organic compounds, most notably fossil fuels such as wood and coal (Committee on Biological Effects of Atmospheric Pollutants, 1972). Benzo[a]pyrene (BAP), a prototypical PAH, is released into the air at an estimated level of 1.8 million pounds per year, primarily derived from coal refuse piles, abandoned coal mines, coke manufacture, and residential combustion of bituminous and anthracite coals. Occupational exposure to PAH is also of concern, particularly for workers at coal tar and aluminum production plants, road and roof tarring operations, and municipal incineration sites that burn carbonaceous materials (National Occupational Exposure Survey 1981–1983). It should be noted that a primary route of human exposure to PAH is also via cigarette and cigar smoke (Mattison and Thorgeirsson, 1978), as levels of BAP and dimethylbenz[a]anthracene (DMBA) have been estimated at 0.1 mg and 4 mg per cigarette, respectively (World Health Organization/International Agency for Research on Cancer, 1973). A link between maternal cigarette smoking during pregnancy and reduced fecundity of female offspring exposed *in utero* has been reported (Weinberg *et al.*, 1989). These findings, taken with data which indicate that women who smoke undergo the menopause at an earlier age (Mattison *et al.*, 1989), strongly suggest that chemicals present in cigarette smoke impair foetal oogenesis and destroy pre-existing germ cells established in the ovary. These epidemiological findings that implicate PAH as ovotoxicants in humans are fully supported by experimental data obtained through analysis of the actions of BAP and DMBA in numerous animal studies. Massive foetal germ cell loss has been reported to occur in pregnant female mice injected with BAP (Mackenzie and Angevine, 1981). These findings have been reinforced by several studies documenting the effects of BAP and related PAHs (such as DMBA and 3-methylcholanthrene) on oocyte loss and disrupted follicular dynamics in postnatal mouse ovaries (Mattison and Thorgeirsson, 1979, Mattison, 1980, Mattison and Nightingale, 1980).

Although a large volume of data has been reported for the ovotoxic properties of PAH over the past twenty years, comparatively little is known of the mechanisms by which PAH cause follicular destruction. Almost two decades ago, Mattison noted that the morphology of degenerating primordial oocytes in mice dosed with PAH *in vivo* resembled that which occurs during prenatal degeneration of the female germ line (Mattison, 1980). Based on the points already discussed, it is probable that apoptosis is the mechanism by which follicle loss occurs in response to PAH exposure (Tilly and Perez, 1997). Nevertheless, it remains uncertain how PAH trigger apoptosis in oocytes. Several lines of evidence have indicated that metabolic conversion of PAH to reactive intermediates may be an important step in the ovotoxic response. In support of this hypothesis, a positive correlation between ovarian metabolic activity (activity of aryl hydrocarbon hydroxylase or AHH, an enzyme encoded by the P4501A1 gene) and ovotoxicity of PAH in several strains of mice has been reported (Mattison and Thorgeirsson, 1979, Mattison and Nightingale, 1980). Based on these data it has been proposed that free radical-like species, such as the diol-epoxides generated by AHH-catalyzed PAH metabolism, damage DNA and other intracellular components, events that culminate in cellular dissolution. However, significant levels of oocyte destruction occur in mice which possess relatively low levels of ovarian AHH activity (Mattison and Nightingale, 1980), suggesting that metabolism of the PAH to a reactive intermediate(s) is not the only mechanism involved in ovotoxicity. Along these lines, data derived from studies using compounds which are thought to inhibit PAH metabolism, such as α-naphthoflavone (ANF), have been put forth to support the concept that metabolic conversion of the parent PAH compound to reactive intermediates is required for ovotoxicity. For example, animals treated with PAH in the presence of ANF, *in vivo*, do not exhibit oocyte destruction (Mattison and Thorgeirsson, 1979). However, interpretation of these data requires a great deal of caution as more recent studies have shown that ANF and related flavone compounds function as potent antagonists of the cytoplasmic binding protein for PAH, the aryl hydrocarbon receptor (AhR) (Lu *et al.*, 1995, Hankinson, 1995). Therefore, the ovotoxic effects of PAH may be linked to their specific interaction with the AhR and the subsequent regulation of genes responsive to the ligand-activated AhR, in addition to/in lieu of metabolism of PAH to reactive intermediates that perturb cellular homeostasis. Future studies to investigate this possibility, with apoptosis-regulatory genes as candidate targets for PAH/AhR-regulated transcription, will be of interest since many cell death gene knockout mice now exist to confirm or refute this hypothesis (see sections 5.2.1 and 5.2.2).



### 5.3.3 Vinylcyclohexene (VCH)

This third example of ovarian toxicity is interesting because VCH and its metabolites are, from all accounts, relatively specific gonadal toxicants (Hoyer and Sipes, 1996). The chemical synthesis of tyres, insecticides, flame retardants and plasticizers releases by-products into the environment such as 4-vinyl-1-cyclohexene (VCH) and its epoxide metabolites, 4-vinyl-1-cyclohexene 1,2-epoxide, 4-vinyl-1-cyclohexene 7,8-epoxide and 4-vinyl-1-cyclohexene diepoxide (VCD) (Hoyer and Sipes, 1996). Daily intraperitoneal injection of mice with VCH for 30 days has been reported to destroy 85% of the oocytes contained within the smallest ovarian follicles (primordial, primary), whereas similar treatment of rats did not result in any apparent toxic effects within the ovary (Hooser *et al.*, 1993,

1994). Three epoxide metabolites of VCH were also tested and were found to be up to 18-fold more potent than VCH in their ability to destroy small ovarian follicles in mice. Moreover, unlike VCH, these epoxide metabolites also caused oocyte loss in rats (Smith *et al.*, 1990b). Since epoxide metabolites were found to be more ovotoxic than the parent compound (VCH), it has been proposed that the conversion of VCH to the diepoxide form (VCD) is required for this compound to cause germ cell destruction. Data from *in vitro* studies suggest that the greater susceptibility of mouse ovaries, as compared to rat ovaries, to VCH is due to a species-related difference in the rate of epoxide formation in the liver (Smith *et al.*, 1990a). Mouse hepatic microsomes incubated with VCH form VCH-1,2-epoxide at a rate 6.8-fold greater than rat hepatic microsomes. Additionally, human liver microsomes convert VCH to the epoxide form at a rate half of that observed for rat hepatic microsomes (Smith and Sipes, 1991). Consequently, if the generation of epoxide metabolites is required for the ovotoxicity of VCH, the rat probably offers a more appropriate experimental model to predict the effects of VCH in humans.

A study of the long-term effects of VCH exposure on the ovary have revealed that treatment of mice with VCH for 30 days results in markedly reduced ovarian and uterine weights, premature ovarian failure, increased circulating levels of follicle-stimulating hormone (FSH) and pre-neoplastic changes within 360 days of the start of VCH treatment, as compared to control animals (Hooser *et al.*, 1994). In contrast to the irreversible toxic effects of VCH on the ovary, however, male mice treated in an identical fashion displayed decreased testicular weight and a loss of sperm production that were rapidly reversed following cessation of VCH exposure (Hooser *et al.*, 1995). To uncover the possible mechanisms underlying VCD-induced ovarian damage, Springer *et al.* (1996a,b) conducted a parallel set of investigations in rats to address the possibility that apoptosis and cell-death regulatory genes, respectively, were components of primordial and primary follicle loss. Following a daily dosing regimen for up to 30 days, morphological assessments of oocyte and granulosa cell health status confirmed the occurrence of apoptosis as a time-dependent response to VCD treatment (Springer *et al.*, 1996a). Moreover, increased expression of the cell-death susceptibility gene, *bax*, was identified in primordial and primary follicles following VCD exposure at a time preceding any evidence of follicular cell apoptosis or follicle loss (Springer *et al.*, 1996b). These findings support the hypothesis that apoptosis in the ovary, occurring naturally or driven by toxicant exposure, involves a Bax-mediated pathway of execution (see Table 5.1).

## 5.3.4 Anti-cancer drugs

Many classes of cancer chemotherapeutic agents damage DNA leading to cell death, and reproductive tissues are highly vulnerable to unfortunate side-effects of such toxicants. Girls and women treated with these chemicals to combat cancer often show fertility problems and premature ovarian failure at some point post-therapy (Horning *et al.*, 1981,

Waxman, 1983, Familiari *et al.*, 1993). In men faced with the long-term consequences of cancer treatment, fertility can be spared by collection and cryopreservation of sperm prior to the commencement of therapy for use in subsequent assisted reproductive technologies (Meirow and Shenker, 1995). However, a similar course of action is essentially out of reach for women since oocytes harvested for cryopreservation have extremely poor rates of post-thaw fertilization and embryo development (Trounson and Bongso, 1996). Recent efforts to show a recovery of oestrous cyclicity in animal models subjected to transplant of cryopreserved ovarian autografts offer promise for the future (Baird *et al.*, 1999). However, there currently exist no clinically-proven interventions that can be used to protect the ovaries, *in vivo*, from the ravages of chemotherapy.

Unfortunately, little is known of the mechanisms underlying this paradigm of clinically-induced ovarian damage. Importantly, however, a recent series of investigations using female mice has established that oocytes respond to at least one widely-prescribed antineoplastic agent, doxorubicin, by undergoing apoptosis (Perez *et al.*, 1997). These findings provide the first insights into how cancer drugs may cause infertility, and raise the possibility that specific manipulation of apoptosis-regulatory molecules in oocytes may one day yield new approaches to combat premature ovarian failure in female cancer patients (Reynolds, 1999). In support of this concept, pharmacologic suppression of ceramide function or caspase activity, or *bax* deficiency through gene knockout, in murine oocytes prevents the apoptotic response to doxorubicin (Perez *et al.*, 1997). By comparison, p53 deficiency had no protective effect in oocytes exposed to chemotherapy, arguing the existence of cell type-specific requirements for various molecules associated with cell death regulation.

The involvement of caspases in ovotoxicity was suggested by findings that showed peptide inhibitors of caspases prevented oocyte fragmentation caused by doxorubicin treatment (Perez *et al.*, 1997). Follow-up studies showed that in caspase-2-deficient female mice, oocytes were resistant to doxorubicin-induced death (Bergeron *et al.*, 1998). At present, there are no reported studies to confirm the role of ceramide in oocyte loss as deduced from the pharmacologic inhibition studies with sphingosine-1-phosphate (Perez *et al.*, 1997). It should be stressed, however, that Perez *et al.* (1997) identified hydrolytic pathways (i.e. sphingomyelinases) as opposed to synthetic pathways (ceramide synthase) as the source of ceramide that may be produced in oocytes exposed to doxorubicin. Therefore, future studies of the ovarian response of acid sphingomyelinase-deficient female mice (Horinouchi *et al.*, 1995) to antineoplastic drugs will certainly be of interest.

## 5.4 SUMMARY

With the growing amount of literature pointing to apoptosis as a principal effector mechanism underlying ovarian toxicity (and probably toxicity in non-gonadal tissues associated with female reproductive function as well), future efforts should continue to focus on elucidation of the intracellular events set in motion by xenobiotics in ovarian germ cells and somatic cells that lead to their untimely demise. Elaboration of the molecular framework through which specific chemicals wreak havoc in the ovary may one day lead to the development of new technologies to protect the precious reserve of female germ cells from premature death following toxicant exposure. One major consideration that must be met is designing approaches that specifically target ovarian cells, and in particular the germ line, for defense against pathologic insults. Although much work is needed to accomplish these goals, the recent identification of specific cell death genes being required for the ovotoxic response to chemotherapy (Perez *et al.*, 1997, Bergeron *et al.*, 1998), coupled with findings that oocyte-specific expression of protective molecules such as Bcl-2 (Morita *et al.*, 1999b) can be achieved, provide strong impetus to continue studies on understanding the impact of manipulating apoptosis in the context of female reprotoxicology.

## ACKNOWLEDGMENTS

The editor would like to thank Dr Sabina Cosulich for critical appraisal of this chapter and Zana Dye for invaluable editorial assistance. Work conducted by the author discussed herein was supported by NIH grants ROI-ES06999, ROI-ES08430, ROI-AG12279 and ROI-HD34226.

## REFERENCES

Baird, D.T., Webb, R., Campbell, B.K., Harkness, L.M. and Gosden, R.G., 1999, Long-term ovarian function in sheep after ovariectomy and transplantation of autografts stored at −196°C, *Endocrinology*, **140**, 462–471

Baker, T.G., 1963, A quantitative and cytological study of germ cells in human ovaries, *Proceedings of the Royal Society of London (Biology)*, **158**, 417–433

Beaumont, H.M. and Mandl, A.M., 1962, A quantitative and cytological study of oogonia and oocytes in the foetal and neonatal rat, *Proceedings of the Royal Society of London (Biology)*, **155**, 557–579

Bergeron, L., Perez, G.I., Mcdonald, G., Shi, L., Sun, Y., Jurisicova, A., *et al.*, 1998, Defects in regulation of apoptosis in caspase-2-deficient mice, *Genes and Development*, **12**, 1304–1314

Bishop, J.B., Witt, K.L. and Sloane, R.A., 1997b, Genetic toxicities of human teratogens, *Mutation Research*, **396**, 9–43

Bishop, J.B., Morris, R.W., Seely, J.C., Hughes, L.A., Cain, K.T. and Generoso, W.M., 1997a, Alterations in the reproductive patterns of female mice exposed to xenobiotics, *Fundamental and Applied Toxicology*, **40**, 191–204

BOLON, B., BUCCI, T.J., WARBRITTON, A.R., CHEN, J.J., MATTISON, D.R. and HEINDEL, J.J., 1997, Differential follicle counts as a screen for chemically induced ovarian toxicity: results from continuous breeding bioassays, *Fundamental and Applied Toxicology*, **39**, 1–10

BOONE, D.L. and TSANG, B.K., 1998, Caspase-3 in the rat ovary: localization and possible role in follicular atresia and luteal regression, *Biology of Reproduction*, **58**, 1533–1539

BORUM, K., 1961, Oogenesis in the mouse: a study of the meiotic prophase, *Experimental Cell Research*, **24**, 495–507

BUCCI, T.J., BOLON, B., WARBRITTON, A.R., CHEN, J.J. and HEINDEL, J.J., 1997, Influence of sampling on the reproducibility of ovarian follicle counts in mouse toxicity studies, *Reproductive Toxicology*, **11**, 689–696

BUEHR, M., 1997, The primordial germ cells of mammals: some current perspectives, *Experimental Cell Research*, **232**, 194–207

BYSKOV, A.G., 1986, Differentiation of the mammalian embryonic gonad, *Physiological Reviews*, **66**, 71–117

CHUN, S.-Y., BILLIG, H., TILLY, J.L., FURUTA, I., TSAFRIRI, A. and HSUEH, A.J.W., 1994, Gonadotropin suppression of apoptosis in cultured preovulatory follicles: mediatory role of endogenous insulin-like growth factor-I, *Endocrinology*, **135**, 1845–1853

COMMITTEE ON BIOLOGICAL EFFECTS OF ATMOSPHERIC POLLUTANTS, 1972, Particulate polycyclic organic matter: sources of polycyclic organic matter, National Academy of Sciences, Washington, DC, 13–35

COUCOUVANIS, E.C., SHERWOOD, S.W., CARSWELL-CRUMPTON, C., SPACK, E.G. and JONES, P.P., 1993, Evidence that the mechanism of prenatal germ cell death in the mouse is apoptosis, *Experimental Cell Research*, **209**, 238–247

DATTA, S.R., DUDEK, H., TAO, X., MASTERS, S., FU, H., GOTOH, Y. and GREENBERG, M.E., 1997, Akt phosphorylation of BAD couples survival signals to the cell-intrinsic death machinery, *Cell*, **91**, 231–241

DE POL, A., VACCINA, F., FORABOSCO, A., CAVAZUTTI, E. and MARZONA, L., 1997, Apoptosis of germ cells during human prenatal oogenesis, *Human Reproduction*, **12**, 2235–2241

DHARMARAJAN, A.M., GOODMAN, S.B., TILLY, K.I. and TILLY, J.L., 1994, Apoptosis during functional corpus luteum regression: evidence of a role for chorionic gonadotropin in promoting luteal cell survival, *Endocrine Journal (Endocrine)*, **2**, 295–303

DHARMARAJAN, A.M., HISHEH, S., SINGH, B., PARKINSON, S., TILLY, K.I. and TILLY, J.L., 1999, Anti-oxidants mimic the ability of chorionic gonadotropin to suppress apoptosis in the rabbit corpus luteum *in vitro*: a novel role for superoxide dismutase in regulating *bax* expression, *Endocrinology*, **140**, 2555–2561

ENAN, E., LASLEY, B., STEWART, D., OVERSTREET, J. and VANDEVOORT, C.A., 1996b, 2,3,7,8-tetrachlorodibenzo-p-dioxin (TCDD) modulates function of human luteinizing granulosa cells via cAMP signaling and early reduction of glucose transporting activity, *Reproductive Toxicology*, **10**, 191–198

ENAN, E., MORAN, F., VANDEVOORT, C.A., STEWART, D.R., OVERSTREET, J.W. and LASLEY, B.L., 1996a, Mechanism of action of 2,3,7,8-tetrachlorodibenzo-p-dioxin (TCDD) in cultured human luteinized granulosa cells, *Reproductive Toxicology*, **10**, 497–508

FAMILIARI, G., CAGGIATI, A., NOTTOLA, S.A., ERMINI, M., DI BENEDETTO, M.R. and MOTTA, P.M., 1993, Ultrastructure of human ovarian primordial follicles after combination chemotherapy for Hodgkin's disease, *Human Reproduction*, **8**, 2080–2087

FLAWS, J.A., KUGU, K., TRBOVICH, A.M., TILLY, K.I., DESANTI, A., HIRSHFIELD, A.N. and TILLY, J.L., 1995, Interleukin-1β-converting enzyme-related proteases (IRPs) and mammalian cell death: dissociation of IRP-induced oligonucleosomal endonuclease activity from morphological apoptosis in granulosa cells of the ovarian follicle, *Endocrinology*, **136**, 5042–5053

FORABOSCO, A., SFORZA, C., DE POL, A., VIZZOTTO, L., MARZONA, L. and FERRARIO, V.F., 1991, Morphometric study of the human neonatal ovary, *Anatomical Record*, **231**, 201–208

GAYTAN, F., BELLIDO, C., MORALES, C. and SANCHEZ-CRIADO, J.E., 1998, Both prolactin and progesterone in proestrus are necessary for the induction of apoptosis in the regressing corpus luteum of the rat, *Biology of Reproduction*, **59**, 1200–1206

GOODMAN, S.B., KUGU, K., CHEN, S.H., PREUTTHIPAN, S., TILLY, K.I., TILLY, J.L. and DHARMARAJAN, A.M., 1998, Estradiol-mediated suppression of apoptosis in the rabbit corpus luteum is associated with a shift in expression of *bcl-2* family members favoring cellular survival, *Biology of Reproduction*, **59**, 820–827

GOUGEON, A., 1996, Regulation of ovarian follicular development in primates: facts and hypotheses, *Endocrine Reviews*, **17**, 121–155

HANKINSON, O., 1995, The aryl hydrocarbon receptor complex., *Annual Review of Pharmacology and Toxicology*, **35**, 307–340

HEIMLER, I., RAWLINS, R.G., OWEN, H. and HUTZ, R.J., 1998b, Dioxin perturbs, in a dose- and time-dependent fashion, steroid secretion, and induces apoptosis of human luteinized granulosa cells, *Endocrinology*, **139**, 4373–4379

HEIMLER, I., TREWIN, A.L., CHAFFIN, C.L., RAWLINS, R.G. and HUTZ, R.J., 1998a, Modulation of ovarian follicle maturation and effects on apoptotic cell death in Holtzman rats exposed to 2,3,7,8-tetrachlorodibenzo-p-dioxin (TCDD) *in utero* and lactationally, *Reproductive Toxicology*, **12**, 69–73

HOOSER, S.B., PAROLA, L.R., VAN ERT, M.D. and SIPES, I.G., 1993, Differential ovotoxicity of 4-vinylcyclohexene and its analog 4-phenylcyclohexene, *Toxicology and Applied Pharmacology*, **119**, 302–305

HOOSER, S.B., DOUDS, D.P., DEMERELL, D.G., HOYER, P.B. and SIPES, I.G., 1994, Long-term ovarian and gonadotropin changes in mice exposed to 4-vinylcyclohexene, *Reproductive Toxicology*, **8**, 315–323

HOOSER, S.B., DEMERELL, D.G., DOUDS, D.A., HOYER, P.B. and SIPES, I.G., 1995, Testicular germ cell toxicity caused by vinylcyclohexene diepoxide in mice, *Reproductive Toxicology*, **9**, 359–367

HORINOUCHI, K., ERLICH, S., PERL, D.P., FERLINZ, K., BISGAIER, C.L., SANDHOFF, K., *et al.*, 1995, Acid sphingomyelinase deficient mice: a model of types A and B Niemann-Pick diseases, *Nature Genetics*, **10**, 288–293

HORNING, S.J., HOPPE, R.T., DAPLAN, H.S. and ROSENBERG, S.A., 1981, Female reproductive potential after treatment for Hodgkin's disease, *New England Journal of Medicine*, **304**, 1377–1381

HOSOKAWA, K., AHARONI, D., DANTES, A., SHAULIAN, E., SCHERE-LEVY, C., ATZMON, R., KOTSUJI, F., OREN, M., VLODAVSKY, I. and AMSTERDAM, A., 1998, Modulation of Mdm2 expression and p53-induced apoptosis in immortalized human ovarian granulosa cells, *Endocrinology*, **139**, 4688–4700

HOYER, P.B. and SIPES, I.G., 1996, Assessment of follicle destruction in chemical-induced ovarian toxicity, *Annual Review of Pharmacology and Toxicology*, **36**, 307–331

HUGHES, JR., F.M. and GOROSPE, W.C., 1991, Biochemical identification of apoptosis (programmed cell death) in granulosa cells: evidence for a potential mechanism underlying follicular atresia, *Endocrinology*, **129**, 2415–2422

JUENGEL, J.L., GARVERICK, H.A., JOHNSON, A.L., YOUNGQUIST, R.S. and SMITH, M.F., 1993, Apoptosis during luteal regression in cattle, *Endocrinology*, **132**, 249–254

JURISICOVA, A., LATHAM, K., CASPER, R.F. and VARMUZA, S.L., 1998, Expression and regulation of genes associated with cell death during murine preimplantation embryo development, *Molecular Reproduction and Development*, **51**, 243–253

KAMRIN, M.A., CARNEY, E.W., CHOU, K., CUMMINGS, A., DOSTAL, L.A., HARRIS, C., *et al.*, 1994, Female reproductive and developmental toxicology: overview and current approaches, *Toxicology Letters*, **74**, 99–119

KENNEDY, S.G., WAGNER, A.J., CONZEN, S.D., JORDAN, J., BELLACOSA, A., TSICHLIS, P.N. and HAY, N., 1997, The PI 3-kinase/Akt signaling pathway delivers an anti-apoptotic signal, *Genes and Development*, **11**, 701–713

KNUDSON, C.M., TUNG, K.S.K., TOURTELLOTTE, W.G., BROWN, G.A.J. and KORSMEYER, S.J., 1995, Bax-deficient mice with lymphoid hyperplasia and male germ cell death, *Science*, **270**, 96–99

KUGU, K., RATTS, V.S., PIQUETTE, G.N., TILLY, K.I., TAO, X.-J., MARTIMBEAU, S., *et al.*, 1998, Analysis of apoptosis and expression of *bcl-2* gene family members in the human and baboon ovary, *Cell Death and Differentiation*, **5**, 67–76

LI, X., JOHNSON, D.C. and ROZMAN, K.K., 1995, Reproductive effects of 2,3,7,8-tetrachlorodibenzo-*p*-dioxin (TCDD) in female rats: ovulation, hormonal regulation, and possible mechanisms, *Toxicology and Applied Pharmacology*, **133**, 321–327

LU, Y.-F., SANTOSTEFANO, M., CUNNINGHAM, B.D.M., THREADGILL, M.D. and SAFE, S., 1995, Identification of 3′-methoxy-4′-nitroflavone as a pure aryl hydrocarbon (Ah) receptor antagonist and evidence for more than one form of the nuclear Ah receptor in MCF-7 human breast cancer cells, *Archives of Biochemistry and Biophysics*, **316**, 470–477

MACKENZIE, K.M. and ANGEVINE, D.M., 1981, Infertility in mice exposed *in utero* to benzo(a)pyrene, *Biology of Reproduction*, **24**, 183–191

MARAVEI, D.V., TRBOVICH, A.M., PEREZ, G.I., TILLY, K.I., TALANIAN, R.V., BANACH, D., *et al.*, 1997, Cleavage of cytoskeletal proteins by caspases during ovarian cell death: evidence that cell-free systems do not always mimic apoptotic events in intact cells, *Cell Death and Differentiation*, **4**, 707–712

MATSUYAMA, S., CHANG, K.T., KANUKA, H., OHNISHI, M., IKEDA, A., NISHIHARA, M. and TAKAHASHI, M., 1996, Occurrence of deoxyribonucleic acid fragmentation during prolactin-induced structural luteolysis in cycling rats, *Biology of Reproduction*, **54**, 1245–1251

MATTISON, D.R., 1980, Morphology of oocyte and follicle destruction by polycyclic aromatic hydrocarbons in mice, *Toxicology and Applied Pharmacology*, **53**, 249–259

MATTISON, D.R., 1993, Sites of female reproductive vulnerability: implications for testing and risk assessment, *Reproductive Toxicology*, **7** (Suppl. 1), 53–62

MATTISON, D.R. and NIGHTINGALE, M.R., 1980, The biochemical and genetic characteristics of murine ovarian aryl hydrocarbon (benzo[a]pyrene) hydroxylase activity and its relationship to primordial oocyte destruction by polycyclic aromatic hydrocarbons, *Toxicology and Applied Pharmacology*, **56**, 399–408

MATTISON, D.R. and THORGEIRSSON, S.S., 1978, Smoking and industrial pollution and their effects on menopause and ovarian cancer, *Lancet*, **1**, 187–188

MATTISON, D.R. and THORGEIRSSON, S.S., 1979, Ovarian aryl hydrocarbon hydroxylase activity and primordial oocyte toxicity of polycyclic aromatic hydrocarbons in mice, *Cancer Research*, **39**, 3471–3475

MATTISON, D.R., PLOWCHALK, D.R., MEADOWS, M.J., MILLER, M.M., MALEK, A. and LONDON, S., 1989, The effect of smoking on oogenesis, fertilization and implantation, *Seminars in Reproductive Health*, **7**, 291–304

MEIROW, D. and SHENKER, J.G., 1995, Cancer and male fertility, *Human Reproduction*, **23**, 102–131

MORITA, Y., and TILLY, J.L., 1999, Oocyte apoptosis: like sand through an hourglass, *Developmental Biology*, **213**, 1–17

MORITA, Y., PEREZ, G.I. and TILLY, J.L., 1999a, Caspases are required for germ cell apoptosis during mouse fetal ovarian development, *Journal of the Society for Gynecologic Investigation*, **6**, 95A

MORITA, Y., PEREZ, G.I., MARAVEI, D.V., TILLY, K.I. and TILLY, J.L., 1999b, Targeted expression of Bcl-2 in mouse oocytes inhibits ovarian follicle atresia and prevents spontaneous and chemotherapy-induced oocyte apoptosis *in vitro*, *Molecular Endocrinology*, **13**, 841–850

MORITA, Y., MANGANARO, T.F., TAO, X.-J., MARTIMBEAU, S., DONAHOE, P.K. and TILLY, J.L., 1999, Requirement for phosphatidylinositol 3'-kinase in cytokine-mediated germ cell survival during fetal oogenesis in the mouse, *Endocrinology*, **140**, 941–949

NATIONAL OCCUPATIONAL EXPOSURE SURVEY 1981–1983, National Institute of Occupational and Safety Hazards (NIOSH), 1984

PATTON, P.E. and STOUFFER, R.L., 1991, Current understanding of the corpus luteum in women and nonhuman primates, *Clinical Obstetrics and Gynecology*, **34**, 127–143

PEREZ, G.I., KNUDSON, C.M., LEYKIN, L., KORSMEYER, S.J. and TILLY, J.L., 1997, Apoptosis-associated signaling pathways are required for chemotherapy-mediated female germ cell destruction, *Nature Medicine*, **3**, 1228–1232

PEREZ, G.I., ROBLES, R., KNUDSON, C.M., FLAWS, J.A., KORSMEYER, S.J. and TILLY, J.L., 1999, Prolongation of ovarian lifespan into advanced chronological age by Bax-deficiency, *Nature Genetics*, **21**, 200–203

POLAND, A. and KNUTSON, J.C., 1982, 2,3,7,8-Tetrachlorodibenzo-*p*-dioxin and related halogenated hydrocarbons: examination of the mechanism of toxicity, *Annual Review of Pharmacology and Toxicology*, **22**, 517–554

RATTS, V.S., FLAWS, J.A., KOLP, R., SORENSON, C.M. and TILLY, J.L., 1995, Ablation of *bcl-2* gene expression decreases the numbers of oocytes and primordial follicles established in the post-natal female mouse gonad, *Endocrinology*, **136**, 3665–3668

REYNOLDS, T., 1999, Cell death genes may hold clues to preserving fertility after chemotherapy, *Journal of the National Cancer Institute*, **91**, 664–666

RICHARDSON, S.J., SENIKAS, V. and NELSON, J.F., 1987, Follicular depletion during the menopausal transition: evidence for accelerated loss and ultimate exhaustion, *Journal of Clinical Endocrinology and Metabolism*, **65**, 1231–1237

ROBLES, R., TAO, X.-J., TRBOVICH, A.M., MARAVEI, D.V., NAHUM, R., PEREZ, G.I., TILLY, K.I. and TILLY, J.L., 1999, Localization, regulation and possible consequences of apoptotic protease-activating factor-1 (Apaf-1) expression in granulosa cells of the mouse ovary, *Endocrinology*, **140**, 2641–2644

RUEDA, B.R., HENDRY, I.R., TILLY, J.L. and HAMERNIK, D.L., 1999, Accumulation of *caspase-3* mRNA and induction of caspase activity in the ovine corpus luteum following prostaglandin-$F_{2\alpha}$ treatment *in vivo*, *Biology of Reproduction*, **60**, 1087–1092

RUEDA, B.R., HOYER, P.B., HAMERNIK, D.L. and TILLY, J.L., 1997a, Potential regulators of physiological cell death in the corpus luteum, in Tilly, J.L., Strauss, J.F. and Tenniswood, M. (eds), *Cell Death in Reproductive Physiology*, New York: Springer-Verlag, pp 161–181

RUEDA, B.R., TILLY, K.I., HANSEN, T.R., HOYER, P.B. and TILLY, J.L., 1995, Expression of superoxide dismutase, catalase and glutathione peroxidase in the bovine corpus luteum: evidence supporting a role for oxidative stress in luteolysis, *Endocrine*, **3**, 227–232

RUEDA, B.R., TILLY, K.I., BOTROS, I., JOLLY, P.D., HANSEN, T.R., HOYER, P.B. and TILLY, J.L., 1997b, Increased *bax* and interleukin-1β-converting enzyme messenger RNA levels coincide with apoptosis in the bovine corpus luteum during structural regression, *Biology of Reproduction*, **56**, 186–193

SASTRY, B.V., 1991, Placental toxicology: tobacco smoke, abused drugs, multiple chemical interactions, and placental function, *Reproduction, Fertility and Development*, **3**, 355–372

SHARARA, F.I., SEIFER, D.B. and FLAWS, J.A., 1998, Environmental toxicants and female reproduction, *Fertility and Sterility*, **70**, 613–622

SHIKONE, T., YAMOTO, M., KOKAWA, K., YAMASHITA, K., NISHIMORI, K. and NAKANO, R., 1996, Apoptosis of human corpora lutea during cyclic luteal regression and early pregnancy, *Journal of Clinical Endocrinology and Metabolism*, **81**, 2376–2380

SILBERGELD, E.K., 1995, Understanding risk: the case of dioxin, *Scientific American*, November–December, 48–57

SMITH, B.J. and SIPES, I.G., 1991, Epoxidation of 4-vinylcyclohexene by human hepatic microsomes, *Toxicology and Applied Pharmacology*, **109**, 367–371

SMITH, B.J., CARTER, D.E. and SIPES, I.G., 1990a, Comparison of the disposition and *in vitro* metabolism of 4-vinylcyclohexene in the female mouse and rat, *Toxicology and Applied Pharmacology*, **105**, 364–371

SMITH, B.J., MATTISON, D.R. and SIPES, I.G., 1990b, The role of epoxidation in 4-vinylcyclohexene-induced ovarian toxicity, *Toxicology and Applied Pharmacology*, **105**, 372–381

SPIELMANN, H., 1998, Reproduction and development, *Environmental Health Perspectives*, **106** (Suppl. 2), 571–576

SPRINGER, L.N., TILLY, J.L., SIPES, I.G. and HOYER, P.B., 1996b, Enhanced expression of *bax* in small preantral follicles during 4-vinylcyclohexene diepoxide-induced ovotoxicity in the rat, *Toxicology and Applied Pharmacology*, **139**, 402–410

SPRINGER, L.N., MCASEY, M.E., FLAWS, J.A., TILLY, J.L., SIPES, I.G. and HOYER, P.B., 1996a, Involvement of apoptosis in 4-vinylcyclohexene diepoxide-induced ovotoxicity in rats, *Toxicology and Applied Pharmacology*, **139**, 394–401

TILLY, J.L., 1996, Apoptosis and ovarian function, *Reviews of Reproduction*, **1**, 162–172

TILLY, J.L., 1998, Cell death and species propagation: molecular and genetic aspects of apoptosis in the vertebrate female gonad, in LOCKSHIN, R.A., ZAKERI, Z. and TILLY, J.L.

(eds), *When Cells Die. A Comprehensive Evaluation of Apoptosis and Programmed Cell Death*, New York: Wiley-Liss, 431–452

Tilly, J.L. and Perez, G.I., 1997, Mechanisms and genes of physiological cell death: a new direction for toxicological risk assessments?, in Sipes, I.G., McQueen, C.A. and Gandolfi, A.J. (eds), *Comprehensive Toxicology*, Oxford: Elsevier Press, **10**, 379–385

Tilly, J.L. and Tilly, K.I., 1995, Inhibitors of oxidative stress mimic the ability of follicle-stimulating hormone to suppress apoptosis in cultured rat ovarian follicles, *Endocrinology*, **136**, 242–252

Tilly, J.L., Tilly, K.I. and Perez, G.I., 1997, The genes of cell death and cellular susceptibility to apoptosis in the ovary: a hypothesis, *Cell Death and Differentiation*, **4**, 180–187

Tilly, J.L., Kowalski, K.I., Johnson, A.L. and Hsueh, A.J.W., 1991, Involvement of apoptosis in ovarian follicular atresia and postovulatory regression, *Endocrinology*, **129**, 2799–2801

Tilly, J.L., Tilly, K.I., Kenton, M.L. and Johnson, A.L., 1995a, Expression of members of the *bcl-2* gene family in the immature rat ovary: equine chorionic gonadotropin-mediated inhibition of apoptosis is associated with decreased *bax* and constitutive *bcl-2* and $bcl\text{-}x_{long}$ messenger ribonucleic acid levels, *Endocrinology*, **136**, 232–241

Tilly, K.I., Banerjee, S., Banerjee, P.P. and Tilly, J.L., 1995b, Expression of the p53 and Wilms' tumor suppressor genes in the rat ovary: gonadotropin repression *in vivo* and immunohistochemical localization of nuclear p53 protein to apoptotic granulosa cells of atretic follicles, *Endocrinology*, **136**, 1394–1402

Trbovich, A.M., Hughes, Jr., F.M., Perez, G.I., Kugu, K., Tilly, K.I., Cidlowski, J.A. and Tilly, J.L., 1998, High and low molecular weight DNA cleavage in ovarian granulosa cells: characterization and protease modulation in intact cells and in cell-free nuclear autodigestion assays, *Cell Death and Differentiation*, **5**, 38–49

Trounson, A.O. and Bongso, A., 1996, Fertilisation and development in humans, *Current Topics in Developmental Biology*, **32**, 59–101

Tsutsumi, O., Uechi, H., Sone, H., Yonemoto, J., Takai, Y., Momoeda, M., *et al.*, 1998, Presence of dioxins in human follicular fluid: their possible stage-specific action on the development of preimplantation mouse embryos, *Biochemical and Biophysical Research Communications*, **250**, 498–501

Wang, L., Miura, M., Bergeron, L., Zhu, H. and Yuan, J., 1994, Ich-1, an Ice/ced-3-related gene, encodes both positive and negative regulators of programmed cell death, *Cell*, **78**, 739–750

Waxman, J., 1983, Chemotherapy and the adult gonad: a review, *Journal of the Royal Society of Medicine*, **76**, 144–148

Witty, J.P., Bridgham, J.T. and Johnson, A.L., 1996, Induction of apoptotic cell death in hen granulosa cells by ceramide, *Endocrinology*, **137**, 5269–5277

World Health Organization (WHO)/International Agency for Research on Cancer (IARC), 1973, Information bulletin on the survey of chemicals being tested for carcinogenicity, volume 3

CHAPTER 6

# Modulation of Apoptosis in Renal Toxicity

**MYRTLE DAVIS**

Department of Pathology, University of Maryland, Baltimore, Maryland 21201-1192, USA

## Contents

## 6.1 INTRODUCTION

Apoptosis is a highly regulated mechanism of cell death that plays a role in pathogenic organ responses following toxicant exposure (see Chapter 1). There are many toxicants that induce apoptosis in kidney or renal cells although the relationship between aberrant triggering of the process by toxicants and renal toxicity remains to be defined. Perhaps the relatively recent interest in apoptosis as a mechanism of toxicant-induced renal damage can be correlated with the need to examine the more subtle effects of low level environmental exposures, minimizing extrapolation from higher doses. This chapter outlines identification of apoptosis in the kidney then explores its role and regulation in normal kidney development and function. Next, several renal toxicants are described followed by a discussion of the molecular mechanisms that may underpin their mode of action.

## 6.2 APOPTOSIS IN THE KIDNEY

Apoptosis is a term first introduced in the 1970s to define the morphology of dying cells (Kerr *et al.*, 1972) although Glucksmann and colleagues described the same morphologic alterations in the kidney during the early 1950s (Glucksmann, 1951). Apoptosis during kidney development and after tubular damage permits proper nephron formation and tubular repair, respectively.

### 6.2.1 Renal apoptosis: morphology and detection

As with other tissues and organs, the diagnosis of kidney apoptosis is not solely by morphology, but also by the frequency, distribution and timecourse of occurrence. Also, apoptotic bodies in the kidney and elsewhere typically undergo apoptotic necrosis (or secondary necrosis; Chapters 1 and 10) within the phagolysosomes of the engulfing cell (Figure 6.1). In contrast, cells not engulfed will proceed to a necrotic phase in the tissue and cellular contents will spill into the extracellular space, often inciting inflammation.

Histopathologic examination of fixed cells or tissue sections is required for a definitive diagnosis of apoptosis in renal tissue or renal cell culture. As described in Chapter 10, various stains make nuclear fragmentation readily apparent in fixed tissues or fixed cells (McGahon *et al.*, 1995) and an acridine orange/ethidium bromide combination is frequently used in cell culture to distinguish early, late or necrotic kidney cells (Darzynkiewicz *et al.*, 1994). Similarly, Hoechst 33258 or DAPI will detect chromatin fragmentation and/or condensation. DNA fragmentation can be detected using agarose gel electrophoresis, the JAM assay, *in situ* end labelling or terminal deoxytidyl transferase-mediated dUTP nick-end labelling (ISEL or TUNEL) or anti-histone antibodies (which detect histone bound DNA fragments (Walker *et al.*, 1993, Wijsman *et al.*, 1993) (Chapter 10). In kidney, there is usually no inflammatory response mounted against apoptotic

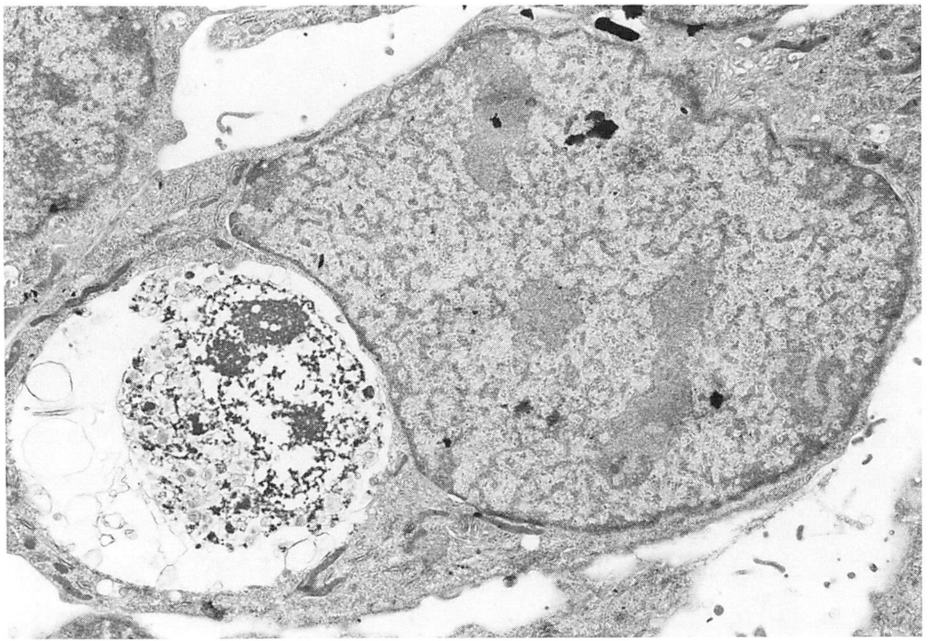

**Figure 6.1** Cell engulfment and secondary necrosis. Primary cultures of rat proximal tubules were exposed to TNFα *in vitro* and processed for electron microscopy. A proximal epithelial cell is shown in the cytoplasm of a neighbouring cell after phagocytosis. The engulfed cell shows morphologic evidence of secondary necrosis. Reproduced with permission from *Toxicologic Pathology*, 1998.



bodies since they are quickly ingested by neighbouring cells or shed into the tubular lumen. However, as mentioned earlier, a combination of apoptosis/oncosis and necrosis can occur *in vivo* which will attract inflammatory cells to the area. Recognition and engulfment of apoptotic cells and cell fragments is associated with externalization of phosphatidylserine (PS) from the inner plasma membrane to the cell surface (Stuart *et al.*, 1998). There are numerous techniques available to detect many of the biochemical changes that occur during apoptosis and some of these are covered in Chapter 10 of this volume. Many of these techniques are also useful in mechanistic studies. The biochemical tests chosen should be used to complement, but not to replace morphologic examination.

### 6.2.2 Apoptosis during kidney development

The ureteric bud is an outgrowth of the wolfian ducts that secretes a variety of factors which protect the mesenchymal cells from apoptosis whereas diffusion-limited basolateral factors trigger the conversion of mesenchyme into epithelium (Barasch *et al.*, 1996) (reviewed in Davies, 1993, Lechner and Dressler, 1997). Generally, apoptosis is the fate of the uninduced, undifferentiated mesenchymal cells and cells residing in the walls of tubular structures (Coles *et al.*, 1993) such as type B intercalated cells of the medullary

**TABLE 6.1**
Roles of apoptosis in renal disease

| Disease | Role of apoptosis in pathogenesis | Sites of apoptosis | References |
|---|---|---|---|
| Congenital dysplastic kidneys | Contributes to renal malformation | Mesenchymal cells surrounding the dysplastic tubules | Kreidberg *et al.*, 1993, Winyard *et al.*, 1996 |
| Diabetic nephropathy | May cause atrophy of tubular epithelium | Tubular epithelium | Ortiz-Arduan *et al.*, 1996 |
| Glomerulonephritis | Resolution of the disease; progression to glomerular sclerosis | Glomeruli | Takemura *et al.*, 1995, Shimizu *et al.*, 1996a,b |
| Glomerular sclerosis | Correlates with loss of renal function | Tubular epithelium; mesangial cells, interstitial cells, endothelial cells | Shimizu *et al.*, 1996 |
| Hemolytic uremic syndrome | Destruction of microvasculature | Endothelial cells Tubular epithelium | Uchida *et al.*, 1997 Williams *et al.*, 1997 |
| HIV nephropathy | Increased apoptosis in tubular epithelium leads to decreased tubular function | Tubular epithelium | Bodi *et al.*, 1995 |
| Obstructive uropathy and hydronephrosis | Contributes to tubular atrophy and renal tissue loss | Tubular epithelium Interstitial cells | Chevalier *et al.*, 1996 Cummings *et al.*, 1996, Truong *et al.*, 1996 |
| Polycystic kidney disease | Loss of normal renal tissue; may account for loss of renal function | Interstitial cells Cystic epithelium Proximal tubular epithelium | Lanoix *et al.*, 1996 Winyard *et al.*, 1996 Woo, 1995 |
| Renal cancer | May modify tumour growth; susceptibility may also determine response to chemotherapeutic agents | Tumour specific | Bardeesy *et al.*, 1995, Chandler *et al.*, 1994, Hofmockel *et al.*, 1996, Paraf *et al.*, 1995, Tannapfel *et al.*, 1997 |

collecting duct (Kim *et al.*, 1996a). Without the inducing effect of the ureteric bud, the entire nephrogenic mesenchyme undergoes widespread apoptosis (Koseki, 1993). The thin ascending limb of the loop of henle is derived from the thick ascending limb by apoptotic deletion of the thick ascending limb followed by cell transformation whereas, the type A intercalated cells are removed by extrusion from the epithelium (Kim *et al.*, 1996b). There is a distinct temporal pattern of apoptotic deletion during rat kidney development. Approximately 2.7% of cells within the nephrogenic zone are apoptotic during the embryonic period which declines to a basal level of 0.15% by the fourteenth postnatal day (Coles *et al.*, 1993). A second peak of apoptosis occurs in the medullary papilla at postnatal days 6–7 when the percentage of apoptotic cells reached 3.2%. Peak apoptosis in the medullary papilla correlates temporally with deletion of the thick ascending limb cell during the formation of the ascending thin limb of the loop of Henle (Kim *et al.*, 1996b). The predicted effect of renal toxicant exposure *in utero* will depend on the cellular events during that developmental period and whether the toxicant induces or inhibits apoptosis. An inducer of apoptosis may kill cell populations that are required for normal tubular function whereas an apoptosis inhibitor may result in malformed tubules of cells with a poorly differentiated phenotype. Many of the renal diseases associated with altered apoptosis are summarized in Table 6.1.

## 6.3 MOLECULAR CONTROL OF RENAL APOPTOSIS

Tremendous progress has been made in understanding the intracellular mechanisms of apoptosis and these pathways are reviewed in detail in Chapter 1. However, certain factors have particular relevance for the kidney; these are discussed below.

### 6.3.1 Tumour necrosis factor α (TNFα) and Fas



The tumour necrosis factor (TNF) receptor family is one of the best characterized of the receptor-mediated cell death pathways (see Chapters 1 and 2). TNFα and the Fas ligand Fas-L) are the cytokine ligands of the TNF superfamily which, upon binding to their receptors, engage the apoptotic pathway via a region of intracellular homology named the 'death domain' (Vandevoorde *et al.*, 1997). Neither receptor has intrinsic catalytic activity, but through the death domain, they induce a cascade of protein interactions that result in activation of apoptotic executioners such as caspases (Muzio *et al.*, 1998, for review see Baker and Reddy, 1996) (Chapter 2). The regulation of Fas and TNFα pathways may also involve the activation of pro-apoptotic protein kinases such as p38 (Goillot *et al.*, 1997) and Jun N terminal Kinase (JNK), sphingomyelinase activation and generation of sphingomyelin metabolites such as ceramide (Tepper *et al.*, 1995, Sweeney *et al.*, 1996, Ichijo *et al.*, 1997). Tumour necrosis factor (TNF) has been shown to cause apoptosis in numerous renal cells in culture including mouse mesangial cells, LLCPK, and mouse proximal tubular cells (Pelletier *et al.*, 1991, Liu *et al.*, 1996, Ortiz-Arduan *et al.*, 1996). Indeed, decreased susceptibility of renal tumour cells to TNF-induced apoptosis is also an important mechanism of chemotherapeutic resistance (Woo *et al.*, 1996). Fas-L expressed on lymphocytes triggers apoptosis in renal cell carcinoma cell lines and murine mesangial and tubular epithelial cells (Ortiz-Arduan *et al.*, 1996, Tomita *et al.*, 1996). The TNF family includes the cell surface receptor CD95 (also called Fas or APO-1), nerve growth factor receptor, CD40 and CD30 (120). CD95 differs from the TNF receptor in that normal human renal cells do not express the CD95 receptor (Kato *et al.*, 1997).

### 6.3.2 Granzyme B

Granzyme B is a serine esterase produced by cytotoxic T lymphocytes and natural killer cells. In most cases, granzyme B enters the targeted cell via perforin mediated pore formation, but it can also cross the cellular membrane independent of perforin (Shi *et al.*, 1997). Within the targeted cell, granzyme B causes proteolysis of intracellular components and apoptosis (Williams and Henkart, 1994). Studies on renal cells in culture indicate that expression of the nuclear kinase Wee1 can protect against granzyme B and perforin induced apoptosis (Chen *et al.*, 1995b). Interestingly, increased levels of granzyme B and perforin correlate with acute rejection of renal transplants (Kummer *et al.*, 1995, Sharma *et al.*, 1996, Strehlau *et al.*, 1997, Suthanthiran, 1997).

### 6.3.3 Cysteine aspartic serine proteases (caspases)

The caspases are a family of cysteine aspartic serine proteases (reviewed in Cohen, 1997) decribed in detail in Chapter 2. Two members of the caspase family, caspase 1 (interleukin 1 converting enzyme; ICE) and Nedd have been localized within the murine foetal kidney (Kumar *et al.*, 1994) and caspase 3 (CPP32) has been localized within the human renal tubular epithelium (Krajewski *et al.*, 1996). The essential role of caspases in apoptosis appears to be in the execution phase although a common feature of the identified caspase substrates (see Chapter 2) is that they all play a vital role in maintenance of cellular homeostasis and function. Recent studies suggest that numerous signal transduction proteins are caspase substrates and that their cleavage contributes to cell death (Krajewski *et al.*, 1996, Widmann *et al.*, 1998a, b). Mitogen activated protein kinase kinase kinase (MEKK) was shown to be cleaved to a 91 kDa fragment that activates caspases. In addition, Cbl, Cblb, Raf, Focal adhesion kinase and AKT appear to be substrates for caspases. Additional studies are certain to identify many additional caspase substrates in the coming years.

### 6.3.4 The Bcl-2 family

The Bcl-2 family of proteins is a closely associated group of intracellular pro- and anti-apoptotic regulators. Six members of the family have been specifically mapped to various locations within the kidney; Bcl-2, Bcl-XL, Bfl-1, Bax, Bak, MCL-1 and BAD (Eguchi *et al.*, 1992, Gonzalez-Garcia *et al.*, 1994, McDonnell *et al.*, 1994, D'Sa *et al.*, 1996, Krajewski *et al.*, 1996, Sorenson *et al.*, 1996). During renal development, Bcl-2 expression has been reported in the murine ureteric bud and tubular structures (Novack and Korsmeyer, 1994) and in induced human mesenchymal cells (McDonnell *et al.*, 1994). Intracellularly, Bcl-2 is localized with mitochondrial, endoplasmic reticulum and nuclear membranes (for review of the Bcl-2 family see Korsmeyer *et al.*, 1995, Kroemer, 1997). Bcl-2 also has unique distribution during mitosis; it appears in early prophase or late G2 and rapidly disappears at telophase. There appeared to be a specific concentration of Bcl-2 at the margins of condensed chromatin (Willingham and Bhalla, 1994).

The anti-apoptotic proteins of the Bcl-2 family are uniquely associated with regulation of apoptosis in the kidney. During murine development, a lack of Bcl-2 results in increased apoptosis and cystogenesis (Sorenson *et al.*, 1995, 1996). In contrast, overexpression of Bcl-2 may contribute to the development of renal neoplasia. The mechanism(s) by which Bcl-2 family members function are presently unclear. There are numerous biological effects of Bcl-2 on intact cells and as a result, various mechanisms have been proposed to explain how the Bcl-2 family members functionally regulate apoptosis. The proposed mechanisms of anti-apoptotic function include protection of membranes from oxidant damage, antagonism of pro-apoptotic proteins such as Bax and protection of organelle membranes by regulating permeability via ion channel formation (Fabisiak *et al.*, 1997, Minn *et al.*, 1997, Schendel *et al.*, 1997). In addition, serine phosphorylation also appears to modulate the function of some Bcl-2 family members such as Bcl-2 and BAD (Haldar *et al.*, 1996, Zha *et al.*, 1996). In a general sense, renal epithelial cells are unique in their physiological relevance to many of the proposed mechanisms of Bcl-2 function since ion channels, response to oxidant stress, and pH regulation are normal functions carried out by renal epithelial cells *in vivo*.



### 6.3.5 p53 and WT-1

P53 is a zinc finger transcription factor which induces the transcription of a variety of proteins including p21, Gadd45, and Gadd153 (Jeong *et al.*, 1997a, b) which are involved in cell cycle arrest or DNA repair. High expression of p53 protein in murine renal development is associated with increased apoptosis and kidney failure in mice (Godley *et al.*, 1996). Similarly, the Wilms' tumour suppression gene (WT-1) is an important regulator of apoptosis in kidney (Englert *et al.*, 1995) and plays a critical role in renal development. Homozygous mutation of murine WT-1 is lethal and results in failure of the development of many organs including the kidney (Kreidberg *et al.*, 1993). Heterozygous mutations have been associated with formation of Wilms' tumour (Little *et al.*, 1992) and urogenital anomalies associated with Denys-Drash syndrome (Pelletier *et al.*, 1991). The mechanism by which WT-1 regulates apoptosis is unclear. However, in human kidney cells, WT-1 has been shown to suppress the synthesis of Bcl-2 and c-myc and to inhibit p53-induced apoptosis in rat kidney (Hewitt *et al.*, 1995, Maheswaran *et al.*, 1995).

### 6.3.6 Growth factors

Growth factors provide a survival signal that suppresses apoptosis in most cell populations; growth factor deprivation induces one of two responses, quiescence or apoptosis. Apoptosis in the metanephric mesenchyme is prevented by epidermal growth factor (EGF) and basic fibroblast growth factor (bFGF). In the adult kidney, acceleration of renal repair following tubular damage can be achieved by exogenous administration of growth factors (Takeda and Endou, 1996). Classically, growth factors were thought to contribute to renal repair by activating cell signalling pathways that induce cell proliferation. However, it is now apparent that growth factors also serve to inhibit apoptosis in the kidney both during development and in renal repair of adult kidneys independent of proliferative effects. The response of the cell to growth factors may depend upon the state of the cell since EGF was shown to abrogate apoptosis of renal epithelial cells without affecting mitotic index (Coles *et al.*, 1993). Similarly, exogenous insulin growth factor (IGF-1), EGF or hepatocyte growth factor (HGF) administered in the early phase post ischemia accelerates renal tubular recovery (Humes *et al.*, 1989). The precise roles of HGF and IGF-1 in apoptosis in the kidney is unclear. However, IGF-1 was shown to prevent apoptosis in serum-deprived PC-12 cells, correlating with an increase in *bcl-x* mRNA and protein (Parrizas and LeRoith, 1997). After ischemic renal injury, IGF-1 expression is localized to the regenerative zones of the renal tubule (Lake and Humes, 1994). In addition, the therapeutic effects of hepatocyte growth factor associated with suppression of apoptosis have been demonstrated in drug-induced nephrotoxicity (Woo *et al.*, 1996).

## 6.3.7 Extracellular matrix interactions

Proper interaction with the extracellular matrix is necessary to prevent spontaneous apoptosis in most epithelial cells. Anoikis (pronounced an-o-É-kis) is a special term used to describe the entry of epithelial or endothelial cells into apoptosis when they lose contact with the extracellular matrix (Frisch, 1994). Activation of apoptosis through detachment is especially important *in vivo* because it insures that detached cells are unable to proliferate or survive in an inappropriate location after detachment inhibiting metastasis. The intracellular signals linking the extracellular matrix to cell response pathways include phosphoinositide 3-OH (PI3) kinase, Ras, focal adhesion kinase and the mitogen activated protein kinase pathways (Frisch and Dolter, 1995, Khan *et al.*, 1995, Cardone *et al.*, 1997). The intracellular signals regulated by extracellular matrix interaction at the cell surface appear to be linked through the integrin family of cell surface proteins (Saelman *et al.*, 1995, Strater *et al.*, 1997). Disrupting the ability alpha 2 beta 1 integrin to bind to the type I and IV collagen caused increased apoptosis in the Madin-Darby canine kidney (MDCK) (Saelman *et al.*, 1995) and inhibition of focal adhesion kinase, a cytoplasmic tyrosine kinase which acts as an integrin transducer, induces apoptosis (Frisch *et al.*, 1996). Cell matrix regulated signal pathways may also mediate the function of anti-apoptotic and pro-apoptotic proteins such as Bcl-2 (Frisch *et al.*, 1996, Zha *et al.*, 1996). Phosphorylation of BAD (a pro-apoptotic member of the Bcl-2 family) by AKT in a PI3-kinase dependent pathway was shown to inactivate BAD and promote cell survival (del Peso *et al.*, 1997). Similarly, cell-to-cell interactions that sensitize cells to anoikis have been shown to down-regulate *bcl-2* mRNA and protein (Frisch *et al.*, 1996).

Alternatively, the cell's connection to the extracellular matrix may influence apoptosis by directly influencing cell spreading. The degree of cell spreading of endothelial cells was key to regulating apoptosis in endothelial cells (Chen *et al.*, 1997). Since the cell shape is dictated by the number of attachments the cell has with the basement membrane, an increase in the area that a cell can occupy will allow the cells to spread and suppress cell death in the remaining cells. An example of a situation in which the area for cell spreading increases can be found in acute tubular necrosis when cells are lost from the basement membrane due to tubular damage.

### 6.3.8 Clusterin

Clusterin (also termed sulphated glycoprotein-2 (SGP2), testosterone repressed prostate message-2 (TRPM-2) or apoplipoprotein) may be linked mechanistically to apoptosis in the kidney due to its role in aggregation of renal epithelial cells (French *et al.*, 1994). However, although the association is quite clear, the precise regulatory role of clusterin in renal apoptosis is still unknown. The increase in clusterin expression may act as a protection mechanism in the damaged tubules, promoting cell interaction and cell signalling in the surviving cells.

### 6.3.9 Protein phosphorylation

Protein phosphorylation is a protein modification that regulates the activity of a number of cellular factors involved in apoptosis. The active state of the substrate protein will ultimately determine the effect of phosphorylation on protein activity and the cellular response to the modification. Multiple phosphorylation based signal transduction cascades such as the mitogen activated protein kinase family (MAPK) and stress activated protein kinases (SAPK) appear to be involved in mediating apoptosis in kidney. A pattern of kinase activation (p38, extracellular regulatory kinase (ERK) and JNK) together with stress response genes (c-jun, ATF-2) correlates with apoptosis in kidney after ischemia reperfusion (Xia *et al.*, 1995). The importance of this correlation is underscored by the improvement in renal function and improved outcome of ischemic renal failure afforded by down-regulation of MAPK (Di Mari, 1997a). The precise role of kinase activation in ischemia reperfusion is still unclear since these kinases can regulate proliferative and apoptotic pathways.

Protein phosphatases, PP1 and PP2A, appear to regulate proliferation in early embryonic kidney with PP1 and PP2A expression highest in E15 kidney (Svennilson *et al.*, 1995). In neurons, a balance between growth factor activated ERK and p38/JNK is proposed to be the switch for apoptosis or cell proliferation (Woolf *et al.*, 1995). A similar balance in kinase and phosphatase activity may be regulatory in renal cells. Kinase signalling pathways are especially relevant to toxicants since they are the major intracellular transducers for endogenous molecules and thus could be employed by a number of exogenous chemicals.



LIVERPOOL JOHN MOORES UNIVERSITY
LEARNING & INFORMATION SERVICES

## 6.4 TOXICANT-INDUCED RENAL APOPTOSIS

A growing list of toxicants and natural toxins are associated with alterations in renal apoptosis (Tables 6.2 and 6.3). Although various heavy metals, pharmacologic agents and naturally occurring toxins cause apoptosis in the kidney, the mechanisms and molecular

**TABLE 6.2**
Pro-apoptotic toxins and toxicants

| Toxin/Toxicant | Class | Source(s) |
|---|---|---|
| *Alternaria alternata lycopersici* toxin | Mycotoxin | Wang *et al.*, 1996<br>Winter *et al.*, 1996 |
| Cadmium | Heavy metal | Hamada *et al.*, 1991,<br>Hamada *et al.*, 1994 |
| Ceramide | Sphingolipid metabolite | Jarvis *et al.*, 1994 |
| Cisplatin | Antineoplastic drug | Leiberthal *et al.*, 1996,<br>Takeda *et al.*, 1996 |
| Curcumin | Dietary pigment/spice | Jiang *et al.*, 1996 |
| Cyclophosphamide | Antineoplastic drug | Cha *et al.*, 1996 |
| Cyclosporin A | Immunosuppressant | Cha *et al.*, 1996 |
| Dexamethasone | Antineoplastic drug | Cha *et al.*, 1996 |
| Doxorubicin | Antineoplastic drug | Zhang *et al.*, 1996 |
| Dithiothreitol | Reducing agent | Van de Water *et al.*, 1996a |
| Fumonisins | Mycotoxin | Tolleson *et al.*, 1996,<br>Voss *et al.*, 1996, Wang *et al.*, 1996 |
| Gliotoxin | Mycotoxin | Waring *et al.*, 1994 |
| Hygromycin | Antibiotic | Chen *et al.*, 1995a |
| Melphalan | Antineoplastic drug | Cha *et al.*, 1996 |
| Mercuric chloride | Heavy metal | Duncan-Achanzar *et al.*, 1996,<br>Nath *et al.*, 1996 |
| Metolazone | Thiazide diuretic | Loffing *et al.*, 1996 |
| Microcystin LR | Cyanobacterial toxin | Khan *et al.*, 1995,<br>MacKintosh *et al.*, 1990,<br>MacKintosh *et al.*, 1995 |
| Ochratoxin A | Mycotoxin | Seegers *et al.*, 1994 |
| Okadaic Acid | Poly ether fatty acid | Davis *et al.*, 1994, 1996 |
| Ricin | Ribosome inactivating protein | Williams *et al.*, 1997 |
| s-(1,2 dichlorovinyl)-L-cysteine | Reducing agent | Van de Water *et al.*, 1996a |
| Verocytotoxin | *E. coli* enterotoxin | Williams *et al.*, 1997 |

**TABLE 6.3**
Pro and anti-apoptotic viral proteins

| Virus/Viral protein | Source(s) |
|---|---|
| Adenovirus E1A protein | Frisch *et al.*, 1995, Mymryk *et al.*, 1994, Rao *et al.*, 1992, Sabbatini *et al.*, 1995a |
| Adenovirus E1A12S protein | Quinlan *et al.*, 1993 |
| Adenovirus E1B protein | Chiou *et al.*, 1994b, Sabbatini *et al.*, 1995b |
| Cowpox (crmA negative) | Ray *et al.*, 1996 |
| Dearing reovirus (type 3) | Rodgers *et al.*, 1997 |
| Epstein Barr virus BHRF 1 protein | Theodorakis *et al.*, 1996 |
| Human papilloma virus E6 protein | Thomas *et al.*, 1996 |
| Influenza A, B and C | Hechtfischer *et al.*, 1997, Hinshaw *et al.*, 1994 |
| Lang reovirus (Type 1) | Rodgers *et al.*, 1997 |
| Sindbis virus | Levine *et al.*, 1993 |

pathways employed are diverse. Exposure to renal toxicants that affect apoptosis in kidney can result in altered renal function and contribute to or cause renal disease.



## 6.4.1 Natural toxins

Many natural toxins such as verocytotoxin and ricin are potent inducers of apoptosis. Verocytotoxin (produced by *E. coli*) induces apoptosis in renal cells which is potentiated by tumour necrosis factor beta (TNFβ) (Uchida *et al.*, 1997). TNFβ increases the number of verocytotoxin receptors on the endothelial cell surface and has been linked to hemolytic uremic syndrome (Richardson *et al.*, 1988, van Setten *et al.*, 1997). Ricin (from the castor bean) has been reported to induce apoptosis of Vero cells (derived from green monkey kidney) by inhibiting ribosomal protein synthesis (Williams *et al.*, 1997). Recently, diisopropyl fluorophosphate, a general serine protease inhibitor, has been shown to inhibit ricin-induced DNA fragmentation and cell death of MDCK cells (Oda *et al.*, 1998). In those studies, protein synthesis was not inhibited suggesting that the pathway of ricin-induced cell death may be independent of the pathway of protein synthesis inhibition.

Toxins such as *Alternaria alternata lycopersici* (AAL) and fumonisin are sphingonine analogue mycotoxins which induce apoptosis in the kidney. These toxins contaminate food and water supplies, increasing their potential for renal exposure (Dutton, 1996, Wang *et al.*, 1996). AAL and fumonisin induce apoptosis in tubular epithelial cells; this is most prevalent in the outer medulla and is associated with development of hydronephrosis

(Lin *et al.*, 1995, Tolleson *et al.*, 1996, Sharma *et al.*, 1997). In rats and rabbits, the kidney is the most sensitive target organ to these toxins (Bucci *et al.*, 1996). The mechanism of AAL or fumonisin-induced apoptosis is not fully understood, but they cause disruption of sphingolipid metabolism (reviewed in Pena *et al.*, 1997) by inhibition of ceramide synthase and/or inhibiting n-acyl transferase (Wang *et al.*, 1991, Norred *et al.*, 1992, Merrill *et al.*, 1993). Exogenous ceramide triggers apoptosis in numerous cell types (Jarvis *et al.*, 1994, Tepper *et al.*, 1995) and since sphingolipids are found in many foods such as beef and cheese, intestinal metabolism of sphingolipids to ceramides could potentially supply the kidney with a dietary source of ceramide (Schmelz *et al.*, 1994). Fumonisin B1 also appears to increase TNF$\alpha$ production by macrophages which may contribute to fumonisin B1-induced apoptosis *in vivo* (Dugyala *et al.*, 1998).

The natural toxins, ochratoxin A, microcystin LR and okadaic acid are important inducers of apoptosis of renal cells. Ochratoxin A is a product of strains of Aspergillus and Penicillium fungi (Veldman *et al.*, 1992) that induces apoptosis in kidney cells (Seegers *et al.*, 1994) via its affect on mitochondrial respiration and total cellular oxygen consumption (Veldman *et al.*, 1992, Seegers *et al.*, 1994). Unlike ochratoxin, microcystin LR and okadaic acid are protein phosphatase inhibitors that induce apoptosis of renal epithelial cells *in vitro* (MacKintosh *et al.*, 1990, Davis *et al.*, 1994, Khan *et al.*, 1995). The regulation of apoptosis by phosphatase inhibitors appears to involve complex kinase signalling pathways. Interestingly, H-ras oncogene expression increases susceptibility to okadaic acid-induced apoptosis (Davis *et al.*, 1996). Other toxins have been specifically associated with apoptosis of transformed renal cells. Curcumin, a yellow dietary pigment found in the spice, tumeric, has been shown to cause apoptosis of renal cancer cells by inhibiting protein kinase C (Jiang *et al.*, 1996) and the mycotoxin, gliotoxin, has been found to cause apoptosis in Wilm's tumour cells (Waring *et al.*, 1994).

Numerous viral toxins have been shown to mediate renal apoptosis (Table 6.3). Viruses or viral proteins which stimulate or inhibit renal apoptosis include the SV40T antigen, adenovirus E1A protein, influenza A virus, influenza B virus, cowpox viruses which lack the crm A gene, sindbis virus, and reovirus (Rao *et al.*, 1992, Levine *et al.*, 1993, Hinshaw *et al.*, 1994, Mymryk *et al.*, 1994, Sabbatini *et al.*, 1995b, Ray *et al.*, 1996, Martel *et al.*, 1997, Rodgers *et al.*, 1997). Viral inhibitors of apoptosis in the kidney include Epstein barr virus BHRF 1 protein, human papilloma virus 18E6 protein, human immunodeficiency virus TAT gene product, adenovirus E1A12S protein and adenovirus E1B 19-kDa protein (Quinlan, 1993, Sabbatini *et al.*, 1995a, Subramanian *et al.*, 1995, Theodorakis *et al.*, 1996, Thomas *et al.*, 1996, Yin *et al.*, 1997). The mechanisms involved in viral associated apoptosis vary. Caspase activation appears to accompany CrmA/SPI-2 mutants of cowpox and rabbitpox viruses (Macen *et al.*, 1998, Nava *et al.*, 1998). Interestingly, the pathogenesis of HIV associated nephropathy, a progressive glomerular and tubular disease that is common in AIDS patients, involves apoptosis directly mediated by HIV-1 expression (Bruggeman *et al.*, 1997).

### 6.4.2 Metals

The kidney is the principle target organ of toxicity for many metals. Many of the toxic effects of metals may be attributable to the induction of apoptosis in renal cells, but the molecular pathways are primarily unclear. For example, inorganic mercuric chloride induces apoptosis of porcine renal proximal tubule cells (LLCPK) although it appears to activate both pro-apoptotic and anti-apoptotic pathways. Exposure of LLCPK1 cells to $HgCl_2$ causes a rise in intracellular hydrogen peroxide level and it induces anti-apoptotic proteins such as Bcl-2 and Bcl-x (Duncan *et al.*, 1996, Nath *et al.*, 1996). In this case, elevation in anti-apoptotic Bcl-2 family members may be part of a generalized elevation of redox-sensitive genes in response to oxidative stress.

Metals such as iron induce apoptosis via an oxidative stress mechanism. Iron-induced apoptosis of mouse renal proximal tubular cells has been reported in the toxicity of iron-nitrilotriacetate. In these studies, iron appears to cause free radical generation (Kawabata *et al.*, 1997). Cadmium induces apoptosis of proximal convoluted tubular cells when administered to dogs or added to primary cultures of renal proximal tubules isolated from canine kidney (Hamada *et al.*, 1991, 1994). Normally, cadmium is rapidly bound to methallothionein (Cd-MT) in the liver and binding is generally considered to be protective to cells. However, it has been hypothesized that the Cd-MT complex may play a mechanistic role in renal apoptosis by increasing nuclear accumulation of Cd and inducing nuclear changes (Hamada *et al.*, 1996). Cd has been localized to the cytoplasm and nucleus of proximal renal epithelial cells and apoptosis may be protective in Cd toxicity since the apoptotic cells loaded with Cd-MT are shed into the urine providing a unique mechanism for cadmium excretion (Tanimoto *et al.*, 1993).



### 6.4.3 Antineoplastic agents and therapies

The toxicologic effects of several antineoplastic agent include the induction of apoptosis in renal epithelial cells. Cisplatin is one of the most well studied renal toxicants whose major cytotoxic effect is apoptosis of the proximal tubule cells. Cisplatin induces apoptosis at low concentrations, resulting in cell loss over the space of several days which appears to be mediated by reactive oxygen species generation (Lieberthal *et al.*, 1996). Other studies have implied that cisplatin-induced apoptosis is mediated by thromboxane A2 (Jariyawat *et al.*, 1997). Several microtubule damaging drugs such as paclitaxel, vincristin and vinblastine have been reported to activate protein kinase A and C-Jun-N-terminal kinase in response to microtubule damage, suggesting that distinct signal transduction cascades may be induced by cytoskeletal damage (Srivastava *et al.*, 1998, Wang *et al.*, 1998). Another group of nephrotoxicants causes apoptosis specifically in association with cytoskeletal disorganization in rat renal proximal tubules. Members of this group include S-(1,2dichlorovinyl) -L-cysteine (DCVC), cytochalasin D and dithiothreitol (Van de Water *et al.*, 1996a). DCVC also results in increased intracellular calcium, mitochondrial damage and oxidative stress, all of which may also play a role in induction of apoptosis (Vamvakas *et al.*, 1990, Van de Water *et al.*, 1996b).

Exposure of kidney cells *in vitro* or *in vivo* to ionizing radiation induces apoptosis. Foetal kidneys exposed to radiation underwent extensive apoptosis, with the highest amount of apoptosis occurring within 24 hours of exposure. In the foetal kidney, the most sensitive cells are the rapidly dividing cells of the developing nephrons while those cells within the more mature nephrons suffered less damage (Gobe *et al.*, 1988). Adult mesangial cells are also susceptible to the apoptosis inducing effects of ionizing radiation (Cha *et al.*, 1996). Local delivery of interferon alpha was shown to increase the radiosensitivity of renal cell carcinoma (Syljuasen *et al.*, 1997). Gamma radiation does not have similar apoptotic inducing effects on the kidney or other radioresistant tissues such as heart, skeletal muscle, brain, liver, lung (Kitada *et al.*, 1996). However, in radiosensitive tissues, gamma radiation induces apoptosis and up-regulation of Bax protein.

## 6.5 MOLECULAR REGULATION OF TOXICANT-INDUCED APOPTOSIS

There are many diverse toxins and toxicants that induce renal apoptosis and these have been outlined above. However, information on the molecular mechanisms underlying this induction of cell death is incomplete. There are numerous instances in which TNF receptor ligands are elevated *in vivo*, increasing susceptibility of renal cells to apoptosis and renal damage. TNF$\alpha$ levels are increased in rat glomeruli during adriamycin or puromycin aminonucleoside-induced hydronephrosis (Gomez *et al.*, 1994). TNF$\alpha$ elevation may also potentiate the apoptotic effects of exogenous toxins or chemicals. Indeed, TNF$\alpha$ increases the apoptotic effects of verocytotoxin producing *E. coli* by increasing the number of verocytotoxin receptors on human renal endothelial cells (van Setten *et al.*, 1997). CD95 receptor mRNA expression is increased in diseased kidney and in human renal allografts undergoing acute rejection (Sharma *et al.*, 1996) and in cultured renal epithelial cells exposed to TNF$\alpha$, *E. coli* lipopolysaccharide, interleukin 1-beta, or gamma interferon (Ortiz-Arduan *et al.*, 1996).

Glucose is an important molecular trigger of apoptosis of the renal cells. Thus, alterations in glucose metabolism that occur in diabetes or toxicant exposure may induce or increase renal susceptibility to apoptosis. The mechanism by which glucose modulates apoptosis appears to involve changes in the Bax : Bcl-2 ratios in murine tubular epithelial cells (Ortiz *et al.*, 1997). By decreasing the levels of Bcl-2 and increasing Bax levels, incubation with 25 mmol glucose engages the apoptotic process (Ortiz *et al.*, 1997).

Inhibition of PP1 and PP2A activity by the toxicant, okadaic acid, inhibited growth, nephron formation and induced apoptosis of developing nephrons. Similarly, okadaic acid-induced apoptosis of renal epithelial cells in culture (Davis *et al.*, 1994). Calcineurin, a calcium-dependent protein phosphatase appears to regulate calcium-induced apoptosis in renal cells deprived of growth factors which is dependent on calcineurin's phosphatase activity (Shibasaki and McKeon, 1995). The inhibition of phosphatase activity by okadaic acid was accompanied by an increase in various kinase activities (Davis *et al.*, 1996). Thus the induction of apoptosis by phosphatase inhibitors may be due to the loss of phosphatase activity and the concomitant activation of kinases.

Tyrosine phosphorylation may link numerous seemingly independent induction pathways of renal apoptosis such as hypoxia, glucose and nitric oxide. Nitric oxide stimulates JNK activity in glomerular mesangial cells which was attenuated by genistein and N-acetylcysteine (DiMari *et al.*, 1997b). Similarly, D-glucose increases apoptosis and decreases replication in cultured metanephroi by perturbing phosphorylation of proteins that regulate perlecan (a mediator of epithelial mesenchymal interaction) (Kanwar *et al.*, 1996). In addition, chemical hypoxia (antimycin A and substrate deprivation) caused cell death and induced an increase in protein tyrosine kinase activity and protein tyrosine phosphorylation in LLCPK1 cells which was also inhibited by genistein. All of these cases support a role for protein phosphorylation as a common link in various mechanisms of apoptosis in renal cells.

## 6.6 SUMMARY

The results discussed in this chapter emphasize the importance of apoptosis in the kidney and in the potential functional consequences of toxicant or pharmacologic regulation of the process. The ongoing challenge is to study toxicants and novel pharmacologic agents to identify induction pathways that will be useful in drug design, preventative therapies and risk assessment. Since many toxicants and pharmacologic agents function during specific points in the cell cycle or have specific intracellular targets, they will provide a direct link to drug design and strategies that enhance or suppress apoptosis. We should look forward to seeing the use of these agents extended to therapeutic control of viral infection, cancer or developmental abnormalities.



## ACKNOWLEDGEMENTS

The editor would like to thank Zana Dye for invaluable assistance in editing this chapter.

## REFERENCES

BAKER, S.J. and REDDY, E.P., 1996, Transducers of life and death: TNF receptor superfamily and associated proteins, *Oncogene*, **12**, 1–9

BARASCH, J., PRESSLER, L., CONNOR, J. and MALIK, A., 1996, A ureteric bud cell line induces nephrogenesis in two steps by two distinct signals, *American Journal of Physiology*, F50–F61

BARASCH, J., QIAO, J., MCWILLIAMS, G., CHEN, D., OLIVER, J.A. and HERZLINGER, D., 1997, Ureteric bud cells secrete multiple factors, including bFGF, which rescue renal progenitors from apoptosis, *American Journal of Physiology*, **273**, F757–F767

BARDEESY, N., BECKWITH, J.B. and PELLETIER, J., 1995, Clonal expansion and attenuated apoptosis inWilms' tumors are associated with p53 gene mutations, *Cancer Research*, **55**, 215–219

BODI, I., ABRAHAM, A.A. and KIMMEL, P.L., 1995, Apoptosis in human immunodeficiency virus-associated nephropathy, *American Journal of Kidney Diseases*, **26**, 286–291

BRUGGEMAN, L.A., DIKMAN, S., MENG, C., QUAGGIN, S.E., COFFMAN, T.M. and KLOTMAN, P.E., 1997, Nephropathy in human immunodeficiency virus-1 transgenic mice is due to renal transgene expression, *Journal of Clinical Investigation*, **100**, 84–92

BUCCI, T.J., HANSEN, D.K. and LABORDE, J.B., 1996, Leukoencephalomalacia and hemorrhage in the brain of rabbits gavaged with mycotoxin fumonisin B1, *Natural Toxins*, **4**, 51–52

CARDONE, M.H., SALVESEN, G.S., WIDMANN, C., JOHNSON, G. and FRISCH, S.M., 1997, The regulation of anoikis: MEKK-1 activation requires cleavage by caspases, *Cell*, **90**, 315–323

CHA, D.R., FELD, S.M., NAST, C., LAPAGE, J. and ADLER, S.G., 1996, Apoptosis in mesangial cells induced by ionizing radiation and cytotoxic drugs, *Kidney International*, **50**, 1565–1571

CHEN, C.S., BRANTON, P.E. and SHORE, G.C., 1995a, Induction of p53-independent apoptosis by hygromysin B: suppression by Bcl-2 and adenovirus E1B 19-kDa protein, *Experimental Cell Research*, **221**, 55–59

CHEN, G., SHI, L., LITCHFIELD, D.W. and GREENBERG, A.H., 1995b, Rescue from granzyme B-induced apoptosis by Wee1 kinase, *Journal of Experimental Medicine*, **181**, 2295–2300

CHEN, C.S., MRKSICH, M., HUANG, S., WHITESIDES, G.M. and INGBER, D.E., 1997, Geometric control of cell life and death, *Science* **276**, 1425–1428

CHEVALIER, R.L., 1996, Growth factors and apoptosis in neonatal ureteral obstruction, *Journal of Americam Society of Nephrology*, **7**, 1098–1105

CHIOU, S.K., RAO, L. and WHITE, E., 1994a, Bcl-2 blocks p53-dependent apoptosis, *Molecular and Cellular Biology* **14**, 2556–2563

CHIOU, S.K., TSENG, C.C., RAO, L. and WHITE, E., 1994b, Functional complementation of the adenovirus E1B 19-kilodaltion protein with Bcl-2 in the inhibition of apoptosis in infected cells, *American Society for Microbiology*, **68**, 6553–6566

COHEN, G.M., 1997, Caspases: the executioners of apoptosis. Biochemical Journal

COLES, H.S., BURNE, J.F. and RAFF, M.C., 1993, Large-scale normal cell death in the developing rat kidney and its reduction by epidermal growth factor, *Development*, **118**, 777–784

CUMMINGS, M.C., 1996, Increased p53 mRNA expression in liver and kidney apoptosis, *Biochimica et Biophysica Acta*, **1315**, 100–104

DARZYNKIEWICZ, Z., LI, X. and GONG, J., 1994, Assays of cell viability: discrimination of cells dying by apoptosis, *Methods in Cell Biology*, **41**, 15–38

DAVIES, J., 1993, How to build a kidney, *Cell Biology* **4**, 213–219

DAVIS, M.A., CHANG, S.H. and TRUMP, B.F., 1996, Differential sensitivity of normal and H-ras oncogene-transformed rat kidney epithelial cells to okadaic acid-induced apoptosis, *Toxicology and Applied Pharmacology*, **141**, 93–101

DAVIS, M.A., SMITH, M.W., CHANG, S.H. and TRUMP, B.F., 1994, Characterization of a renal epithelial cell model of apoptosis using okadaic acid and the NRK-52E cell line, *Toxicologic Pathology*, **22**, 595–605

DEL PESO, L., GONZALEZ, G.M., PAGE, C., HERRERA, R. and NUNEZ, G., 1997, Interleukin-3-induced phosphorylation of BAD through the protein kinase Akt, *Science*, **278**, 687–689

DIMARI, J.F. and SAFIRSTEIN, R.L., 1997a, MAPK activation determines cellular survival during oxidative injury in renal epithelial cells, *Journal of the American Society of Nephrology*, **9**, 585A only

DIMARI, J., MEGYESI, J., UDVARHELYI, N., PRICE, P., DAVIS, R. and SAFIRSTEIN, R., 1997b, N-acetyl cysteine ameliorates ischemic renal failure, *American Journal of Physiology*, 41, F292–F298

D'SA, E.C., SUBRAMANIAN, T. and CHINNADURAI, G., 1996, bfl-1, a bcl-2 homologue, suppresses p53-induced apoptosis and exhibits potent cooperative transforming activity, *Cancer Research*, **56**, 3879–3882

DUGYALA, R.R., SHARMA, R.P., TSUNODA, M. and RILEY, R.T., 1998, Tumor necrosis factor-alpha as a contributor in fumonisin B1 toxicity, *Journal of Pharmacology and Experimental Theraputics*, **285**, 317–324

DUNCAN, A.K., JONES, J.T., BURKE, M.F., CARTER, D.E. and LAIRD, H.N., 1996, Inorganic mercury chloride-induced apoptosis in the cultured porcine renal cell line LLC-PK1, *Journal of Pharmacology and Experimental Therapeutics*, **277**, 1726–1732

DUTTON, M.F., 1996, Fumonisins, mycotoxins of increasing importance: their nature and their effects, *Journal of Pharmacology and Experimental Therapeutics*, **70**, 137–161

EGUCHI, Y., EWERT, D.L. and TSUJIMOTO, Y., 1992, Isolation and characterization of the chicken bcl-2 gene: expression in a variety of tissues including lymphoid and neuronal organs in adult and embryo, *Nucleic Acids Research*, **20**, 4187–4192

ENGLERT, C., HOU, X., MAHESWARAN, S., BENNETT, P., NGWU, C., RE, G.G., *et al.*, 1995, WT1 suppresses synthesis of the epidermal growth factor receptor and induces apoptosis, *EMBO Journal*, **14**, 4662–4675

FABISIAK, J.P., KAGAN, V.E., RITOV, V.B., JOHNSON, D.E. and LAZO, J.S., 1997, Bcl-2 inhibits selective oxidation and externalization of phosphatidylserine during paraquat-induced apoptosis, *American Journal of Physiology*, **41**, C675–C684

FRENCH, L.E., WOHLWEND, A., SAPPINO, A.P., TSCHOPP, J. and SCHIFFERLI, J.A., 1994, Human clusterin gene expression is confined to surviving cells during in vitro programmed cell death, *Journal of Clinical Investigation*, **93**, 877–884

FRISCH, S.M., 1994, E1a induces the expression of epithelial characteristics, *Journal of Cell Biology*, **127**, 1085–1096

FRISCH, S.M. and DOLTER, K.E., 1995, Adenovirus E1a-mediated tumor suppression by a c-erbB-2/neu-independent mechanism, *Cancer Research*, **55**, 5551–5555

FRISCH, S.M., VUORI, K., RUOSLAHTI, E. and CHAN, H.P., 1996, Control of adhesion-dependent cell survival by focal adhesion kinase, *Journal of Cell Biology*, **134**, 793–799

GLUCKSMANN, A., 1951, Cell deaths in normal vertebrate ontogeny, *Biological Reviews*, **26**, 59–86

GOBE, G.C., AXELSEN, R.A., HARMON, B.V. and ALLAN, D.J., 1988, Cell death by apoptosis following X-irradiation of the foetal and neonatal rat kidney, *International Journal of Radiation Biology*, **54**, 567–576

GODLEY, L.A., KOPP, J.B., ECKHAUS, M., PAGLINO, J.J., OWENS, J. and VARMUS, H.E., 1996, Wild-type p53 transgenic mice exhibit altered differentiation of the ureteric bud and possess small kidneys, *Genes and Development* **10**, 836–850

GOILLOT, E., RAINGEAUD, J., RANGER, A., TEPPER, R.I., DAVIS, R.J., HARLOW, E. and SANCHEZ, I., 1997, Mitogen-activated protein kinase-mediated Fas apoptotic signalling pathway, *Proceedings of the National Academy of Sciences of the United States of America*, **94**, 3302–3307

GOMEZ, C.M., ORTIZ, A., LERMA, J.L., LOPEZ, A.M., MAMPASO, F., GONZALEZ, E. and EGIDO, J., 1994, Involvement of tumor necrosis factor and platelet-activating factor in the pathogenesis of experimental nephrosis in rats, *Laboratory Investigations*, **70**, 449–459

GONZALEZ-GARCIA, M., PEREZ, B.R., DING, L., DUAN, L., BOISE, L.H., THOMPSON, C.B. and NUNEZ, G., 1994, Bcl-XL is the major Bcl-x mRNA form expressed during murine development and its product localizes to mitochondria, *Development*, **120**, 3033–3042

HALDAR, S., CHINTAPALLI, J. and CROCE, C.M., 1996, Taxol induces bcl-2 phosphorylation and death of prostate cancer cells, *Cancer Research*, **56**, 1253–1255

HAMADA, T., NAKANO, S., IWAI, S., TANIMOTO, A., ARIYOSHI, K. and KOIDE, O., 1991, Pathological study on beagles after long-term oral administration of cadmium, *Toxicological Pathology*, **19**, 138–147

HAMADA, T., SASAGURI, T., TANIMOTO, A., ARIMA, N., SHIMAJIRI, S., ABE, T. and SASAGURI, Y., 1996, Apoptosis of human kidney 293 cells is promoted by polymerized cadmium-metallothionein, *Biochemical and Biophysical Research Communications*, **219**, 829–834

HAMADA, T., TANIMOTO, A., IWAI, S., FUJIWARA, H. and SASAGURI, Y., 1994, Cytopathological changes induced by cadmium-exposure in canine proximal tubular cells: a cytochemical and ultrastructural study, *Nephron*, **68**, 104–111

HECHTFISCHER, A., MARSCHALL, M., HELTEN, A., BOESWALD, C. and MEIER-EWERT, H., 1997, A highly cytopathogenic influenza C virus variant induces apoptosis in cell culture, *Journal of General Virology*, **78**, 1327–1330

HEWITT, S.M., HAMADA, S., MCDONNELL, T.J., RAUSCHER, F.R. and SAUNDERS, G.F., 1995, Regulation of the proto-oncogenes bcl-2 and c-myc by the Wilms' tumor suppressor gene WT1, *Cancer Research*, **55**, 5386–5389

HINSHAW, V.S., OLSEN, C.W., DYBDAHL, S.N. and EVANS, D., 1994, Apoptosis: a mechanism of cell killing by influenza A and B viruses, *Journal of Virology*, **68**, 3667–3673

HOFMOCKEL, G., WITTMANN, A., DAEMMRICH, J. and BASSUKAS, I.D., 1996, Expression of p53 and bcl-2 in primary locally confined renal cell carcinomas: no evidence for prognostic significance, *Anticancer Research*, **16**, 3807–3812

HUMES, H.D., CIESLINSKI, D.A., COIMBRA, T.M., MESSANA, J.M. and GALVAO, C., 1989, Epidermal growth factor enhances renal tubule cell regeneration and repair and accelerates the recovery of renal function in postischemic acute renal failure, *Journal of Clinical Investigation*, **84**, 1757–1761

ICHIJO, H., NISHIDA, E., IRIE, K., TEN, D.P., SAITOH, M., MORIGUCHI, T., *et al.*, 1997, Induction of apoptosis by ASK1, a mammalian MAPKKK that activates SAPK/JNK and p38 signalling pathways, *Science*, **275**, 90–94

JARIYAWAT, S., TAKEDA, M., KOBAYASHI, M. and ENDOU, H., 1997, Thromboxane A2 mediates cisplatin-induced apoptosis of renal tubule cells, *Biochemical and Molecular Biology Interactions*, **42**, 113–121

JARVIS, W.D., KOLESNICK, R.N., FORNARI, F.A., TRAYLOR, R.S., GEWIRTZ, D.A. and GRANT, S., 1994, Induction of apoptotic DNA damage and cell death by activation of the sphingomyelin pathway, *Proceedings of the National Academy of Sciences of the United States of America*, **91**, 73–77

JEONG, J.K., DYBING, E., SODERLUND, E., BRUNBORG, G., HOLME, J.A., LAU, S.S. and MONKS, T.J., 1997a, DNA damage, gadd153 expression, and cytotoxicity in plateau-phase renal proximal tubular epithelial cells treated with a quinol thioether, *Archives of Biochemistry and Biophysics*, **341**, 300–308

JEONG, J.K., HUANG, Q., LAU, S.S. and MONKS, T.J., 1997b, The response of renal tubular epithelial cells to physiologically and chemically induced growth arrest, *Journal of Biological Chemistry*, **272**, 7511–7518

JIANG, M.C., YANG, Y.H., YEN, J.J. and LIN, J.K., 1996, Curcumin induces apoptosis in immortalized NIH 3T3 and malignant cancer cell lines, *Nutrition and Cancer*, **26**, 111–120

KANWAR, Y.S., LIU, Z.Z., KUMAR, A., USMAN, M.I., WADA, J. and WALLNER, E.I., 1996, D-glucose-induced dysmorphogenesis of embryonic kidney, *Journal of Clinical Investigation*, **98**, 2478–2488

KATO, S., AKASAKA, Y. and KAWAMURA, S., 1997, Fas antigen expression and its relationship with apoptosis in transplanted kidney, *Pathology International*, **47**, 230–237

KAWABATA, T., MA, Y., YAMADOR, I. and OKADA, S., 1997, Iron-induced apoptosis in mouse renal proximal tubules after an injection of a renal carcinogen, iron-nitrilotriacetate, *Carcinogenesis*, **18**, 1389–1394

KERR, J., WYLLIE, A. and CURRIE, A., 1972, Apoptosis: a basic biological phenomenon with wide-ranging implications in tissue kinetics, *British Journal of Cancer*, **26**, 239–257

KHAN, S.A., GHOSH, S., WICKSTROM, M., MILLER, L.A., HESS, R., HASCHEK, W.M. and BEASLEY, V.R., 1995, Comparative pathology of microcystin-LR in cultured hepatocytes, fibroblasts, and renal epithelial cells, *Natural Toxins*, **3**, 119–128

KIM, J., CHA, J.H., TISHER, C.C. and MADSEN, K.M., 1996a, Role of apoptotic and non-apoptotic cell death in removal of intercalated cells from developing rat kidney, *American Journal of Physiology*, 270, F575–F592

KIM, J., LEE, G.S., TISHER, C.C. and MADSEN, K.M., 1996b, Role of apoptosis in development of the ascending thin limb of the loop of Henle in rat kidney, *American Journal of Physiology*, **271**, F831–F845

KITADA, S., KRAJEWSKI, S., MIYASHITA, T., KRAJEWSKA, M. and REED, J.C., 1996, Gamma-radiation induces upregulation of Bax protein and apoptosis in radiosensitive cells in vivo, *Oncogene* **12**, 187–192

KORSMEYER, S.J., YIN, X.M., OLTVAI, Z.N., VEIS, N.D. and LINETTE, G.P., 1995, Reactive oxygen species and the regulation of cell death by the Bcl-2 gene family, *Biochimica et Biophysica Acta*, **1271**, 63–66

KOSEKI, C., 1993, Cell death programmed in uninduced metanephric mesenchymal cells, *Pediatric Nephrology*, **7**, 609–611

KRAJEWSKI, S., KRAJEWSKA, M. and REED, J.C., 1996, Immunohistochemical analysis of in vivo patterns of Bak expression, a proapoptotic member of the Bcl-2 protein family, *Cancer Research*, **56**, 2849–2855

KREIDBERG, J.A., SARIOLA, H., LORING, J.M., MAEDA, M., PELLETIER, J., HOUSMAN, D. and JAENISCH, R., 1993, WT-1 is required for early kidney development, *Cell*, **74**, 679–691

KROEMER, G., 1997, The proto-oncogene Bcl-2 and its role in regulating apoptosis, *Nature Medicine*, **3**, 614–620

KUMAR, S., KINOSHITA, M., NODA, M., COPELAND, N.G. and JENKINS, N.A., 1994, Induction of apoptosis by the mouse Nedd2 gene, which encodes a protein similar to the product of the Caenorhabditis elegans cell death gene ced-3 and the mammalian IL-1 beta-converting enzyme, *Genes and Development*, **8**, 1613–1626

KUMMER, J.A., WEVER, P.C., KAMP, A.M., TEN, B.I., HACK, C.E. and WEENING, J.J., 1995, Expression of granzyme A and B proteins by cytotoxic lymphocytes involved in acute renal allograft rejection, *Kidney International*, **47**, 70–77

LAKE, E.W. and HUMES, H.D., 1994, Acute renal failure: directed therapy to enhance renal tubular regeneration, *Seminars in Nephrology*, **14**, 83–97

LANOIX, J., D'AGATI, V., SZABOLCS, M. and TRUDEL, M., 1996, Dysregulation of cellular proliferation and apoptosis mediates human autosomaldominant polycystic kidney disease (ADPKD), *Oncogene*, **13**, 1153–1160

LECHNER, M.S. and DRESSLER, G.R., 1997, The molecular basis of embryonic kidney development, *Mechanisms of Development*, **62**, 105–120

LEVINE, B., HUANG, Q., ISAACS, J.T., REED, J.C., GRIFFIN, D.E. and HARDWICK, J.M., 1993, Conversion of lytic to persistent alphavirus infection by the bcl-2 cellular oncogene, *Nature*, **361**, 739–742

LIEBERTHAL, W., TRIACA, V. and LEVINE, J., 1996, Mechanisms of death induced by cisplatin in proximal tubular epithelial cells: apoptosis vs. Necrosis, *American Journal of Physiology*, **271**, F700–708

LIN, H.J., EVINER, V., PRENDERGAST, G.C. and WHITE, E., 1995, Activated H-ras rescues E1A-induced apoptosis and cooperates with E1A to overcome p53-dependent growth arrest, *Molecular and Cellular Biology*, **15**, 4536–4544

LITTLE, M.H., PROSSER, J., CONDIE, A., SMITH, P.J., VAN, H.V. and HASTIE, N.D., 1992, Zinc finger point mutations within the WT1 gene in Wilms tumor patients, *Proceedings of the National Academy of Sciences of the United States of America*, **89**, 4791–4795

LIU, Z.H., STRIKER, G.E., STETLER, S.M., FUKUSHIMA, P., PATEL, A. and STRIKER, L.J., 1996, TNF-alpha and IL-1 alpha induce mannose receptors and apoptosis in glomerular mesangial but not endothelial cells, *American Journal of Physiology*, **39**, C1595–1601

LOFFING, J., LOFFINGCUENI, D., HEGI, I., KAPLAN, M.R., HERBERT, S.C., LEHIR, M., *et al.*, 1996, Thiazide treatment of rats provokes apoptosis in disteetubule cells, *Kidney International*, **50**, 1180–1190

MACEN, J., TAKAHASHI, A., MOON, K.B., NATHANIEL, R., TURNER, P.C. and MOYER, R.W., 1998, Activation of caspases in pig kidney cells infected with wild-type and CrmA/SPI-2 mutants of cowpox and rabbitpox viruses, *Journal of Virology*, **72**, 3524–3533

MACKINTOSH, C., BEATTIE, K.A., KLUMPP, S., COHEN, P. and CODD, G.A., 1990, Cyanobacterial microcystin-LR is a potent and specific inhibitor of protein phosphatases 1 and 2A from both mammals and higher plants, *FEBS Letters*, **264**, 187–192

MacKintosh, C., Dalby, K.N., Campbell, D.G. and Cohen, P.T.W., 1995, The cyanobacterial toxin microcystin binds covalently to cysteine-273 on protein phosphatase 1, *FEBS Letters*, **371**, 236–240

Maheswaran, S., Englert, C., Bennett, P., Heinrich, G. and Haber, D.A., 1995, The WT1 gene product stabilizes p53 and inhibits p53-mediated apoptosis, *Genes and Development*, **9**, 2143–2156

Martel, C., Harper, F., Cereghini, S., Noe, V., Mareel, M. and Cremisi, C., 1997, Inactivation of retinoblastoma family proteins by SV40 T antigen results in creation of a hepatocyte growth factor/scatter factor autocrine loop associated with an epithelial-fibroblastoid conversion and invasiveness, *Cell Growth and Differentiation*, **8**, 165–178

McDonnell, T.J., El-Naggar, A.K. and Chandler, D., 1994, Apoptosis and expression of Bcl-2 in the developing human Kidney and during renal carcinogenesis, *Modern Pathology*, **7**, 142A only

McGahon, A.J., Martin, S.J., Bissonnette, R.P., Mahboubi, A., Shi, Y., Mogil, R.J., Nishioka, W.K. and Green, D.R., 1995, The end of the (cell) line: methods for the study of apoptosis in vitro, *Methods in Cell Biology*, **46**, 153–185

Merrill, A.J., Van Echten, G., Wang, E. and Sandhoff, K., 1993, Fumonisin B1 inhibits sphingosine (sphinganine) N-acyltransferase and de novo sphingolipid biosynthesis in cultured neurons in situ, *Journal of Biological Chemistry*, **268**, 27299–27306

Minn, A.J., Velez, P., Schendel, S.L., Liang, H., Muchmore, S.W., Fesik, S.W., *et al.*, 1997, Bcl-x(L) forms an ion channel in synthetic lipid membranes, *Nature*, **385**, 353–357

Muzio, M., Stockwell, B.R., Stennicke, H.R., Salvesen, G.S. and Dixit, V.M., 1998, An induced proximity model for caspase-8 activation, *Journal of Biological Chemistry*, **273**, 2926–2930

Mymryk, J.S., Shire, K. and Bayley, S., 1994, Induction of apoptosis by adenovirus type 5 E1A in rat cells requires a proliferation block, *Oncogene*, **9**, 1187–1193

Nath, K.A., Croatt, A.J., Likely, S., Behrens, T.W. and Warden, D., 1996, Renal oxidant injury and oxidant response induced by mercury, *Kidney International*, **50**, 1032–1043

Nava, V.E., Rosen, A., Veliuona, M.A., Clem, R.J., Levine, B. and Hardwick, J.M., 1998, Sindbis virus induces apoptosis through a caspase-dependent, CrmA- sensitive pathway, *Journal of Virology*, **72**, 452–459

Norred, W.P., Wang, E., Yoo, H., Riley, R.T. and Merrill, A.J., 1992, In vitro toxicology of fumonisins and the mechanistic implications, *Mycopathologia*, **117**, 73–78

Novack, D.V. and Korsmeyer, S.J., 1994, Bcl-2 protein expression during murine development, *American Journal of Pathology*, **145**, 61–73

ODA, T., KOMATSU, N. and MURAMATSU, T., 1998, Diisopropylfluorophosphate (DFP) inhibits ricin-induced apoptosis of MDCK cells, *Bioscience, Biotechnology and Biochemistry*, **62**, 325–333

ORTIZ, A., ZIYADEH, F.N. and NEILSON, E.G., 1997, Expression of apoptosis-regulatory genes in renal proximal tubular epithelial cells exposed to high ambient glucose and in diabetic kidneys, *Journal of Investigative Medicine*, **45**, 50–56

ORTIZ-ARDUAN, A., DANOFF, T.M., KALLURI, R., GONZALEZ, C.S., KARP, S.L., ELKON, K., *et al.*, 1996, Regulation of Fas and Fas ligand expression in cultured murine renal cells and in the kidney during endotoxemia, *American Journal of Physiology*, **271**, F1193–F1201

PARAF, F., GOGUSEV, J., CHRETIEN, Y. and DROZ, D., 1995, Expression of bcl-2 oncoprotein in renal cell tumours, *Journal of Pathology*, **177**, 247–252

PARRIZAS, M. and LEROITH, D., 1997, Insulin-like growth factor-1 inhibition of apoptosis is associated with increased expression of the bcl-xL gene product, *Endocrinology*, **138**, 1355–1358

PELLETIER, J., BRUENING, W., KASHTAN, C.E., MAUER, S.M., MANIVEL, J.C., STRIEGEL, J.E., *et al.*, 1991, Germline mutations in the Wilms' tumor suppressor gene are associated with abnormal urogenital development in Denys-Drash syndrome, *Cell*, **67**, 437–447

PENA, L.A., FUKS, Z. and KOLESNICK, R., 1997, Stress-induced apoptosis and the sphingomyelin pathway, *Biochemical Pharmacology*, **53**, 615–621

QUINLAN, M.P., 1993, E1A 12S in the absence of E1B or other cooperating oncogenes enables cells to overcome apoptosis, *Oncogene*, **8**, 3289–3296

RAO, L., DEBBAS, M., SABBATINI, P., HOCKENBERY, D., KORSMEYER, S. and WHITE, E., 1992, The adenovirus E1A proteins induce apoptosis, which is inhibited by the E1B 19-kDa and Bcl-2 proteins, *Proceedings of the National Academy of Sciences of the United States of America*, **89**, 7742–7746

RAY, C.A. and PICKUP, D.J., 1996, The mode of death of pig kidney cells infected with cowpox virus is governed by the expression of the crmA gene, *Virology*, **217**, 384–391

RAY, S.K., PUTTERMAN, C. and DIAMOND, B., 1996, Pathogenic autoantibodies are routinely generated during the response to foreign antigen: a paradigm for autoimmune disease, *Proceedings of the National Academy of Sciences of the United States of America*, **93**, 2019–2024

RICHARDSON, S.E., KARMALI, M.A., BECKER, L.E. and SMITH, C.R., 1988, The histopathology of the hemolytic uremic syndrome associated with verocytotoxin-producing Escherichia coli infections, *Human Pathology*, **19**, 1102–1108

RODGERS, S.E., BARTON, E.S., OBERHAUS, S.M., PIKE, B., GIBSON, C.A., TYLER, K.L. and DERMODY, T.S., 1997, Reovirus-induced apoptosis of MDCK cells is not linked to viral yield and is blocked by Bcl-2, *Journal of Virology*, **71**, 2540–2546

SABBATINI, P., CHIOU, S.K., RAO, L. and WHITE, E., 1995a, Modulation of p53-mediated transcriptional repression and apoptosis by the adenovirus E1B 19K protein, *Molecular and Cellular Biology*, **15**, 1060–1070

SABBATINI, P., LIN, J., LEVINE, A.J. and WHITE, E., 1995b, Essential role for p53-mediated transcription in E1A-induced apoptosis, *Genes and Development*, **9**, 2184–2192

SAELMAN, E.U., KEELY, P.J. and SANTORO, S.A., 1995, Loss of MDCK cell alpha 2 beta 1 integrin expression results in reduced cyst formation, failure of hepatocyte growth factor/scatter factor-induced branching morphogenesis, and increased apoptosis, *Journal of Cell Science*, **108**, 3531–3540

SCHENDEL, S.L., XIE, Z., MONTAL, M.O., MATSUYAMA, S., MONTAL, M. and REED, J.C., 1997, Channel formation by antiapoptotic protein Bcl-2, *Proceedings of the National Academy of Sciences of the United States of America*, **94**, 5113–5118

SCHMELZ, E.M., CRALL, K.J., LAROCQUE, R., DILLEHAY, D.L. and MERRILL, A.J., 1994, Uptake and metabolism of sphingolipids in isolated intestinal loops of mice, *Journal of Nutrition*, **124**, 702–712

SEEGERS, J.C., BOHMER, L.H., KRUGER, M.C., LOTTERING, M.L. and DE KOCK, M., 1994, A comparative study of ochratoxin A-induced apoptosis in hamster kidney and HeLa cells, *Toxicology and Applied Pharmacology*, **129**, 1–11

SHARMA, R.P., DUGYALA, R.R. and VOSS, K.A., 1997, Demonstration of in-situ apoptosis in mouse liver and kidney after short-term repeated exposure to fumonisin B1, *Journal of Comparative Pathology*, **117**, 371–381

SHARMA, V.K., BOLOGA, R.M., LI, B., XU, G.P., LAGMAN, M., HISCOCK, W., *et al.*, 1996, Molecular executors of cell death-differential intrarenal expression of Fas ligand, Fas, granzyme B, and perforin during acute and/or chronic rejection of human renal allografts, *Transplantation*, **62**, 1860–1866

SHI, L., MAI, S., ISRAELS, S., BROWNE, K., TRAPANI, J.A. and GREENBERG, A.H., 1997, Granzyme B (GraB) autonomously crosses the cell membrane and perforin initiates apoptosis and GraB nuclear localization, *Journal of Experimental Medicine*, **185**, 855–866

SHIBASAKI, F. and MCKEON, F., 1995, Calcineurin functions in Ca(2+)-activated cell death in mammalian cells, *Journal of Cell Biology*, **131**, 735–743

SHIMIZU, A., KITAMURA, H., MASUDA, Y., ISHIZAKI, M., SUGISAKI, Y. and YAMANAKA, N., 1996a, Glomerular cappillary regeneration and endothelia cell apoptosis in both reversable and progressive models of glomerulonephritis, *Contributions to Nephrology*, **118**, 29–40

SHIMIZU, A., MASUDFA, Y., KITAMURA, H., ISHIZAKI, M., SUGISAKI, Y. and YAMANAKA, N., 1996b, Apoptosis in progressive anescentic glomerulonephritis, *Laboratory Investigation*, **74**, 941–952

SORENSON, C.M., PADANILAM, B.J. and HAMMERMAN, M.R., 1996, Abnormal postpartum renal development and cystogenesis in the bcl-2 (-/-) mouse, *American Journal of Physiology*, **40**, F184–F193

SORENSON, C.M., ROGERS, S.A., KORSMEYER, S.J. and HAMMERMAN, M.R., 1995, Fulminant metanephric apoptosis and abnormal kidney development in bcl-2-deficient mice, *Molecular Biology of the Cell*, **5**, F73–F81

SRIVASTAVA, R.K., SRIVASTAVA, A.R., KORSMEYER, S.J., NESTEROVA, M., CHO, C.Y. and LONGO, D.L., 1998, Involvement of microtubules in the regulation of Bcl2 phosphorylation and apoptosis through cyclic AMP-dependent protein kinase, *Molecular and Cellular Biology*, **18**, 3509–3517

STRATER, J., WELLISCH, I., RIEDL, S., WALCZAK, H., KORETZ, K., TANDARA, A., *et al.*, 1997, CD95 (APO-1/Fas)-mediated apoptosis in colon epithelial cells: a possible role in ulcerative colitis, *Gastroenterology*, **113**, 160–167

STREHLAU, J., PAVLAKIS, M., LIPMAN, M., SHAPIRO, M., VASCONCELLOS, L., HARMON, W. and STROM, T.B., 1997, Quantitative detection of immune activation transcripts as a diagnostic tool in kidney transplantation, *Proceedings of the National Academy of Sciences of the United States of America*, **94**, 695–700

STUART, M.C., DAMOISEAUX, J.G., FREDERIK, P.M., ARENDS, J.W. and REUTELINGSPERGER, C.P., 1998, Surface exposure of phosphatidylserine during apoptosis of rat thymocytes precedes nuclear changes, *European Journal of Cell Biology*, **76**, 77–83

SUBRAMANIAN, T., TARODI, B. and CHINNADURAI, G., 1995, p53-independent apoptotic and necrotic cell deaths induced by adenovirus infection: suppression by E1B 19K and Bcl-2 proteins, *Cell Growth and Differentiation*, **6**, 131–137

SUTHANTHIRAN, M., 1997, Molecular analyses of human renal allografts: differential intragraft gene expression during rejection, *Kidney International Supplement*, S15–S21

SVENNILSON, J., DURBEEJ, M., CELSI, G., LAESTADIUS, A., DA CRUZ, E., SILVA, E., *et al.*, 1995, Evidence for a role of protein phosphatases 1 and 2A during early nephrogenesis, *Kidney International*, **48**, 103–110

SWEENEY, E.A., SAKAKURA, C., SHIRAHAMA, T., MASAMUNE, A., OHTA, H., HAKOMORI, S. and IGARASHI, Y., 1996, Sphingosine and its methylated derivative N,N-dimethylsphingosine (DMS) induce apoptosis in a variety of human cancer cell lines, *International Journal of Cancer*, **66**, 358–366

SYLJUASEN, R.G., BELLDEGRUN, A., TSO, C.L., WITHERS, H.R. and MCBRIDE, W.H., 1997, Sensitization of renal carcinoma to radiation using alpha interferon (IFNA) gene transfection, *Radiation Research*, **148**, 443–448

TAKEDA, M. and ENDOU, H., 1996, Drug-induced nephrotoxicity, *Folia Pharmacologica Japonica*, **107**, 1–8

TAKEDA, M., FUKUCKA, K. and ENDOU, H., 1996, Cisplatin-induced apoptosis in mouse proximal tubular cell line, *Contributions to Nephrology*, **118**, 24–28

TAKEMURA, T., MURAKAMI, K., MIYAZATO, H., YAGI, K. and YOSHIOKA, K., 1995, Expression of Fas antigen and bcl in human glomerulonephritis, *Kidney International*, **48**, 1886–1892

TANIMOTO, A., HAMADA, T. and KOIDE, O., 1993, Cell death and regeneration of renal proximal tubular cells in rats with subchronic cadmium intoxication, *Toxicological Pathology*, **21**, 341–352

TANNAPFEL, A., HANN, H.A., KATALINIC, A., FIETKAU, R.J., KUENN, R. and WITTEKIND, C.W., 1997, Incidence of apoptosis, cell proliferation and p53 expression in renal cell carcinomas, *Anticancer Research*, **17**, 1155–1162

TEPPER, C.G., JAYADEV, S., LIU, B., BIELAWSKA, A., WOLFF, R., YONEHARA, S., *et al.*, 1995, Role for ceramide as an endogenous mediator of Fas-induced cytotoxicity, *Proceedings of the National Academy of Sciences of the United States of America*, **92**, 8443–8447

THEODORAKIS, P., D'SA, E.C., SUBRAMANIAN, T. and CHINNADURAI, G., 1996, Unmasking of a proliferation-restraining activity of the anti-apoptosis protein EBV BHRF1, *Oncogene*, **12**, 1707–1713

THOMAS, M., MATLASHEWSKI, G., PIM, D. and BANKS, L., 1996, Induction of apoptosis by p53 is independent of its oligomeric state and can be abolished by HPV-18 E6 through ubiquitin mediated degradation, *Oncogene*, **13**, 265–273

TOLLESON, W.H., DOOLEY, K.L., SHELDON, W.G., THURMAN, J.D., BUCCI, T.J. and HOWARD, P.C., 1996, The mycotoxin fumonisin induces apoptosis in cultured human cells and in livers and kidneys of rats, *Advances in Experimental Medicine and Biology*, **392**, 237–250

TOMITA, Y., KAWASAKI, T., BILIM, V., TAKEDA, M. and TAKAHASHI, K., 1996, Tetrapeptide DEVD-aldehyde or YVAD-chloromethylketone inhibits Fas/Apo- 1(CD95)-mediated apoptosis in renal-cell-cancer cells, *International Journal of Cancer*, **68**, 132–135

TRUONG, L.D., PETRUSEVSKA, G., GURPINAR, T., SHAPPELL, S., LECHAGE, J. and ROUSE, D., 1996, Cell apoptosis and proliferation in experimental chrinic obstructive uropathy, *Kidney International*, **50**, 200–207

UCHIDA, H., FUJIMOTO, J. and TAKEDA, T., 1997, Primary tubular impairment by verocytotoxin in hemolytic uremic syndrome, *Nippon Rinsho*, **55**, 216–220

VAMVAKAS, S., SHARMA, V.K., SHEU, S.S. and ANDERS, M.W., 1990, Perturbations of intracellular calcium distribution in kidney cells by nephrotoxic haloalkenyl cysteine S-conjugates, *Molecular Pharmacologyl*, **38**, 455–461

VAN DE WATER, B., KRUIDERING, M. and NAGELKERKE, J.F., 1996a, F-actin disorganization in apoptotic cell death of cultured rat renal proximal tubular cells, *American Journal of Physiology*, F593–F603

VAN DE WATER, B., ZOETEWEIJ, J.P. and NAGELKERKE, J.F., 1996b, Alkylation-induced oxidative cell injury of renal proximal tubular cells: involovment of glutathione redox-cycle inhibition, *Archives of Biochemistry and Biophysics*, **327**, 71–80

VANDEVOORDE, V., HAEGEMAN, G. and FIERS, W., 1997, Induced expression of trimerized intracellular domains of the human tumor necrosis factor (TNF) p55 receptor elicits TNF effects, *Journal of Cell Biology*, **137**, 1627–1638

VAN SETTEN, P.A., VAN HINSBERGH, V.W., VAN DER VELDEN, T.J., VAN DE KAR, N.C., VERMEER, M., MAHAN, J.D., *et al.*, 1997, Effects of TNF alpha on verocytotoxin cytotoxicity in purified human glomerular microvascular endothelial cells, *Kidney International*, **51**, 1245–1256

VELDMAN, A., BORGGREVE, G.J., MULDERS, E.J., VAN DE LAGEMAAT, D., 1992, Occurrence of the mycotoxins ochratoxin A, zearalenone and deoxynivalenol in feed components, *Food Additives and Contaminants*, **9**, 647–655

VOSS, K.A., RILEY, R.T., BACON, C.W., CHAMBERLAIN, W.J. and NORRED, W.P., 1996, Subchronic toxic effects of fusarium moniliforme and fumonisin B1 in rats and mice, *Natural Toxins*, **4**, 16–23

WALKER, P.R., KOKILEVA, L., LEBLANC, J. and SIKORSKA, M., 1993, Detection of the initial stages of DNA fragmentation in apoptosis, *Biotechniques*, **15**, 1032–1040

WANG, E., NORRED, W.P., BACON, C.W., RILEY, R.T. and MERRILL, A.J., 1991, Inhibition of sphingolipid biosynthesis by fumonisins. Implications for diseases associated with Fusarium moniliforme, *Journal of Biological Chemistry*, **266**, 14486–14490

WANG, T.H., WANG, H.S., ICHIJO, H., GIANNAKAKOU, P., FOSTER, J.S., FOJO, T. and WIMALASENA, J., 1998, Microtubule-interfering agents activate c-Jun N-terminal kinase/stress-activated protein kinase through both Ras and apoptosis signal-regulating kinase pathways, *Journal of Biological Chemistry*, **273**, 4928–4936

WANG, W., JONES, C., CIACCI, Z.J., HOLT, T., GILCHRIST, D.G. and DICKMAN, M.B., 1996, Fumonisins and *Alternaria alternata* lycopersici toxins: sphinganine analog mycotoxins induce apoptosis in monkey kidney cells, *Proceedings of the National Academy of Sciences of the United States of America*, **93**, 3461–3465

WARING, P., NEWCOMBE, N., EDEL, M., LIN, Q.H., JIANG, H., SJAARDA, A., *et al.*, 1994, Cellular uptake and release of the immunomodulating fungal toxin gliotoxin, *Toxicon*, **32**, 491–504

WIDMANN, C., GIBSON, S. and JOHNSON, G.L., 1998b, Caspase-dependent cleavage of signalling proteins during apoptosis. A turn-off mechanism for anti-apoptotic signals, *Journal of Biological Chemistry*, **273**, 7141–7147

WIDMANN, C., GERWINS, P., JOHNSON, N.L., JARPE, M.B. and JOHNSON, G.L., 1998a, MEK kinase 1, a substrate for DEVD-directed caspases, is involved in genotoxin-induced apoptosis, *Molecular and Cellular Biology*, **18**, 2416–2429

WIJSMAN, J.H., JONKER, R.R., KEIJZER, R., VAN DE VELDE, C.J.H., CORNELISSE, C.J. and VAN DIERENDONCK, J.H., 1993, A new method to detect apoptosis in paraffin sections: in situ end-labeling of fragmented DNA, *Journal of Histochemistry and Cytochemistry*, **41**, 7–12

WILLIAMS, J.M., LEA, N., LORD, J.M., ROBERTS, L.M., MILFORD, D.V. and TAYLOR, C.M., 1997, Comparison of ribosome-inactivating proteins in the induction of apoptosis, *Toxicology Letters*, **91**, 121–127

WILLIAMS, M.S. and HENKART, P.A., 1994, Apoptotic cell death induced by intracellular proteolysis, *Journal of Immunology*, **153**, 4247–4255

WILLINGHAM, M.C. and BHALLA, K., 1994, Transient mitotic phase localization of bcl-2 oncoprotein in human carcinoma cells and its possible role in prevention of apoptosis, *Journal of Histochemistry and Cytochemistry*, **42**, 441–450

WINTER, C.K., GILCHRIST, D.G., DICKMAN, M.B. and JONES, C., 1996, Chemistry and biological activity of AAL toxins, *Advances in Experimental Medicine and Biology*, **392**, 307–316

WINYARD, P.J.D., NAUTA, J., LIVENMAN, D.S., HARDMAN, P., SAMS, V.R., RISDON, R.A. and WOOLF, A.S., 1996, Deregulation of cell survival in cystix and dysplastic renal development, *Kidney International*, **49**, 135–146

WOO, D., 1995, Apoptosis and loss of renal tissue in polycystic kidney diseases, *New England Journal of Medicine*, **333**, 18–25

WOO, K.R., SHU, W.P., KONG, L. and LIU, B.C., 1996, Tumor necrosis factor mediates apoptosis via Ca++/Mg++ dependent endonuclease with protein kinase C as a possible mechanism for cytokine resistance in human renal carcinoma cells, *Journal of Urology*, **155**, 1779–1783

WOOLF, A.S., KOLATSI, J.M., HARDMAN, P., ANDERMARCHER, E., MOORBY, C., FINE, L.G., *et al.*, 1995, Roles of hepatocyte growth factor/scatter factor and the met receptor in the early development of the metanephros, *Journal of Cell Biology*, **128**, 171–184

XIA, Z., DICKENS, M., RAINGEAUD, J., DAVIS, R.J. and GREENBERG, M.E., 1995, Opposing effects of ERK and JNK-p38 MAP kinases on apoptosis, *Science*, **270**, 1326–1331

YIN, T., SANDHU, G., WOLFGANG, C.D., BURRIER, A., WEBB, R.L., RIGEL, D.F., *et al.*, 1997, Tissue-specific pattern of stress kinase activation in ischemic/reperfused heart and kidney, *Journal of Biological Chemistry*, **272**, 19943–19950

ZHA, J., HARADA, H., YANG, E., JOCKEL, J. and KORSMEYER, S.J., 1996, Serine phosphorylation of death agonist BAD in response to survival factor results in binding to 14-3-3 not Bcl-X(L), *Cell*, **87**, 619–628

ZHANG, J., CLARK, J.R., HERMAN, E.H. and FERRANS, V.J., 1996, Doxrubin-induced apoptosis in spontaneously hypersensitive rats: differential effects in heart, kidney and intestine and inhibiton by ICRF-187, *Journal of Molecular and Cellular Cardiology*, **28**, 1931–1944

CHAPTER 7

# *Role of Apoptosis in Neuronal Toxicology*

**ROSEMARY M. GIBSON**

School of Biological Sciences, University of Manchester, Manchester, M13 9PT, UK

## *Contents*

## 7.1 INTRODUCTION

Cell death by apoptosis plays a fundamental role in the nervous system. It is essential for establishing neuronal networks during development since neurones initially are overproduced and the excess subsequently eliminated by apoptosis (reviewed in Oppenheim, 1991). Neurones require trophic support from several sources including their target cells and the Schwann cells that myelinate their processes (Snider, 1994) (Figure 7.1), and the blockade or removal of these signals induces cell death. The molecular mechanisms underlying the response of cells to such survival signals involves the Bcl-2 family of proteins.

Certain features of neurones such as their postmitotic phenotype and longevity make them susceptible to particular types of toxic insult. Despite this, there are few well documented examples of toxicants inducing neuronal cell death. This review will describe the occurrence and molecular mechanisms of neuronal apoptosis under normal physiological conditions and then will consider the role that apoptotic cell death plays in degenerative diseases. Finally, some examples of the role of apoptosis in mediating the effects of neurotoxicants will be considered.

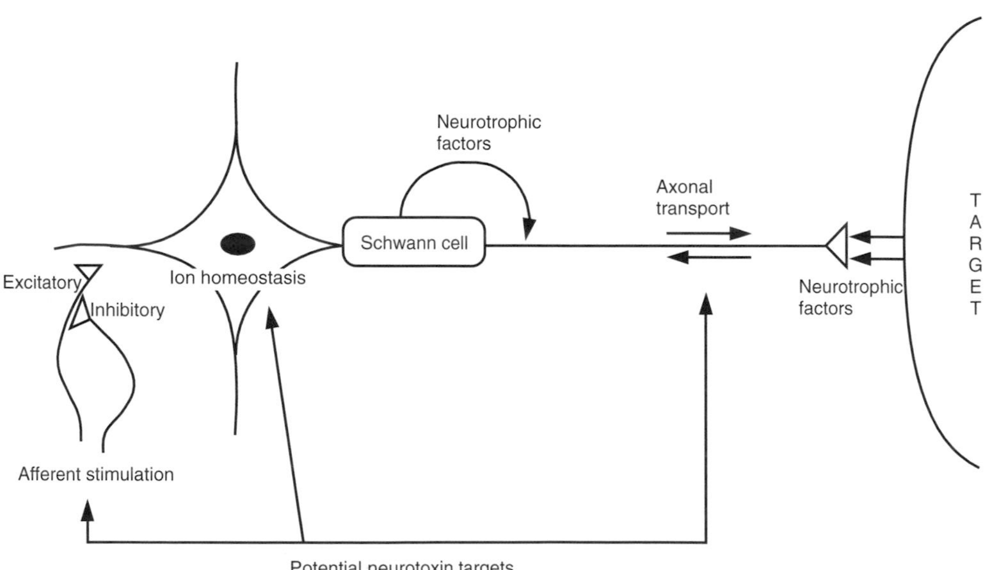


**Figure 7.1** Neurons depend on multiple signals to survive.

## 7.2 NEURONAL APOPTOSIS

### 7.2.1 Detection and quantitation of neuronal apoptosis

Techniques are well developed for detecting apoptosis in proliferating non-neuronal cells, using for examples, chromatin stains, viability assays, detection of cleavage of cellular proteins by caspases and DNA fragmentation (reviewed in Chapter 10). Neuronal cells *in vitro* can be similarly assessed. Methods and reagents for looking for caspase activation in tissue samples and sections are being developed; it is possible to homogenise brain tissue, for example, and assay for DNA fragmentation and active caspases (Cheng *et al.*, 1998), but the TUNEL (or terminal deoxynucleotidyl transferase-mediated dUTP-biotin nick end labelling) technique, which detects DNA strand breaks, remains one of the most widely used methods for detecting apoptosis *in situ* (Ben-Sasson *et al.*, 1995). Since DNA strand breaks are not confined to apoptosis, caution is needed when interpreting the results (Charriaut-Marlangue and Ben-Ari, 1995). However, the distinctions between neuronal apoptosis and necrosis may be blurred; neuronal cell death *in vivo* may not be so readily compartmentalized into either type (for discussion, see Baringa, 1998b).



### 7.2.2 Neuronal apoptosis during development

Neurones are over-produced during development and the excess are subsequently eliminated by apoptosis (reviewed in Oppenheim, 1991). It was generally thought that the deletion reflects competition for contact with targets and target-derived neurotrophic factors, so that neurones whose processes fail to reach their target die by apoptosis. In this way, the number of neurones is carefully matched with the targets which require innervation. Exogenous application of neurotrophic factors can prevent this neuronal death (Hofer and Barde, 1988) and antibodies to nerve growth factor (NGF) can induce neuronal death (Levi-Montalcini and Booker, 1960). In the sympathetic nervous system, for example, survival of neuronal precursor cells is initially independent of neurotrophic factors, allowing the cells to grow and proliferate. As the cells differentiate, they become increasingly dependent on NGF and neurones that fail to reach their target-derived supply of NGF die by apoptosis; in the rat, this begins at approximately day 19.5 (Birren *et al.*, 1993). Once connections are established, the extent of cell death declines and the neurones become less acutely dependent on neurotrophic factors. In fact, many neurones require trophic support from several sources, such as both their targets and the Schwann cells myelinating their processes (Snider, 1994). Furthermore, it is now clear that neurotrophic factors constitute just one of a complex array of signals upon which neurones depend for survival (Figure 7.1). They also depend upon afferent stimulation, blockade or removal of which

LIVERPOOL
JOHN MOORES UNIVERSITY
AVRIL ROBARTS LRC
TITHEBARN STREET
LIVERPOOL L2 2ER
TEL. 0151 231 4022

can induce cell death (Okado and Oppenheim, 1984, Galli-Resta *et al.*, 1993). These mechanisms lead to careful regulation of target innervation and stabilization of the adult neuronal networks.

## 7.3 GENETIC REGULATION OF NEURONAL APOPTOSIS

The molecular mechanisms underlying the response of cells to such survival signals involves the Bcl-2 family of proteins. The Bcl-2 protein itself, as would be predicted on the basis of data from non-neuronal systems (Chapter 1; Adams and Cory, 1998), can protect neuronal cells from cell death induced by a diverse array of stimuli (reviewed in Merry and Korsmeyer, 1997). Overexpression of Bcl-2 in neuronal cells *in vitro* can inhibit apoptosis: it can protect neuronally differentiated PC12 cells and primary cultures of sympathetic neurones from cell death induced by NGF withdrawal (Garcia *et al.*, 1992, Zhong *et al.*, 1993). Some neuronal cells however are not protected by overexpression of Bcl-2 protein. For example, ciliary ganglion neurones overexpressing Bcl-2 die by apoptosis when ciliary neurotrophic factor is withdrawn (Allsopp *et al.*, 1993). *In vivo*, overexpression of Bcl-2 leads to decreased cell death in response to diverse stimuli including anti-tumour drugs or focal ischaemia induced by middle cerebral artery occlusion (Martinou *et al.*, 1994, Lawrence *et al.*, 1996). Likewise, transgenic mice expressing Bcl-2 targeted to the peripheral and central nervous systems by the neurone-specific enolase promoter, have more neurones than their wild-type counterparts and fewer cells die following axotomy (Farlie *et al.*, 1995). However, loss of function analyzes suggest that another member of the Bcl-2 family, Bcl-XL, may be more important than Bcl-2 in neuronal survival. Mice lacking Bcl-XL die at embryonic day 13, displaying extensive death of neurones throughout the developing brain and spinal cord (Motoyama *et al.*, 1995), whereas mice lacking Bcl-2 can live to one month of age albeit with progressive postnatal loss of sensory and motoneurones (Michaelidis *et al.*, 1996).

The Bcl-2 family also contains pro-apoptotic members, and Bax has been shown to play a role in neuronal cell death. Overexpression of Bax leads to neuronal cell death (Vekrellis *et al.*, 1997) and expression of both Bax and Bcl-2 (but not Bcl-XL) is developmentally regulated such that their levels decrease in the cortex and cerebellum during postnatal rat brain development (Vekrellis *et al.*, 1997). Mice lacking Bax have more facial motoneurones and more neurones in the superior cervical ganglia at birth than their wild-type siblings and the neurones show increased resistance to apoptosis in response to insults such as axotomy (Knudson *et al.*, 1995, Deckwerth *et al.*, 1996). In addition to these roles of the Bcl-2 family in modulating death responses, Bcl-2 plays a role in axonal growth and neuronal differentiation. When overexpressed in several neuronal cell lines, Bcl-2 enhances neurite outgrowth (Zhang *et al.*, 1996) and neurones isolated from mice lacking Bcl-2 show slower neurite outgrowth and maturation than neurones from wild-type mice (Hilton *et al.*, 1997, Middleton *et al.*, 1998).

**TABLE 7.1**
Characteristics of mice lacking caspases

| Caspase deficiency | Life span of mice | Defects of -/- neurones | References |
|---|---|---|---|
| Caspase-1 | Normal | None reported | Kuida *et al.*, 1995, Li *et al.*, 1995 |
| Caspase-2 | Normal | Facial motoneurone apoptosis during development accelerated, but final number of neurones equals w.t. Isolated sympathetic neurones died after NGF withdrawal faster than w.t. neurones | Bergeron *et al.*, 1998 |
| Caspase-3 | Maximum age 1–3 weeks | Reduced developmental neuronal apoptosis; increased numbers of neurones (and glia) giving ectopic masses between cerebral cortex, hippocampus and striatum | Kuida *et al.*, 1996 |
| Caspase-9 | Die before day 3 | Reduced developmental neuronal apoptosis; expansion of ventricular zone and cortex | Hakem *et al.*, 1998, Kuida *et al.*, 1998 |
| Caspase-11 | Normal | None reported | Wang *et al.*, 1998 |



The growing family of enzymes called the caspases are responsible for cleaving key components of cells so that cellular processes are disabled and the cell is packaged for phagocytosis (reviewed in Cryns and Yuan, 1998, Thornberry and Lazebnik, 1998). Activation of caspases has been demonstrated during neuronal apoptosis and cell death can be prevented by caspase inhibitors such as *crm*A and z-Val-Ala-DL-Asp-fluoromethyl ketone (Gagliardini *et al.*, 1994, Deshmukh *et al.*, 1996) (see Chapter 2). Inhibition of caspases can protect sympathetic neurones from apoptosis induced by growth factor withdrawal (Deshmukh *et al.*, 1997) and can attenuate neuronal cell death *in vivo*, induced, for example, by ischaemic injury (Loddick *et al.*, 1996, Cheng *et al.*, 1998). This will be discussed in more detail in section 7.5.1. Strong evidence for a role for caspases in neuronal cell death comes from gene knockout mice (summarized in Table 7.1). Although mice lacking caspase-1 show no detectable neuronal deficits, neurones isolated either from these mice or from transgenic mice expressing a mutant form of caspase-1 are resistant to trophic factor withdrawal-induced apoptosis and the mice show reduced brain injury after ischaemic damage (Li *et al.*, 1995, Friedlander *et al.*, 1997). Interestingly, mice lacking caspase-2 display accelerated neuronal apoptosis suggesting that this enzyme has some neuroprotective properties. Furthermore, caspase inhibitors cannot block all neuronal apoptosis: apoptosis of cerebellar granule cells in response to potassium deprivation is unaffected by such inhibitors (Miller *et al.*, 1997, Taylor *et al.*, 1997). This may indicate the existence of a caspase-independent pathway to cell death. Alternatively, it may be that in some cells the decision to die (or commitment point) is made upstream of the caspases so that inhibiting the enzymes does not promote long-term survival. These concepts of commitment and execution have been reviewed in Chapter 1.

One potent method to induce apoptosis in many cells is via activation of the tumour necrosis factor (TNF)/Fas receptors. These receptors contain death domains in their cytoplasmic tails. Upon binding ligand, the receptors trimerize and form a death inducing signalling complex (DISC) that directly leads to activation of caspase-8 (reviewed in Nagata, 1997). Another member of this receptor superfamily is the low-affinity neurotrophin receptor p75, which together with the Trk receptors Trk A, B and C, mediates cellular responses to neurotrophins. The mode of action of p75 is rather unusual in that, in contrast with the TNF receptor and Fas, it initiates a death signal when no ligand is bound. Thus when NGF is withdrawn from NGF-dependent cells, the cells die by apoptosis. Down-regulation of p75 expression can lead to resistance to apoptosis in response to NGF withdrawal. However, the existence of multiple neurotrophin receptors complicates this picture. In fact, the response to neurotrophic factors critically depends on which receptors are expressed and in what ratio (reviewed in Bredesen and Rabizadeh, 1997). When p75 and a Trk receptor are co-expressed and can both bind the neurotrophin present, co-operation between the two receptors generates binding sites of higher affinity than that of the Trk receptor alone. However, conversely, mutual repression can occur when the neurotrophin can bind only p75 and not the Trk receptor expressed on the cells, the signalling from one receptor then inhibits that from the other receptor. For example, the neuronal cell line PC12 expresses both p75 and Trk A. When serum is withdrawn, the cells die by apoptosis and whilst NGF can rescue the cells, brain-derived neurotrophic factor (BDNF) cannot. If the expression of Trk A is decreased, however, BDNF becomes anti-apoptotic (Taglialatela *et al.*, 1996). Similarly, introduction of Trk A into an oligodendrocyte line can suppress activation of the apoptotic signal from p75 (Yoon *et al.*, 1998). Thus these receptors appear to initiate complex interacting and merging signals that modulate the cellular response to neurotrophins.

## 7.4 WHAT MAKES NEURONES SUSCEPTIBLE TO TOXICANTS

Certain features of neurones make them susceptible to particular types of toxic insult either during the onset of disease or following administration of a toxicant. The first feature is their postmitotic phenotype and longevity. Because neurones are formed largely when mammals are born and retain very limited regenerative capacity into adulthood, they are susceptible to agents which can gradually accumulate to toxic levels.

The second feature is their structure. Neurones are typically 10,000 times longer than other cell types and have 1000 times more cell volume, most of which is in the axon. As a result, distal parts of the cell can be up to one metre from the cell body, yet protein synthesis is restricted to the cell body and some dendrites. Transport of proteins, vesicles and organelles is therefore essential for neuronal viability and axonal integrity. Axonal transport depends on the cytoskeleton and can be broadly classified into slow and rapid transport. Rapid transport uses the molecular motors, kinesin and dynein, which require ATP. There are emerging connections between defects in axonal transport and some neurodegenerative disorders, suggesting that agents which interfere with transport can with time lead to neuronal apoptosis (see section 7.6).

The third cellular feature unique to neurones is the synapse. Functioning of the nervous system depends on information transfer within and between cells. Information transfer between cells occurs via chemical and electrical synapses and since neurones have been shown to depend on stimulation for survival, it follows that toxicants which interfere with synaptic function can ultimately lead to cell death. Conversely, over-stimulation of glutamate receptors can also lead to cell death (reviewed in Choi, 1992) although the mode of death remains controversial. The morphology of neurones dying by glutamate excitotoxicity is consistent with a necrotic cell death: the cell bodies and dendrites swell and organelles disintegrate. This leads in turn to further release of glutamate and propagation of the toxic signal, leading to the development of so-called secondary lesions.

## 7.5 NEURONAL APOPTOSIS AND DISORDERS OF THE CENTRAL NERVOUS SYSTEM



Despite controversies on the mode of neuronal cell death (see 7.2.1), it is generally accepted that neurons die by apoptosis during development. In contrast, the view that acute injury and neurodegeneration lead to necrosis has predominated. However, this has been challenged by observations of apparent apoptosis in these types of pathological neuronal death. Furthermore, it appears that genes implicated in neurodegeneration may actually participate in fundamental aspects of apoptosis, possibly even in developmental cell deaths. This is leading to exciting possibilities for the development of new intervention therapies for the treatment of a wider variety of neurological diseases, described below.

### 7.5.1 Ischaemic brain injury

Focal cerebral ischaemia can result from stroke, haemorrhage, cardiac or respiratory failure. It leads to neuronal injury, activation of glia and extravasation of leukocytes from brain microvessels. Although ischaemia was thought to induce necrosis rather than apoptosis on the basis of the morphology of the dying neurones, some experimental evidence now supports a role for apoptosis in ischaemic injuries (see Baringa, 1998b). For example, DNA fragmentation, a hallmark feature of apoptosis, has been observed in cells isolated from the ischaemic cortex (Linnik *et al.*, 1993). One neurotoxic stimulus during ischaemia is thought to be the accumulation of high local concentrations of excitatory amino acids, such as glutamate. Following ischaemia, cytokines of the interleukin-1 family are induced, triggering an inflammatory response. IL-1ß is quickly expressed upon injury and although itself is not neurotoxic, it acts to mediate and exacerbate the injury caused by ischaemia (Loddick *et al.*, 1996). IL-1ß is synthesized in a pro-form; activation requires the enzyme interleukin-1ß converting enzyme (ICE) or caspase-1. Thus, agents which inhibit production of IL-1ß can decrease the extent of damage resulting from ischaemia; administration of inhibitors of caspase-1 can reduce the damage in response to ischaemia (Loddick *et al.*, 1996). Transgenic mice expressing a mutated form of caspase-1, which is incapable of cleaving pro-IL-1ß, show significantly reduced injury in terms of both infarct volume and neurological score of the recovering mice compared to wild-type controls (Friedlander *et al.*, 1997). Mice deficient in caspase-1 also show reduced brain injury after middle cerebral artery occlusion (Schielke *et al.*, 1998). It is as yet unclear, however, what are the relative contributions of caspase-mediated production of proinflammatory IL-1ß and caspase-induced apoptosis (Baringa, 1998b). Caspases other than caspase-1 have also been implicated in ischaemic neuronal injury. Ischaemia induces activation of caspase-3: the active p20 subunit can be detected in neuronal cells at the time of reperfusion (Namura *et al.*, 1998), and expression of the messenger RNA and protein is up-regulated several hours later (Chen *et al.*, 1998). Caspase inhibitors can reduce the infarct size, brain swelling and neurological deficits (Hara *et al.*, 1997), even when administration is delayed until after the ischaemic insult (Cheng *et al.*, 1998), and caspase inhibitors given in conjunction with an excitatory amino acid antagonist can act synergistically to give greater neuroprotection after ischaemia (Ma *et al.*, 1998). Caspase-11 is required for activation of caspase-1 and mice with mutant caspase-11 are more resistant to ischaemic damage than wild-type mice (Wang *et al.*, 1998). In addition, overexpression of the anti-apoptotic proteins Bcl-2 or Bcl-XL can reduce the amount of damage following cerebral ischaemia (Martinou *et al.*, 1994, Parsadanian *et al.*, 1998).

These results suggest that IL-1ß, caspases and excitatory amino acids all contribute to neuronal damage in response to ischaemia and that successful therapeutic intervention may be possible using a multiple hit strategy.

### 7.5.2 Chronic neurodegenerative diseases

Extensive neuronal death is a hallmark of neurodegenerative diseases as well as acute injury. Although most cases are sporadic, studies of families with hereditary forms of these diseases have yielded insight into some of the genes that potentially regulate the disorders and have given clues as to the nature of the neurotoxic stimulus as well as avenues to explore for therapies. Protein aggregation is a common feature of chronic neurological disorders. The mutated proteins form abnormal aggregates (Figure 7.2); the extent to which these contribute to the aetiology of the disorders is not completely clear.

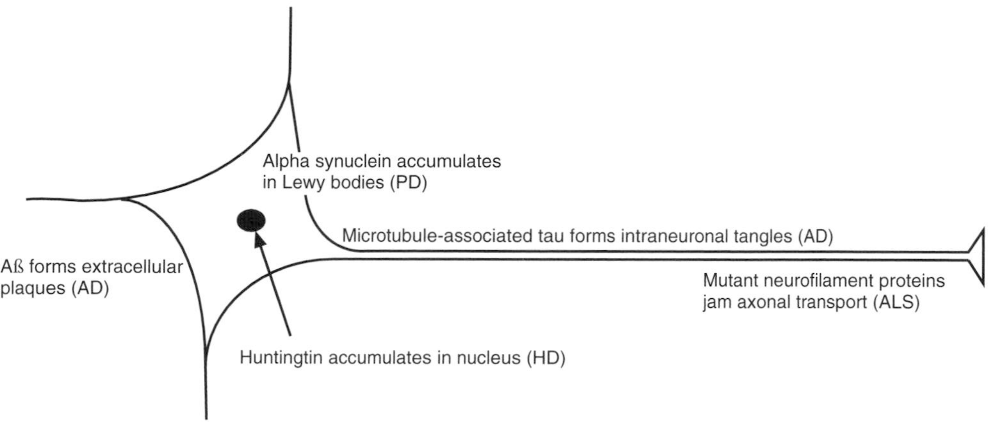


**Figure 7.2** Protein aggregation is a common feature of chronic neurological disorders.

### 7.5.2.1 Alzheimer's disease

Alzheimer's disease (AD) leads to debilitating memory and cognitive losses. Patients develop extracellular deposits of a 40–42 amino acid amyloid-ß peptide (Aß) and intraneuronal tangles of highly phosphorylated tau, a microtubule associated protein. The Aß peptides are produced from ß-amyloid precursor protein (ßAPP) by specific proteolytic cleavages and accumulate in deposits and tangles in areas of the brain which function in memory and cognition. The Aß deposits take several forms: Aß filaments, predominantly consisting of the abundant Aß(1–40) peptide, and non-filamentous deposits or diffuse plaques of Aß(1–42).

Three other genes have been found to be mutated in AD. Presenilins 1 and 2 are homologous proteins with multiple transmembrane segments. Mutations in presenilins may be neurotoxic in several ways: they may disrupt calcium homeostasis, lead to increased levels of neurotoxic Aß and/or increase caspase-mediated fragmentation of the proteins themselves. These mechanisms may then make cells more vulnerable to death either by the loss of a neuroprotective presenilin protein or generation of neurotoxic presenilin fragments.

The third gene found to be mutated in familial AD is apolipoprotein E (apoE). ApoE4 alleles lead to elevation of the density of the Aß deposits by an unknown mechanism. Some light may be shed on the mechanism however by the discovery of another protein implicated in AD pathology: a yeast two-hybrid screen for proteins that interact with Aß isolated a novel protein called endoplasmic-reticulum-associated binding protein (ERAB) (Yan *et al.*, 1997). ERAB is found predominantly in AD near Aß plaques but not in normal brain; overexpression of ERAB increases Aß toxicity whilst blocking ERAB decreases it. It may therefore be that it is the ERAB-Aß complex which is neurotoxic in AD. Interestingly, the protein shares homology with the short chain alcohol dehydrogenases, which are involved in cholesterol biosynthesis, suggesting a connection between ApoE, cholesterol and AD.

Inflammation also plays a role in AD and anti-inflammatory drugs can delay the onset or reduce the rate of progression of the disease (Rogers *et al.*, 1996). Elevation of proinflammatory cytokines such as IL-1ß has been observed (Patterson, 1995), and activated microglia, which can produce cytokines and other toxic factors such as free radicals, cluster near the Aß plaques. One secreted form of ßAPP can indeed induce such an inflammatory response in microglia. This can be blocked by ApoE3 but not the mutant ApoE4 alleles associated with AD, providing another potential link between this ApoE4 allele and the disease (Barger and Harmon, 1997).

### 7.5.2.2 Motoneurone disorders

Amyotrophic lateral sclerosis (ALS), or Lou Gehrig's disease, is an age-dependent neurodegenerative disease, leading to progressive paralysis and death typically within 3 to 5 years of diagnosis. The chief pathological feature is the degeneration of motoneurones in the brain, brainstem and spinal cord. The gene encoding the cytosolic copper–zinc superoxide dismutase (SOD1) was found to be mutated in several cases of familial ALS. The function of SOD is believed to be the detoxification of superoxide anions to form hydrogen peroxide which is subsequently converted to water by catalase or glutathione peroxidase. Mutation of SOD in ALS could thus increase reactive oxygen species to neurotoxic levels.

The mutations of SOD associated with ALS do not map to the active site of the enzyme (Deng *et al.*, 1993), however, but decrease its activity by decreasing the protein's stability more than its enzyme activity *per se* (Borchelt *et al.*, 1994). This suggests that the mutant SOD1 proteins are neurotoxic via a mechanism independent of the wild-type function of the enzyme. It is also possible that the altered enzymes have new substrate affinities which increase production of toxic reaction products.

The mutant forms of SOD generated may themselves be neurotoxic since transgenic mice overexpressing mutant SOD1 develop a neurodegenerative disorder closely resembling ALS, whilst mice overexpressing the wild-type form of the enzyme appear normal (Gurney *et al.*, 1994). Furthermore, it has been shown that whilst wild-type SOD1 inhibits apoptosis, mutated forms of SOD1, which retain enzymatic activity, induce apoptosis of neuronal cells in culture (Rabizadeh *et al.*, 1995).

The neuronal apoptosis which occurs in response to SOD1 down-regulation can be inhibited by peptide inhibitors of caspase-1, as well as by IL-1ra and blocking antibodies to IL-1ß (Troy *et al.*, 1997). Further evidence for a role of apoptosis in ALS comes from the demonstration that overexpressing the neuroprotective Bcl-2 protein in SOD1 mutant mice delays the onset of the disease and prolongs survival (Kostic *et al.*, 1997).

Impaired axonal transport is also a feature of ALS, and this may partly explain the acute susceptibility of the long motoneurones. Neuronal aggregates of neurofilament proteins occur in ALS patients and these could impair the transport critical for distal neuronal viability. Mutation of neurofilament proteins have been reported in this disease (Figlewicz *et al.*, 1994) and mice overexpressing either wild-type or mutant neurofilament proteins develop motoneurone disorders (Cote *et al.*, 1993). Furthermore, transgenic mice expressing SOD1 mutants develop defects in axonal transport before any other obvious pathology (Williamson *et al.*, 1997).

Another motoneurone disorder is spinal muscular atrophy (SMA), which is classified into types I to III according to the severity of the disease and age of onset, with type I patients dying within their first two years, whilst type III patients do not present with symptoms until this age. SMA is characterized by loss of motoneurones and accompanied by progressive weakness and wasting of muscles and finally paralysis. Two genes have been linked to the disease: evidence suggests that mutation of either gene could be neurotoxic and lead to increased apoptosis and motoneurone loss. The gene product of Smn, or survival motoneurone (Lefebvre *et al.*, 1995), has been proposed to be involved in maturation of nuclear RNA (Liu and Dreyfuss, 1996) but, until quite recently, the mechanism whereby Smn affects motoneurone survival was unclear. Smn was then reported to interact with Bcl-2 and enhance its anti-apoptotic effect, although Smn alone is not anti-apoptotic (Iwahashi *et al.*, 1997). In contrast, the identity of the second gene implicated in SMA immediately suggested how mutation can lead to aberrant apoptosis: neuronal apoptosis inhibitor protein (NAIP) (Roy *et al.*, 1995) was the first mammalian homologue of the viral IAP or inhibitor of apoptosis proteins to be described (reviewed in Clem and Duckett, 1997). NAIP is related to the viral IAPs since it shares a similar protein domain structure and it has been shown to have anti-apoptotic activity in mammalian cells (Liston *et al.*, 1996). Furthermore, it is located in motor but not sensory neurones which is consistent with the fact that motoneurones are selectively targeted in SMA.



### 7.5.2.3 Huntington's disease

Huntington's disease (HD) is an autosomal dominant genetic disorder, characterized by progressive movement disturbances and devastating cognitive and psychiatric deterioration, leading to death within 10 to 15 years of disease onset. The major targets are the neurones of the basal ganglia and striatum, especially the γ-aminobutyric acid producing medium spiny neurones.

The genetic alteration associated with the disorder is expansion of an unstable trinucleotide (CAG) repeat within the HD gene (Huntington's Disease Collaborative Research Group, 1993). The size of the CAG repeat varies from 8 to 35 in normal individuals and from 40 to 70 or more in affected individuals and the longer the repeated region, the earlier the onset of the disease. The function of the normal huntingtin protein and the basis for the neurotoxicity of the mutant forms with their elongated stretches of glutamines are unclear. Like SOD1, huntingtin is widely expressed in the body, leaving us to wonder at the basis for the selective neurotoxicity of the disease. It may be that proteins which interact with huntingtin or SOD1 lead to specific targeting. In the brain, huntingtin is present in the cytoplasm and associated with vesicle membranes in neurones, suggesting a potential role in vesicle trafficking (Difiglia *et al.*, 1995).

For all these neurodegenerative disorders, animal models which faithfully reproduce the human features of the disease are invaluable for studying mechanisms of neurotoxicity as well as for evaluating potential therapies. Disruption of the HD gene in mice is lethal beyond embryonic day 8.5 demonstrating that huntingtin is essential for post-implantation development. The heterozygous mice have a phenotype resembling human HD to some extent, including loss of neurones in the basal ganglia and loss of cognitive flexibility (Nasir *et al.*, 1995). These results suggest that there is no functional redundancy for the HD gene. Since humans with expanded CAG repeats in the HD gene survive, this also suggests that the poly-glutamine region does not obliterate the wild-type function of the protein, but the neurotoxicity may rather reflect a novel, selectively lethal gain of function.

Huntingtin is cleaved by caspase-3 and the rate of cleavage increases with expansion of the poly-glutamine region (Goldberg *et al.*, 1996). Since mice lacking huntingtin show higher than normal levels of apoptosis in the embryonic ectoderm (Zeitlin *et al.*, 1996), it may be that the story resembles that of the presenilins in AD: wild-type huntingtin may be anti-apoptotic, but following poly-glutamine expansion, the protein may become more susceptible to caspase-mediated cleavage, the fragments aggregating and becoming neurotoxic.

### 7.5.2.4 Parkinson's disease

Parkinson's disease (PD) is a neurodegenerative disorder characterized by death of the dopaminergic neurones of the substantia nigra. Onset of the disease is rare before the age of 40 and is accompanied by symptoms which include resting tremor, muscular rigidity, postural instability and bradykinesia. These motor symptoms result from an imbalance of inputs into the striatal neurones; loss of the neurones of the substantia nigra denies the striatal neurones of inhibitory input and they therefore respond with increased activity to excitatory cortical inputs. As with the other neurodegenerative disorders discussed, the exact nature of the neurotoxic stimulus in PD remains unresolved. Experimental evidence suggests a role for oxidative stress in many of the diseases (reviewed in Simonian and Coyle, 1996).

Some cases of PD are associated with mutation in a gene for α-synuclein (Polymeropoulos *et al.*, 1997). Alpha synuclein is a protein of unknown function found in presynaptic nerve termini. It has been found in Lewy bodies of the substantia nigra in patients with PD (Spillantini *et al.*, 1997); Lewy bodies are intracellular inclusion bodies characteristic of but not unique to PD. They also contain neurofilaments, ubiquitin and other components of the protein degradation pathway (Mezey *et al.*, 1998). Mutation of α-synuclein may facilitate its aggregation and play a role in formation of Lewy bodies, but how these events lead to neurotoxicity of dopaminergic neurones is unclear; the aggregates may be neurotoxic and/or disrupt the normal function of the protein. However it is important to note that α-synuclein is probably only a minor locus for familial forms of PD (French Parkinson's Disease Genetics Study Group, 1998). Indeed, α-synuclein-containing inclusions are found in other neurodegenerative disorders (Mezey *et al.*, 1998, Clayton and George, 1998) whether they are causative or occur in response to neuronal damage is as yet unclear.

Connections between α-synuclein and neuronal cell death are so far lacking. However, experiments have shown that the neurotoxins MPP+ and 6-hydroxydopamine, which also generates PD-like symptoms in animals, can induce apoptosis in neuronal types of cells *in vitro* (Dipasquale *et al.*, 1991, Walkinshaw and Waters, 1994). Furthermore, striatal neurones undergo apoptosis in animal models of PD and the cell death is blocked by glutamate antagonists suggesting these cells die in response to the excitotoxic stimulus (Mitchell *et al.*, 1994).

## 7.6 APOPTOSIS IN NEUROTOXICOLOGY

As discussed previously, neurones are particularly susceptible to cell death but the response of neuronal cells to neurotoxic stimuli is complex, reflecting the environment of the cells and the ability of the array of proteins expressed to respond to these environmental signals. For example, *in vivo*, overexpression of Bcl-2 leads to decreased cell death in response to diverse stimuli including the anti-tumour drug adriamycin or focal ischaemia induced by middle cerebral artery occlusion (Martinou *et al.*, 1994, Lawrence *et al.*, 1996). Similarly, most caspases can act as potent neurotoxins whereas inhibition of caspases can protect sympathetic neurones from apoptosis induced by growth factor withdrawal; these cells can recover fully their response to nicotine, when the growth factor levels are restored (Deshmukh *et al.*, 1997). Furthermore, there is evidence to suggest that inhibiting caspases can attenuate neuronal cell death *in vivo*, induced, for example, by ischaemic injury (Loddick *et al.*, 1996, Cheng *et al.*, 1998).



### 7.6.1 Glutamate

Information transfer between cells occurs via chemical and electrical synapses and since neurones have been shown to depend on stimulation for survival, it follows that toxicants which interfere with synaptic function can ultimately lead to cell death. Conversely, over-stimulation can also lead to cell death. Thus, the accumulation of high local concentrations of excitatory amino acids can serve as a neurotoxic stimulus (reviewed in Martin *et al.*, 1998). Glutamate can function as a potent excitotoxin for neuronal cells and glutamate-induced excitotoxicity has been linked to both acute and chronic neurological disorders such as ischaemia or HD (reviewed in Martin *et al.*, 1998). The over-stimulation of glutamate receptors leads to an excessive rise in intracellular calcium which contributes to the neurotoxicity (reviewed in Choi, 1992). The mode of cell death remains controversial. The morphology of neurones dying by glutamate excitotoxicity is consistent with a necrotic cell death: the cell bodies and dendrites swell and organelles disintegrate. This leads in turn to further release of glutamate and propagation of the toxic signal and development of so-called secondary lesions.

Evidence suggests that when cerebellar granule cells are treated with glutamate, two phases of cell death are observed; first, there is a rapid phase of necrosis but, subsequently, those neurones which have survived this phase and recovered their mitochondrial membrane potential and energy levels, then go on to die by apoptosis (Ankarcrona *et al.*, 1995). A role for apoptosis in glutamate-induced cell death is also suggested by the observation that it is accompanied by activation of caspase-3 and can be blocked by caspase inhibitors (Du *et al.*, 1997, Tenneti *et al.*, 1998).

### 7.6.2 Other excitatory amino acids

As discussed above, toxicants which interfere with synaptic function can ultimately lead to cell death (reviewed in Martin *et al.*, 1998). Interestingly, the pathology associated with HD can be reproduced in animal models by intrastriatal injection of the glutamate receptor agonist kainic acid (McGeer and McGeer, 1976) or injection of the NMDA receptor agonist quinolinic acid. The latter is reported to give rise to a pathology which more closely resembles that of HD (Beal *et al.*, 1986). The contribution of apoptosis to the neuronal cell death seen in HD is not completely clear and may, like other examples of excitotoxic death, have both necrotic and apoptotic components.

### 7.6.3 Aß

Aß is a peptide deposited in the brain of patients suffering from Alzheimer's disease, specifically in regions which function in memory and cognition. Aß has been shown to be neurotoxic since aggregates of the peptide induce primary cultures of neuronal cells to become dystrophic and to die by apoptosis (Loo *et al.*, 1993). Similarly, primates display neuronal degeneration in response to intracerebral injection of the Aß peptides (Geula *et al.*, 1998). This is both species- and most importantly age-dependent (Geula *et al.*, 1998). Curiously, transgenic mice overexpressing the precursor protein develop Aß aggregates, but do not show the neuronal death associated with human AD (Quon *et al.*, 1991). This may be explained since the toxicity of Aß is significantly greater in the aged brain (Geula *et al.*, 1998). The molecular mechanisms by which Aß induces neuronal apoptosis remain elusive (Baringa, 1998a). However, sublethal doses of Aß can sensitize neurones to other challenges such as exposure to excitotoxic amino acids or glucose deprivation.

### 7.6.4 1-methyl-4-phenyl-1,2,3,6-tetrahydropyridine (MPTP)

1-methyl-4-phenyl-1,2,3,6-tetrahydropyridine or MPTP is a piperidine derivative that causes irreversible symptoms of parkinsonism in humans and monkeys. It has been identified as an impurity in isomeperidine, a designer drug also known as synthetic heroin since people using isomeperidine developed PD-like symptoms. In glial cells, MPTP is converted by monoamine oxidase B into the neurotoxin 1-methyl-4-phenylpyridinium (MPP+) and MPP+ is subsequently taken up into the nerve termini of dopaminergic neurones via their dopamine transporter, explaining the selective targeting to dopaminergic neurones. Once inside the cells, MPP+ inhibits the mitochondrial electron transport chain, leading to cell death. The clinical effects of MPTP closely resemble the symptoms of PD and the drug can be used to generate animal models of the disease (Burns *et al.*, 1983). Inhibition of electron transport has been reported in idiopathic PD (Schapira *et al.*, 1990) and the effects of MPTP appear age-related explaining the late onset of PD (reviewed in Langston, 1996).

### 7.6.5 Toxicants that target axonal transport

Axonal transport is dependent upon a functional cytoskeleton and upon the activity of molecular motors. These molecular motors, kinesin and dynein are dependent upon ATP and drive axonal transport along cytoskeletal microtubules. Thus, axonal transport is susceptible to toxicants that are metabolic inhibitors and prevent the formation of ATP such as cyanide. Similarly, toxicants that effect microtubule function such as the chemotherapeutic drug taxol are able to inhibit the function of the molecular motors. There are emerging connections between defects in axonal transport and some neurodegenerative disorders, suggesting that agents which interfere with transport can, with time, lead to neuronal apoptosis.

## 7.7 SUMMARY

Understanding neurotoxic stimuli and the cellular responses underlying neuronal cell death is critical if we are to protect humans from potentially harmful toxicants whilst developing effective therapies for such devastating disorders as Parkinson's and Alzheimer's. More knowledge is required on the aetiology of these diseases and on the functions of the proteins implicated. Although the distinction between apoptosis and necrosis may be blurred in CNS cells, the surge of interest in apoptosis and the mechanisms of cell death will hopefully help us answer at least some questions and develop effective strategies for intervention.

## ACKNOWLEDGMENTS

The author would like to thank Professor Nancy Rothwell for her comments on this chapter. The editor would like to thank Dr Sabina Cosulich for critical appraisal of and assistance with editing this chapter and Zana Dye for additional editorial assistance.

## REFERENCES

ADAMS, J.M. and CORY, S., 1998, The Bcl-2 protein family: arbiters of cell survival, *Science*, **281**, 1322–1326

ALLSOPP, T.E., WYATT, S., PATERSEN, H.F. and DAVIES, A.M., 1993, The proto-oncogene bcl-2 can selectively rescue neurotrophic factor-dependent neurons from apoptosis, *Cell*, **73**, 295–307

ANKARCRONA, M., DYPBUKT, J.M., BONFOCO, E., ZHIVOTOVSKY, B., ORRENIUS, S., LIPTON, S.A. and NICOTERA, P., 1995, Glutamate-induced neuronal death: a succession of necrosis and apoptosis depending on mitochondrial function, *Neuron*, **15**, 961–973

BARGER, S.W. and HARMON, A.D., 1997, Microglial activation by Alzheimer amyloid precursor protein and modulation by apolipoprotein E, *Nature*, **388**, 878–881

BARINGA, M., 1998a, Is apoptosis key in Alzheimer's disease? *Science*, **281**, 1303–1304

BARINGA, M., 1998b, Stroke-damaged neurons may commit cellular suicide, *Science*, **281**, 1302–1303

BEAL, M.F., KOWALL, N.W., ELLISON, D.W., MAZUREK, M.F., SWARTZ, K.J. and MARTIN, J.B., 1986, Replication of the neurochemical characteristics of Huntington's disease by quinolinic acid, *Nature*, **321**, 168–171

BEN-SASSON, S.A., SHERMAN, Y. and GAVRIELI, Y., 1995, Identification of dying cells – *in situ* staining, in SCHWARTZ, L.M. and OSBORNE, B.A. (eds), *Methods in Cell Biology*. San Diego: Academic Press, Vol. 46, pp. 29–39

BERGERON, L., PEREZ, G.I., MACDONALD, G., SHI, L., SUN, Y., JURISICOVA, A., *et al.*, 1998, Defects in regulation of apoptosis in caspase-2-deficient mice, *Genes and Development*, **12**, 1304–1314

BIRREN, S.J., LICHING, L. and ANDERSEN, D.J., 1993, Sympathetic neuroblasts undergo a developmental switch in trophic dependence, *Development*, **119**, 597–610

BORCHELT, D.R., LEE, M.K., SLUNT, H.S., GUARNIERI, M., XU, Z.-S., WONG, P.-C., *et al.*, 1994, Superoxide dismutase 1 with mutations linked to familial amyotrophic lateral sclerosis possesses significant activity, *Proceedings of the National Academy of Sciences of the United States of America*, **91**, 8292–8296

BREDESEN, D.E. and RABIZADEH, S., 1997, p75NTR and apoptosis: Trk-dependent and Trk-independent effects, *Trends in Neuroscience*, **20**, 287–290

Burns, R.S., Chiueh, C.C., Markey, S.P., Ebert, M.H., Jacobwitz, D.M. and Koplin, I.J., 1983, A primate model of parkinsonism: selective destruction of dopaminergic neurons in the pars compacta of the substantia nigra by N-methyl-4-phenyl-1,2,3,6-tetrahydropyridine, *Proceedings of the National Academy of Sciences of the United States of America*, **80**, 4546–4550

Charriaut-Marlangue, C. and Ben-Ari, Y., 1995, A cautionary note on the use of TUNEL stain to determine apoptosis, *Neuroreport*, **7**, 61–64

Chen, J., Nagayama, T., Jin, K., Stetler, R.A., Zhu, R.L., Graham, S.H. and Simon, R.P., 1998, Induction of caspase-3-like protease may mediate delayed neuronal death in the hippocampus after transient cerebral ischemia, *Journal of Neuroscience*, **18**, 4914–4928

Cheng, Y., Deshmukh, M., D'Costa, A., Demaro, J.A., Gidday, J.M., Shah, A., *et al.*, 1998, Caspase inhibitor affords neuroprotection with delayed administration in a rat model of neonatal hypoxic-ischemic brain injury, *Journal of Clinical Investigation*, **101**, 1992–1999

Choi, D.W., 1992, Excitotoxic cell death, *Journal of Neurobiology*, **23**, 1261–1276

Clayton, D.F. and George, J.M., 1998, The synucleins: a family of proteins involved in synaptic function, plasticity, neurodegeneration and disease, *Trends in Neuroscience*, **21**, 249–254

Clem, R.J. and Duckett, C.S., 1997, The iap genes: unique arbitrators of cell death, *Trends in Cell Biology*, **7**, 337–339

Cote, F., Collard, J.-F. and Julien, J.P., 1993, Progressive neuropathy in transgenic mice expressing the human neurofilament heavy gene: a mouse model of amyotrophic lateral sclerosis, *Cell*, **73**, 35–46

Cryns, V. and Yuan, J., 1998, Proteases to die for, *Genes and Development*, **12**, 1551–1570

Deckwerth, T.L., Elliott, J.L., Knudson, C.M., Johnson, E.M., Snider, W.D. and Korsmeyer, S.J., 1996, BAX is required for neuronal death after trophic factor deprivation and during development, *Neuron*, **17**, 401–411

Deng, H.-X., Hentati, A., Tainer, J.A., Cayabyab, A., Hung, W.-Y., Getzoff, E.D., *et al.*, 1993, Amyotrophic lateral sclerosis and structural defects in Cu, Zn superoxide dismutase, *Science*, **261**, 1047–1051

Deshmukh, M., Cheng, Y., Werth, J., Cocabo, J., Rothman, S., Holtzman, D.M., *et al.*, 1997, Caspase inhibition: site of action in the neuronal death pathway and therapeutic utility in ischemic brain injury, *Molecular Biology of the Cell*, **8** (Suppl.), 351A

Deshmukh, M., Vasilakos, J., Deckwerth, T.L., Lampe, P.A., Shivers, B.D. and Johnson Jr, E.M., 1996, Genetic and metabolic status of NGF-deprived sympathetic neurons saved by an inhibitor of ICE family proteases, *Journal of Cell Biology*, **135**, 1341–1354

DIFIGLIA, M., SAPP, E., CHASE, K., SCHWARZ, C., MELONI, A., YOUNG, C., *et al.*, 1995, Huntingtin is a cytoplasmic protein associated with vesicles in human and rat brain neurons, *Neuron*, **14**, 1075–1081

DIPASQUALE, B., MARINI, A.M. and YOULE, R.J., 1991, Apoptosis and DNA degradation induced by 1-methyl-4-phenylpyridinium in neurons, *Biochemical and Biophysical Research Communications*, **181**, 1442–1448

DU, Y.S., BALES, K.R., DODEL, R.C., HAMILTONBYRD, E., HORN, J.W., CZILLI, D.L., *et al.*, 1997, Activation of a caspase 3-related cysteine protease is required for glutamate-mediated apoptosis of cultured cerebellar granule neurons, *Proceedings of the National Academy of Sciences of the United States of America*, **94**, 11657–11662

FARLIE, P.G., DRINGEN, R., REES, S.M., KANNOURAKIS, G. and BERNARD, O., 1995, Bcl-2 transgene expression can protect neurons against developmental and induced cell death, *Proceedings of the National Academy of Sciences of the United States of America*, **92**, 4397–4401

FIGLEWICZ, D.A., KRIZUS, A., MARTINOLI, M.G., MEININGER, V., DIB, M., ROULEAU, G.A. and JULIEN, J.-P., 1994, Variants of the heavy neurofilament subunit are associated with the development of amyotrophic lateral sclerosis, *Human Molecular Genetics*, **3**, 1757–1761

French Parkinson's Disease Genetics Study Group, 1998, Alpha-synuclein gene and Parkinson's Disease, *Science*, **279**, 1116–1117

FRIEDLANDER, R.M., GAGLIARDII, V., HARA, H., FINK, K.B., LI, W., MACDONALD, G., *et al.*, 1997, Expression of a dominant negative mutant of interleukin-1ß converting enzyme in transgenic mice prevents neuronal cell death induced by trophic factor withdrawal and ischemic brain injury, *Journal of Experimental Medicine*, **185**, 933–940

GAGLIARDINI, V., FERNANDEZ, P.A., LEE, R.K.K., DREXLER, H.C.A., ROTELLO, R.J., FISHMAN, M.C. and YUAN, J., 1994, Prevention of vertebrate neuronal death by the crmA gene, *Science*, **263**, 826–828

GALLI-RESTA, L., ENSINI, M., FUSCO, E., GRAVINA, A. and MARGHERITTI, B., 1993, Afferent spontaneous activity promotes the survival of target cells in the developing retinotectal system of the rat, *Journal of Neuroscience*, **13**, 243–250

GARCIA, I., MARTINOU, I., TSUJIMOTO, Y. and MARTINOU, J.C., 1992, Prevention of programmed cell death of sympathetic neurons by the bcl-2 proto-oncogene, *Science*, **258**, 302–304

GEULA, C., WU, C.-K., SAROFF, D., LORENZO, A., YUAN, M. and YANKNER, B.A., 1998, Aging renders the brain vulnerable to amyloid ß-protein toxicity, *Nature Medicine*, **4**, 827–831

GOLDBERG, Y.P., NICHOLSON, D.W., RASPER, D.M., KALCHMAN, M.A., KOIDE, H.B., GRAHAM, R.K. *et al.*, 1996, Cleavage of huntingtin by apopain, a proapoptotic cysteine protease, is modulated by the polyglutamine tract, *Nature Genetics*, **13**, 442–449

GURNEY, M.E., PU, H., CHIU, A.Y., CANTO, M.C.D., POLCHOW, C.Y., ALEXANDER, D.D., *et al.*, 1994, Motor neuron degeneration in mice that express a human Cu, Zn superoxide dismutase mutation, *Science*, **264**, 1772–1775

HAKEM, R., HAKEM, A., DUNCAN, G.S., HENDERSON, J.T., WOO, M., SOENGAS, M.S., *et al.*, 1998, Differential requirement for Caspase-9 in apoptotic pathways *in vivo*, *Cell*, **94**, 339–352

HARA, H., FRIEDLANDER, R.M., GAGLIARDINI, V., AYATA, C., FINK, K., HUANG, Z., SHIMIZU-SASAMATA, M., YUAN, J. and MOSKOWITZ, M.A., 1997. Inhibition of interleukin-1ss converting enzyme family proteases reduces ischemic and excitotoxic neuronal damage. *Proc. Natl. Acad. Sci.* (USA), **94**, 2007–2012

HILTON, M., MIDDLETON, G. and DAVIES, A.M., 1997, Bcl-2 influences axonal growth rate in embryonic sensory neurons, *Current Biology*, **7**, 798–800

HOFER, M.M. and BARDE, Y.A., 1988, Brain-derived neurotrophic factor prevents neuronal death in vivo, *Nature*, **331**, 261–262

Huntington's Disease Collaborative Research Group, 1993, A novel gene containing a trinucleotide repeat that is expanded and unstable on Huntington's disease chromosomes, *Cell*, **72**, 971–983

IWAHASHI, H., EGUCHI, Y., YASUHARA, N., HANAFUSA, T., MATSUZAWA, Y. and TSUJIMOTO, Y., 1997, Synergistic anti-apoptotic activity between Bcl-2 and SMN implicated in spinal muscular atrophy, *Nature*, **390**, 413–417

KNUDSON, C.M., TUNG, K.S., TOURTELLOTTE, W.G., BROWN, G.A. and KORSMEYER, S.J., 1995, Bax-deficient mice with lymphoid hyperplasia and male germ cell death, *Science*, **270**, 96–99

KOSTIC, V., JACKSON-LEWIS, V., DE BILBAO, F., DUBOIS-DAUPHIN, M. and PRZEDBORSKI, S., 1997, Bcl-2: prolonging life in a transgenic mouse model of familial amyotrophic lateral sclerosis, *Science*, **277**, 559–562

KUIDA, K., HAYDAR, T.F., KUAN, C.Y., GU, Y., TAYA, C., KARASUYAMA, H., *et al.*, 1998, Reduced apoptosis and cytochrome c-mediated caspase activation in mice lacking caspase-9, *Cell*, **94**, 325–337

KUIDA, K., LIPPKE, J.A., KU, G., HARDURG, M.W., LIVINGSTON, D.J., SU, M.S. and FLAVELL, R.A., 1995, Altered cytokine export and apoptosis in mice deficient in interleukin-1 beta converting enzyme, *Science*, **267**, 2000–2003

KUIDA, K., ZHENG, T.S., NA, T., KUAN, C.Y., YANG, D., KARASUYAMA, H., *et al.*, 1996, Decreased apoptosis in the brain and premature lethality in CPP32 deficient mice, *Nature*, **384**, 368–372

LANGSTON, J.W., 1996, The etiology of Parkinson's disease with emphasis on the MPTP story, *Neurology*, **47** (Suppl. 3) S153–S160

LAWRENCE, M.S., HO, D.Y., SUN, G.H., STEINBERG, G.K. and SAPOLSKY, R.M., 1996, Overexpression of Bcl-2 with herpes simplex virus vectors protects CNS neurons against neurological insults in vitro and in vivo, *Journal of Neuroscience*, **16**, 486–496

LEFEBVRE, S., BURGLEN, L., REBOULLET, S., CLERMONT, O., BURLET, P., VIOLLET, L., *et al.*, 1995, Identification and characterization of a spinal muscular atrophy-determining gene, *Cell*, **80**, 155–165

LEVI-MONTALCINI, R. and BOOKER, B., 1960, Destruction of sympathetic ganglia in mammals by an antiserum to a nerve growth protein, *Proceedings of the National Academy of Sciences of the United States of America*, **46**, 384–391

LI, P., ALLEN, H., BANERJEE, S., FRANKLIN, S., HERZOG, L., JOHNSTON, C., *et al.*, 1995, Mice deficient in IL-1ß-converting enzyme are defective in production of mature IL-1ß and resistant to endotoxic shock, *Cell*, **80**, 401–411

LINNIK, M.D., ZOBRIST, R.H. and HATFIELD, M.D., 1993, Evidence supporting a role for programmed cell death in focal cerebral ishemia in rats, *Stroke*, **24**, 2002–2009

LISTON, P., ROY, N., TAMAI, K., LEFEBVRE, C., BAIRD, S., CHERTON-HORVAT, G., *et al.*, 1996, Suppression of apoptosis in mammalian cells by NAIP and a related family of IAP genes, *Nature*, **379**, 349–353

LIU, Q. and DREYFUSS, G., 1996, A novel nuclear structure containing the survival of motor neurons protein, *EMBO Journal*, **15**, 3555–3565

LODDICK, S.A., MACKENZIE, A. and ROTHWELL, N.J., 1996, An ICE inhibitor, (z)-VAD-DCB attenuates ischemic brain damage in the rat, *Neuroreport*, **7**, 1465–1468

LOO, D.T., COPANI, A., PIKE, C.J., WHITTEMORE, E.R., WALENCEWICZ, A.J. and COTMAN, C.W., 1993, Apoptosis is induced by ß-amyloid in cultured central nervous system neurons, *Proceedings of the National Academy of Sciences of the United States of America*, **90**, 7951–7955

MA, J.Y., ENDRES, M. and MOSKOWITZ, M.A., 1998, Synergistic effects of caspase inhibitors and MK-801 in brain injury after transient focal cerebral ischaemia in mice, *British Journal of Pharmacology*, **124**, 756–762

MARTIN, L.J., AL-ABDULLA, N.A., BRAMBRINK, A.M., KIRSCH, J.R., SIEBER, F.E. and PORTERA-CAILLIAU, C., 1998, Neurodegeneration in excitotoxicity, global cerebral ischemia, and target deprivation: a perspective on the contributions of apoptosis and necrosis, *Brain Research Bulletin*, **46**, 281–309

MARTINOU, J.C., DUBOIS-DAUPHIN, M., STAPLE, J.K., RODRIGUEZ, I., FRANKOWSKY, H., MISSOTTEN, M., *et al.*, 1994, Overexpression of bcl-2 in transgenic mice protects neurons from naturally occurring cell death and experimental ischaemia, *Neuron*, **13**, 1017–1030

MCGEER, E.G. and MCGEER, P.L., 1976, Duplication of biochemical changes of Huntington's chorea by intrastriatal injections of glutamic acid, *Nature*, **263**, 517–519

MERRY, D.E. and KORSMEYER, S.J., 1997, Bcl-2 gene family in the nervous system, *Annual Reviews of Neuroscience*, **20**, 245–267

MEZEY, E., DEHEJIA, A., HARTA, G., PAPP, M.I., POLYMEROPOULOS, M.H. and BROWNSTEIN, M.J., 1998, Alpha synuclein in neurodegenerative disorders: murderer or accomplice? *Nature Medicine*, **4**, 755–757

MICHAELIDIS, T.M., SENDTNER, M., COOPER, J.D., AIRAKSINEN, M.S., HOLTMANN, B., MEYER, M. and THOENEN, H., 1996, Inactivation of bcl-2 results in progressive degeneration of motor neurons, sympathetic neurons and sensory neurons during early postnatal development, *Neuron*, **17**, 75–89

MIDDLETON, G., PINON, L.G.P., WYATT, S. and DAVIES, A.M., 1998, Bcl-2 accelerates the maturation of early sensory neurons, *Journal of Neuroscience*, **18**, 3344–3350

MILLER, T.M., MOULDER, K.L., KNUDSON, C.M., CREEDON, D.J., DESHMUKH, M., KORSMEYER, S.J. and JOHNSON, E.M., 1997, Bax deletion further orders the cell death pathway in cerebellar granule cells and suggests a caspase-independent pathway to cell death, *Journal of Cell Biology*, **139**, 205–217

MITCHELL, I.J., LAWSON, S., MOSER, B., LAIDLAW, S.M., COOPER, A.J., WALKINSHAW, G. and WATERS, C.M., 1994, Glutamate-induced apoptosis results in a loss of striatal neurons in the Parkinsonian rat, *Neuroscience*, **63**, 1–5

MOTOYAMA, N., WANG, F., ROTH, K.A., SAWA, H., NAKAYAMA, K., NAKAYAMA, K., *et al.*, 1995, Massive cell death of immature hematopoietic cells and neurons in Bcl-x-deficient mice, *Science*, **267**, 1506–1510

NAGATA, S., 1997, Apoptosis by death factor. *Cell*, **88**, 355–365

NAMURA, S., ZHU, J., FINK, K., ENDRES, M., SRINIVASAN, A., TOMASELLI, K.J., *et al.*, 1998, Activation and cleavage of caspase-3 in apoptosis induced by experimental cerebral ischemia, *Journal of Neuroscience*, **18**, 3659–3668

NASIR, J., FLORESCO, S.B., O'KUSKY, J.R., DIEWERT, V.M., RICHMAN, J.M., ZEISLER, J., *et al.*, 1995, Targeted disruption of the Huntington's Disease gene results in embryonic lethality and behavioral and morphological changes in heterozygotes, *Cell*, **81**, 811–823

OKADO, N. and OPPENHEIM, R.W., 1984, Cell death of motor neurons in the chick embryo spinal cord. IX. The loss of motor neurons following removal of afferent inputs, *Journal of Neuroscience*, **4**, 1639–1652

OPPENHEIM, R.W., 1991, Cell death during development of the nervous system, *Annual Reviews of Neuroscience*, **14**, 453–501

PARSADANIAN, A.S., CHENG, Y., KELLER-PECK, C.R., HOLTZMAN, D.M. and SNIDER, W.D., 1998, Bcl-x(L) is an antiapoptotic regulator for postnatal CNS neurons, *Journal of Neuroscience*, **18**, 1009–1019

PATTERSON, P.H., 1995, Cytokines in Alzheimers-Disease and Multiple-Sclerosis, *Current Opinion in Neurobiology*, **5**, 642–646

POLYMEROPOULOS, M.H., LAVEDAN, C., LEROY, E., IDE, S.E., DEHEJIA, A., DUTRA, A., *et al.*, 1997, Mutation in the alpha-synuclein gene identified in families with Parkinson's disease, *Science*, **276**, 2045–2047

QUON, D., WANG, Y., CATALANO, R., SCARDINA, J.M., MURAKAMI, K. and CORDELL, B., 1991, Formation of ß-amyloid protein deposits in brains of transgenic mice, *Nature*, **352**, 239–241

RABIZADEH, S., GRALLA, E.B., BORCHELT, D.R., GWINN, R., VALENTINE, J.S., SISODIA, S., *et al.*, 1995, Mutations associated with amyotrophic lateral sclerosis convert superoxide dismutase from an antiapoptotic gene to a proapoptotic gene: studies in yeast and neural cell, *Proceedings of the National Academy of Sciences of the United States of America*, **92**, 3024–3028

ROGERS, J., WEBSTER, S., LUE, L.F., BRACHOVA, L., CIVIN, W.H., EMMERLING, M., *et al.*, 1996, Inflammation and Alzheimers-disease pathogenesis, *Neurobiology of Aging*, **17**, 681–686

ROY, N., MAHADEVAN, M.S., MCLEAN, M., SHUTLER, G., YARAGHI, Z., FARAHANI, R., *et al.*, 1995, The gene for neuronal apoptosis inhibitory protein is partially deleted in individuals with spinal muscular atrophy, *Cell*, **80**, 167–178

SCHAPIRA, A.H.V., COOPER, J., DEXTER, D., JENNER, P. and MARSDEN, C.D., 1990, Mitochondrial complex I deficiency in Parkinson's disease, *Journal of Neurochemistry*, **54**, 823–827

SCHIELKE, G.P., YANG, G.-Y., SHOVERS, B.D. and BETZ, A.L., 1998, Reduced ischemic brain injury in interleukin-1ß converting enzyme-deficient mice, *Journal of Cerebral Blood Flow and Metabolism*, **18**, 180–185

SIMONIAN, N.A. and COYLE, J.T., 1996, Oxidative stress in neurodegenerative diseases, *Annual Reviews of Pharmacology and Toxicology*, **36**, 83–106

SNIDER, W.D., 1994, Functions of the neurotrophins during nervous system development: what the knockouts are teaching us, *Cell*, **77**, 627–638

SPILLANTINI, M.G., SCHMIDT, M.L., LEE, V.M.-Y., TROJANOWSKI, J.Q., JAKES, R. and GOEDERT, M., 1997, Alpha-synuclein in Lewy bodies, *Nature*, **388**, 839–840

TAGLIALATELA, G., HIBBERT, C.J., HUTTON, L.A., WERRBACHPEREZ, K. and PEREZPOLO, J.R., 1996, Suppression of p140(trkA) does not abolish nerve growth factor-mediated rescue of serum-free PC12 cells, *Journal of Neurochemistry*, **66**, 1826–1835

TAYLOR, J., GATCHALIAN, C.L., KEEN, G. and RUBIN, L.L., 1997, Apoptosis in cerebellar granule neurones: involvement of interleukin-1ß converting enzyme-like proteases, *Journal of Neurochemistry*, **68**, 1598–1605

TENNETI, L., DEMILIA, D.M., TROY, C.M. and LIPTON, S.A., 1998, Role of caspases in N-methyl-D-aspartate-induced apoptosis in cerebrocortical neurons, *Journal of Neurochemistry*, **71**, 946–959

THORNBERRY, N.A. and LAZEBNIK, Y., 1998, Caspases: enemies within, *Science*, **281**, 1312–1316

TROY, C.M., STEFANIS, L., GREENE, L.A. and SHELANSKI, M.L., 1997, Nedd2 is required for apoptosis after trophic factor withdrawal, but not superoxide dismutase (SOD1) downregulation, in sympathetic neurons and PC12 cells, *Journal of Neuroscience*, **17**, 1911–1918

VEKRELLIS, K., MCCARTHY, M.J., WATSON, A., WHITFIELD, J., RUBIN, L.L. and HAM, J., 1997, Bax promotes neuronal cell death and is downregulated during the development of the nervous system, *Development*, **124**, 1239–1249

WALKINSHAW, G. and WATERS, C.M., 1994, Neurotoxin-induced cell death in neuronal PC12 cells is mediated by induction of apoptosis, *Neuroscience*, **63**, 975–987

WANG, S., MIURA, M., JUNG, Y., ZHU, H., LI, E. and YUAN, J., 1998, Murine caspase-11, an ICE-interacting protease, is essential for the activation of ICE, *Cell*, **92**, 501–509

WILLIAMSON, T.L., ANDERSON, K.L. and CLEVELAND, D.W., 1997, Slowing of axonal tubulin transport in mice expressing two different familial ALS-linked SOD1 mutations, *Molecular Biology of the Cell*, **8** (Suppl.), 123A

YAN, S.D., FU, C., SOTO, C., CHEN, X., ZHU, H., AL-MOHANNA, F., *et al.*, 1997, An intracellular protein that binds amyloid-ß peptide and mediates neurotoxicity in Alzheimer's disease, *Nature*, **389**, 689–695

YOON, S.O., CASACCIA-BONNEFIL, P., CARTER, B. and CHAO, M.V., 1998, Competitive signalling between TrkA and p75 nerve growth factor receptors determines cell survival, *Journal of Neuroscience*, **18**, 3273–3281

ZEITLIN, S., LIU, J.P., CHAPMAN, D.L., PAPAIOANNOU, V.E. and EFSTRATIADIS, A., 1996, Increased apoptosis and early embryonic lethality in mice nullizygous for the Huntington's disease gene homolog, *Nature Genetics*, **11**, 155–163

ZHANG, K., WESTBERG, J.A., HOLTTA, E. and ANDERSSON, L.C., 1996, BCL2 regulates neural differentiation, *Proceedings of the National Academy of Sciences of the United States of America*, **93**, 4504–4508

ZHONG, L.T., SARAFIAN, T., KANE, D.J., CHARLES, A.C., MAH, S.P., EDWARDS, R.H. and BREDESEN, D.E., 1993, Bcl-2 inhibits death of central neural cells induced by multiple agents, *Proceedings of the National Academy of Sciences of the United States of America*, **90**, 4533–4537

CHAPTER

8

# *Perturbation of Apoptosis as a Mechanism of Action of Nongenotoxic Carcinogens*

**SABINA C. COSULICH and RUTH A. ROBERTS**

AstraZeneca Central Toxicology Laboratory, Alderley Park, Macclesfield, SK10 4TJ, UK

## *Contents*

## 8.1 INTRODUCTION

Nongenotoxic carcinogens are a group of structurally diverse compounds that cause cancer in rodents without damaging DNA (reviewed in Ashby *et al.*, 1994). A number of mechanisms have been proposed to explain their mode of action. Generation of reactive oxygen species, induction of DNA synthesis and suppression of apoptosis have been suggested as possible mechanisms for the carcinogenicity observed in rodent bioassays (Ashby *et al.*, 1994).

The liver is the major target organ for carcinogenicity (Ashby and Tennant, 1991) and there are several classes of nongenotoxic carcinogen that can cause hepatocellular adenomas and carcinomas. These include barbiturate drugs such as phenobarbitone (PB) (Butler, 1978), dioxins such as 2,3,7,8-tetradichlrobenzo-b,e-1,4-dioxin (TCDD) (Poland and Knutson, 1982) and the largest group, the peroxisome proliferators (reviewed in Ashby *et al.*, 1994). Despite the prevalence of liver tumours, neoplasia is observed in other tissues. For example, tetradecanoylphorbol ester causes skin carcinoma (Balmain and Pragnell, 1983, Bremner *et al.*, 1994), trimethylpentane is the chemical implicated in the nephrocarcinogenicity of unleaded petrol (Lock, 1990) and monoethylhexylphthalate (MEHP), a metabolite of the plasticizer DEHP, can cause tumours in rat testes (Richburg and Boekelheide, 1996).

## 8.2 NONGENOTOXIC CARCINOGENS AND CANCER

The carcinogenic potential of different nongenotoxic carcinogens differs considerably. Peroxisome proliferators (PPs) are a group of chemically distinct compounds which include chemicals of therapeutic, industrial and environmental significance such as hypolipidaemic fibrate drugs, leukotriene antagonists, clingwrap/medical tubing, plasticizers, herbicides and de-greasing solvents (Moody *et al.*, 1991, Green, 1992, Ashby *et al.*, 1994) (Figure 8.1). Continued treatment of rodents with PPs results in the development of hepatic adenomas and carcinomas (Reddy *et al.*, 1980). The potent non-fibrate hypolipidaemic drug Wyeth-14,643 and the leukotriene antagonist LY-171883 cause 100% tumours at 2 years in rats (0.1% w/w) in the diet (Marsman *et al.*, 1988). Weaker chemicals, such as DEHP and trichloroethylene yield hepatocellular carcinomas at 2 years when administered at 1.2%–2.5% (w/w) and 1200 mg/kg, respectively (National Toxicology Program, 1983). Although proliferation of peroxisomes is observed upon administration of PPs, it has become apparent that peroxisomal proliferation alone cannot account for tumourigenesis. Weak hepatocarcinogens such as DEHP only cause 25% less peroxisome proliferation when compared with potent carcinogens such as Wyeth-14,643 (Moody *et al.*, 1991). In addition, trichloroacetic acid (TCA) stimulates peroxisome proliferation in both rats and mice but only induces DNA synthesis and is only carcinogenic in mouse (Herren-Freund *et al.*, 1987).

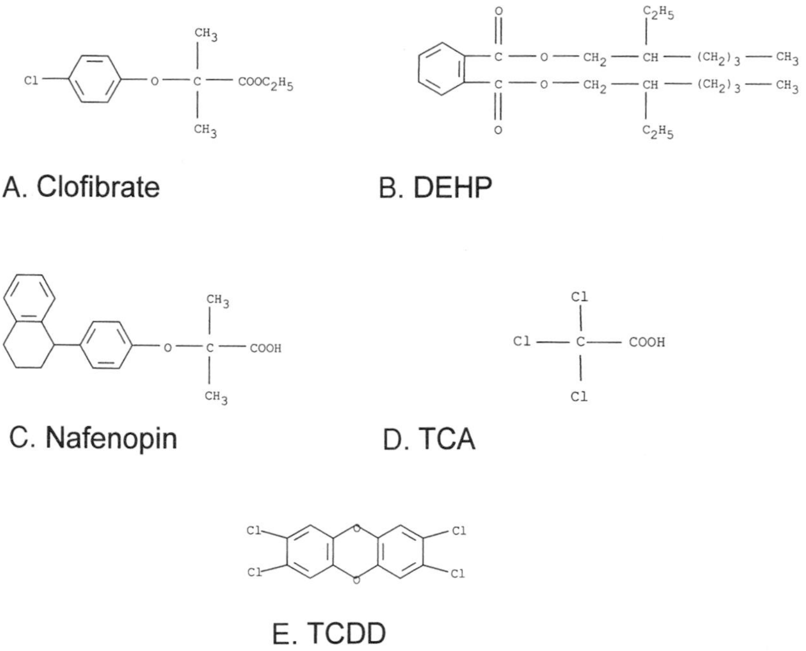

A. Clofibrate

B. DEHP

C. Nafenopin

D. TCA

E. TCDD

**Figure 8.1** Structures of some members of the peroxisome proliferator class of rodent nongenotoxic hepatocarcinogens.



The nongenotoxic liver carcinogen phenobarbital has also been studied extensively (Schulte-Hermann *et al.*, 1990, Phillips *et al.*, 1997). At doses between 400 ppm and 700 ppm, this compound causes more than 30% liver enlargement (Berman *et al.*, 1983, Edwards and Lucas, 1985, Mansbach *et al.*, 1996) probably due to an induction of hepatic DNA synthesis (Berman *et al.*, 1983, Edwards and Lucas, 1985) coupled with suppression of apoptosis (Oberhammer and Qin, 1995, James and Roberts, 1996, Worner, 1996). There are numerous studies showing that PB is hepatocarcinogenic in the rat liver (Rossi *et al.*, 1977, Phillips *et al.*, 1997). However, some authors suggest that PB is not a complete carcinogen but merely acts as a promoter of genotoxic damage caused by compounds such as diethylnitrosamine (DEN) (Peraino *et al.*, 1975, Butler, 1978, Neveu *et al.*, 1990, Worner, 1996). Treatment with PB and other barbiturates causes a number of changes in the expression of different classes of enzymes, such as aldehyde dehydrogenases and the cytochrome p450 mono-oxygenases as well as changes in the levels of a number of peroxisomal, microsomal, cytosolic and mitochondrial enzymes in hepatocytes *in vitro* and *in vivo* (Baron *et al.*, 1981, Liang *et al.*, 1995).

The biochemical and molecular mechanisms by which nongenotoxic carcinogens such as PB or PPs cause liver cancer is still unclear. The data available to date suggest that changes in cell proliferation and apoptosis are likely to play an important role. However, other mechanisms cannot be ruled out and may act alongside hepatic growth perturbation.

## 8.3 SPECIES DIFFERENCES IN RESPONSE TO NONGENOTOXIC CARCINOGENS

There are marked species differences in the response to nongenotoxic carcinogens (Lake *et al.*, 1989, James and Roberts, 1995, Latruffe, 1997). In rat or mouse, hepatocarcinogens such as the PPs or PB cause liver enlargement, hepatocyte replication and induction of detoxification enzymes such as cytochrome p4504A1 and 2B1/2, respectively (Moody *et al.*, 1977, 1991, Berman *et al.*, 1983, Edwards and Lucas, 1985, Mansbach *et al.*, 1996, Bell *et al.*, 1998). However, experiments using cultured human hepatocytes and data from patients receiving fibrate drug therapy suggest that humans do not display the adverse effects of PPs observed in rats and mice (Elcombe and Styles, 1989, Perrone and Williams, 1998). Specifically, human hepatocytes *in vitro* show no peroxisome proliferation nor hepatocyte proliferation and there appears to be no increased risk of developing hepatoadenomas in patients receiving peroxisome proliferators such as fenofibrate and bezafibrate (Ashby *et al.*, 1994). However, humans do respond to the fibrate PPs by altered expression of enzymes that regulate serum cholesterol and lipid homeostasis (Devchand *et al.*, 1996, Schoonjans *et al.*, 1996b). These PP-mediated changes form the basis of the clinical benefit to patients at risk of ischaemic heart disease. Such marked species differences in the nature and magnitude of the response to nongenotoxic carcinogens suggest that, despite their obvious carcinogenic potential in rodents, nongenotoxic carcinogens such as the PPs pose no risk to human health (Cattley *et al.*, 1998).

## 8.4 MECHANISM OF ACTION OF NONGENOTOXIC CARCINOGENS

### 8.4.1 Transcriptional activation by nongenotoxic carcinogens

The biological effects of many nongenotoxic carcinogens appear to be receptor-mediated (Green, 1992), where the toxicant binds to an intracellular receptor and perturbs growth regulation via a direct or indirect effect on gene expression. For example, TPA mimics diacylglyerol (DAG) by binding to and stimulating protein kinase C to switch on gene expression in the skin via the transcription factor AP1 (Mills and Smart, 1989, Sassone-Corsi *et al.*, 1990). PPs activate the peroxisome proliferator activated receptor (PPARα) (Issemann and Green, 1990) which has been shown to be responsible for the pleiotropic effects of these compounds (Lee *et al.*, 1995, Gonzalez, 1997a, Peters *et al.*, 1997). PPARα is expressed highly in hepatocytes, cardiomyocytes, enterocytes and the proximal tubules of the kidney (Issemann and Green, 1990, Braissant *et al.*, 1996). In the liver, activation of PPARα mediates the transcription of a number of genes, including CYP4A1 which encodes the microsomal enzyme cytochrome P4504A1 (Aldridge *et al.*, 1995) and the peroxisomal enzymes of β-oxidation such and acyl-CoA oxidase, enoyl-CoA hydratase and 3-hydroxyacyl-CoA dehydrogenase (Tugwood *et al.*, 1992, Aldridge *et al.*, 1995) (Table 8.1). In addition,

**TABLE 8.1**
Examples of PPARα inducible genes

| Gene | Function |
|---|---|
| Acyl-CoA oxidase | Fatty acid β-oxidation |
| Enoyl-CoA hydratase-3-hydrixyacyl-CoA dehydrogenase | Fatty acid β-oxidation |
| C-acyl-CoA synthetase | Conversion of fatty acids into acyl-CoA derivatives |
| CYP4A6 | Formation of dicarboxylic acids by ω-oxidation |
| aP2 (adipocyte fatty acid binding protein) | Adipose tissue fatty acid metabolism |
| Lipoprotein lipase | Hydrolysis of triglyceride-rich particles |
| Apolipoprotein A-I | Protein component of High Density Lipoprotein |
| Apolipoprotein A-II | Protein component of High Density Lipoprotein |

activation of PPARα modulates high density lipoprotein (HDL) cholesterol levels by increased transcription of the major HDL apolipoproteins, apoAI and apoAII, and reduces plasma triglycerides via induction of LPL and reduction of apoCIII (Auwerx, 1992, Schoonjans *et al.*, 1996a, Latruffe and Vamecq, 1997). The ability to up-regulate expression of the P450 family of genes is common to other nongenotoxic carcinogens such as dioxin and PB (Baron *et al.*, 1981, Bains *et al.*, 1985, Bell *et al.*, 1991, Liang *et al.*, 1995), probably because of the key role played by the cytochrome P450 gene family in oxidative metabolism and detoxification.



PPARα activation by its ligands is essential for PP-mediated transcription of known target genes (Table 8.1), as well as for the induction of cellular proliferation. Transcriptional regulation by PPARα is achieved through PPAR-RXR (where RXR is the receptor for 9-*cis* retinoic acid) heterodimers which bind DNA motifs called PPAR response elements (PPREs) within the promoters of their target genes (Tugwood *et al.*, 1992). Studies using PPARα null mice have indicated a complete lack of transcriptional response to chronic treatment with PPs (Lee *et al.*, 1995). In addition, PPARα null mice do not exhibit the increase in DNA synthesis normally observed upon administration of PPs (Lee *et al.*, 1995, Gonzalez, 1997a). This indicates that the adverse effects of PPs such as liver growth and cancer are mediated by activation of PPARα and are likely to involve transcription of new genes (Lee *et al.*, 1995, Gonzalez, 1997b, Peters *et al.*, 1997).

## 8.4.2 Generation of oxidative damage by nongenotoxic carcinogens

PB and PPs have been suggested to exert their tumour promoting activities by increasing cellular oxidative damage (Reddy, 1989, Chu *et al.*, 1995). Active oxygen, in the form of superoxide or hydrogen peroxide can be produced by an increase in peroxisomal fatty acid β-oxidation coupled with an increase in catalase levels (Nilakantan *et al.*, 1998). The resultant oxidative stress can lead to a number of diverse effects within the cell, including DNA damage, lipid peroxidation and new gene expression, which may lead to tumourigenesis (Cerutti and Trump, 1991). Although an increase in hydroxyl radicals is observed in rats given PPs (Reddy, 1989), conflicting data exist on the generation of DNA damage as an indirect mode of PP action (Huber *et al.*, 1997, Soliman *et al.*, 1997). Active oxygen species can directly modulate transcription through antioxidant response elements (AREs) found in the promoters of certain genes (Pinkus *et al.*, 1993). For example, the aryl hydrocarbon (Ah) gene battery comprises a family of at least six genes that respond to oxidative cytotoxicity (Nebert *et al.*, 1993). Other nongenotoxic carcinogens such as PB can also activate the promoters for glutathione-S-transferase Ya subunit and NAD(P)H: oxidoreductase genes through AREs (Pinkus *et al.*, 1993). Active oxygen species, in turn, can induce transcription through members of the nuclear factor-κB (NF-κB) family of transcriptional regulators (Nilakantan *et al.*, 1998). Thus, oxygen radicals may play a role as secondary messengers for a variety of agents.

### 8.4.3 Perturbation of growth by nongenotoxic carcinogens

It has been proposed that the tumourigenic potential of nongenotoxic carcinogens arises from their ability to promote the survival and growth of spontaneously initiated hepatocytes (Schulte-Hermann, 1983, Cattley *et al.*, 1991). This hypothesis arises from the observation that PPs and PB can cause cellular proliferation in the liver (Berman *et al.*, 1983). The majority of hepatocarcinogenic PPs cause measurable induction of DNA synthesis during the first 2–7 days of administration (Ashby *et al.*, 1994) but, frequently, this is not sustained and may return to control levels by around 10–20 days. However, the highly carcinogenic PP Wyeth-14,643 causes both acute and sustained hepatocyte proliferation in rat livers (Marsman *et al.*, 1988) leading to suggestions that carcinogenesis arises from prolonged mitogenic stimulation. The relative contribution of acute versus sustained proliferation remains to be determined.

The tightly controlled balance exerted in normal tissues between cell growth and cell death is disrupted during chemical-induced carcinogenesis (Roberts *et al.*, 1997). Thus, as well as stimulating proliferation, nongenotoxic carcinogens may alter rates of cell death in their target organs. This has been demonstrated for the PPs class of nongenotoxic hepatocarcinogens and also for PB since these chemicals suppress apoptosis in rodent hepatocytes (Bursch *et al.*, 1984, Schulte-Hermann *et al.*, 1990, Bayly *et al.*, 1994, Gill *et al.*, 1998, Roberts *et al.*, 1995). Induction of apoptosis by a number of stimuli, including the physiological liver factor TGF$\beta_1$, the DNA damaging agent etoposide or activation of the Fas receptor, can be suppressed in cultured rodent hepatocytes by PPs (Gill *et al.*, 1998). In addition to a reduction in the number of cells exhibiting apoptotic morphology, incubation with PPs or with PB results in maintenance of viability over several weeks, in contrast to eight days in control cultures (Bayly *et al.*, 1994). *In vivo* studies using other nongenotoxic carcinogens have shown that although liver hyperplasia is observed upon administration of compounds such as cyproterone acetate or phenobarbital, withdrawal of these compounds causes regression of the hyperplastic liver with a concomitant increase in apoptosis (Schulte-Hermann *et al.*, 1980, Bursch *et al.*, 1984, 1985, Roberts *et al.*, 1995). This liver regression is reversible and can be inhibited by re-administration of the relevant compound.

## 8.4.4 Molecular mechanism for the suppression of apoptosis by nongenotoxic carcinogens

### 8.4.4.1 Cytokine networks

The molecular mechanism of the perturbation of growth by nongenotoxic hepatocarcinogens is unknown, but recent evidence suggests the involvement of cytokine signalling between different cell populations within the liver. In particular, tumour necrosis factor α (TNF-α) (Beutler and Cerami, 1988) has been implicated in mediating the growth perturbation effects of nongenotoxic hepatocarcinogens (James, 1998, Roberts and Kimber, 1999). Although TNF-α is known to be an inflammatory mediator proximally associated with necrotic injury (Deviere *et al.*, 1990), it may also have a role as a positive mediator of hepatocyte growth (Taub, 1996, Roberts and Kimber, 1999). The observation that TNFRI null mice show reduced liver regeneration after partial hepatectomy (Yamada *et al.*, 1997) has lead to the hypothesis that TNF-α may be involved in mediating the effects of nongenotoxic carcinogens (Rolfe *et al.*, 1997). Further studies have shown that TNF-α could mimic the effects of PPs in rat hepatocytes and that antibodies to TNF-α and TNFR1 could attenuate the effects of PPs on DNA synthesis (Bojes *et al.*, 1997) and apoptosis (Rolfe *et al.*, 1997, West *et al.*, 1999). Thus, low levels of TNF-α produced from the resident liver macrophages, the Kupffer cells, may be mediating the effects of PPs on liver homeostasis in a paracrine signalling mechanism (Rose *et al.*, 1997). This hypothesis is supported by the observation that Kupffer cells are required for stimulation of mitosis by PPs (Rose *et al.*, 1997) and that PPs stimulate Kupffer cells (Bojes and Thurman, 1996, Rusyn, 1998). The molecular mechanism of putative TNF-α production and/or release in response to PPs is unclear at present.

TNF-α signalling involves the activation of both pro- and anti-apoptotic processes (Heller and Kronke, 1994, Smith *et al.*, 1994, Wallach, 1997). The anti-apoptotic effects of TNF-α are thought to be mediated by activation of novel gene expression, perhaps through activation of NF-κB (DiDonato *et al.*, 1997, May and Ghosh, 1998, West *et al.*, 1999). NF-κB, a critical regulator of cytokine-inducible gene expression, is activated by TNF-α and by pro-inflammatory cytokines such as IL-1 (Heller and Kronke, 1994, DiDonato *et al.*, 1997, Malinin *et al.*, 1997) (Figure 8.2). Normally, NF-κB is held in an inactive state in the cytoplasm by IκB inhibitory proteins (DiDonato *et al.*, 1997). When cells

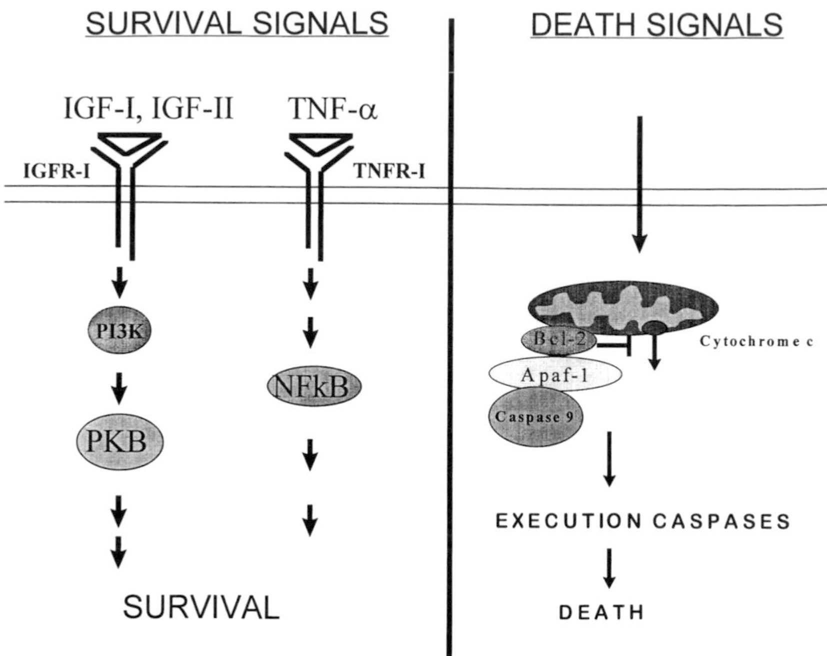


**Figure 8.2** Cell survival and cell death signalling pathways.

are treated with TNF-α or IL-1, protein kinase cascades are activated that lead to the phosphorylation of IκB proteins. This targets IκB for ubiquitination and proteasome mediated degradation, allowing nuclear transport and activation of NF-κB (DiDonato *et al.*, 1997). NF-κB has been shown to be a pivotal transcription factor in liver regeneration (Cressman *et al.*, 1994, Beg *et al.*, 1995). NF-κB activity is induced during liver regeneration after partial hepatectomy, and inhibition of NF-κB by a dominant mutant of IκB results in the induction of massive hepatocyte apoptosis (Hellerbrand *et al.*, 1998, Iimuro *et al.*, 1998). To further support the role of NF-κB as an anti-apoptotic factor, disruption of the NF-κB p65 (RelA) subunit leads to embryonic lethality at 15–16 days of gestation, concomitant with extensive apoptosis of the liver (Beg *et al.*, 1995).

The NF-κB inducible cytokine IL-6 has also been shown to be essential for liver regeneration after partial hepatectomy, indicating that IL-6 might be a critical component of the regenerative response (Cressman *et al.*, 1996). Therefore, direct or indirect activation of cytokines such as TNF-α, IL-1 or IL-6 may be a common mechanism for the suppression of apoptosis by nongenotoxic carcinogens.

### 8.4.4.2 Activation of survival pathways

Hepatocyte survival is likely to be dependent on a number of factors. The regulation of cytokine networks may allow interactions with a number of other pathways which in turn may cause suppression of apoptosis (Figure 8.2). Recent evidence suggests that the basic mechanisms for regulation of apoptosis are tied to cellular proliferation and transformation pathways (Harrington *et al.*, 1994). This co-ordination is illustrated by the observation that serum specific mitogens and growth factors inhibit apoptosis and promote survival (Harrington *et al.*, 1994). A number of survival factors have been described, including insulin-like growth factors I and II (IGF I and II) (Christofori *et al.*, 1994, Resnicoff *et al.*, 1995a) (Figure 8.2). IGFs are mitogenic polypeptides which are thought to be required in several cell types for the establishment and maintenance of transformed phenotypes and tumourigenesis (Resnicoff *et al.*, 1995b). IGFs are synthesized primarily in the liver and are thought to play an important role in the development of certain tumours (Daughaday and Rotwein, 1989). IGF-I mRNA levels have been shown to be high in the adult rat liver, whereas levels of IGF-II are low (Daughaday and Rotwein, 1989). However, hepatoma tissues frequently up-regulate expression of IGFs. In hepatocytes infected with virus, expression of IGF-II increases cell survival and promotes tumourigenesis (Ueda and Ganem, 1996). Furthermore, tumours develop with reduced malignancy and with a higher incidence of apoptosis in mice that are deficient in IGF-II (Christofori *et al.*, 1994). Thus, it is possible that nongenotoxic carcinogens may act by regulating IGF survival pathways in the liver either directly or indirectly via a number of cytokine networks.

The mechanism of action of IGFs has been discussed in Chapter 1 (section 1.3.1) and involves interaction with specific receptors and activation of downstream signalling pathways via PI3 kinase (Kennedy *et al.*, 1997, Marte and Downward, 1997). Recent research from several laboratories has indicated that a signalling pathway from PI3 kinase to the serine/threonine protein kinase PKB (Akt) may mediate protection from apoptosis (Franke *et al.*, 1997, Songyang *et al.*, 1997). Activation of this pathway appears to be able to prevent apoptosis induced by a variety of cellular challenges. The exact mechanisms for the suppression of apoptosis by cell survival factors are unknown, but recent evidence suggests PKB may cause modifications of Bcl-2 proteins. In particular, Bad, a pro-apoptotic member of the family (Kelekar *et al.*, 1997), is phosphorylated by PKB at two sites (Franke *et al.*, 1997). This modification has been shown to prevent Bad from binding to Bcl-XL, thereby inactivating an anti-apoptotic family member (Kelekar *et al.*, 1997, Reed, 1997).



### 8.4.4.3 Regulation of Bcl-2 family members

The regulation of Bcl-2 family members and its role in cell survival has been discussed in Chapter 1 (section 1.3.3). However, little is known about the regulation of Bcl-2 family members or other components of the apoptosis pathway by nongenotoxic carcinogens. In particular, it is not known whether post-translational modifications of Bcl-2 family members may occur in response to treatment with nongenotoxic carcinogens. Although differential regulation of Bcl-2 family members has been observed during liver regeneration, normally the steady state transcript abundance of different family members is regulated post-transcriptionally and in part by mRNA stability (Kren *et al.*, 1996, Tzung *et al.*, 1997). The uncoupling of mRNA and protein levels for different Bcl-2 family members observed in hepatocytes treated with the PP clofibrate, suggests that translational or post-translational controls of different family members may regulate survival in the liver (Kren *et al.*, 1996). Post-translational modifications of Bcl-2 family members may be a mechanism by which cytokine-regulated pathways and survival pathways communicate with the apoptosis execution machinery.

## 8.5 SUMMARY

Nongenotoxic carcinogens cause tumours in rodent bioassays without damaging DNA. Although tumours of the thyroid, skin and kidney can occur, the liver is the major target organ in rats and mice for nongenotoxic carcinogenesis. Dioxin and PB are potent promoters of rat and mouse liver tumours but the largest and most chemically diverse family of nongenotoxic hepatocarcinogens are the PPs. The carcinogenic potential of nongenotoxic carcinogens arises from their ability to perturb growth regulation in their target tissue. PPs mediate their biological responses via activation of the nuclear hormone receptor PPARα which mediates transcriptional regulation of the genes associated with response to PPs. However, the genes that regulate the growth changes associated with hepatocarcinogenesis are unknown. Although suppression of apoptosis and induction of S-phase are likely to play a key role in the carcinogenicity process, recent data support a role for hepatic cytokines such as TNF-α in modulating hepatocyte survival and apoptosis pathways.

## REFERENCES

ALDRIDGE, T., TUGWOOD, J. and GREEN, S., 1995, Identification and characterization of DNA elements implicated in the regulation of CYP4A1 transcription, *Biochemical Journal*, **306** (Pt 2), 473–479

ASHBY, J. and TENNANT, R., 1991, Definitive relationships among chemical structure, carcinogenicity and mutagenicity for 301 chemicals tested by the U.S. NTP, *Mutation Research*, **257**, 229–306

ASHBY, J., BRADY, A., ELCOMBE, C., ELLIOTT, B., ISHMAEL, J., ODUM, J., *et al.*, 1994, Mechanistically-based human hazard assessment of peroxisome proliferator-induced hepatocarcinogenesis, *Human Experimental Toxicology*, **13** (Suppl. 2), S1–S117

AUWERX, J., 1992, Regulation of gene expression by fatty acids and fibric acid derivatives: an integrative role for peroxisome proliferator activated receptors. The Belgian Endocrine Society Lecture, *Hormone Research*, **38**, 269–277

BAINS, S., GARDINER, S., MANNWEILER, K., GILLETT, D. and GIBSON, G., 1985, Immunochemical study on the contribution of hypolipidaemic-induced cytochrome

P-452 to the metabolism of lauric acid and arachidonic acid, *Biochemical Pharmacology*, **34**, 3221–3229

Balmain, A. and Pragnell, I., 1983, Mouse skin carcinomas induced *in vivo* by chemical carcinogens have a transforming Harvey-ras oncogene, *Nature*, **303**, 72–74

Baron, J., Redick, J. and Guengerich, F., 1981, An immunohistochemical study on the localization and distributions of phenobarbital- and 3-methylcholanthrene-inducible cytochromes P-450 within the livers of untreated rats, *Journal of Biological Chemistry*, **256**, 5931–5937

Bayly, A., Roberts, R. and Dive, C., 1994, Suppression of liver cell apoptosis *in vitro* by the non-genotoxic hepatocarcinogen and peroxisome proliferator nafenopin, *Journal of Cell Biology*, **125**, 197–203

Beg, A., Sha, W., Bronson, R., Ghosh, S. and Baltimore, D., 1995, Embryonic lethality and liver degeneration in mice lacking the RelA component of NF-kappa B, *Nature*, **376**, 167–170

Bell, A.S., Savory, R., Horley, N.J., Choudhury, A.I., Dickins, M., Gray, T.J.B., *et al.*, 1998, Molecular basis of non-responsiveness to peroxisome proliferators: the guinea pig PPAR-alpha is functional and mediates peroxisome proliferator-induced hypolipidaemia, *Biochemical Journal*, **332**, 689–693

Bell, D., Bars, R., Gibson, G. and Elcombe, C., 1991, Localization and differential induction of cytochrome P450IVA and acyl-CoA oxidase in rat liver, *Biochemical Journal*, **275** (Pt 1), 247–252

Berman, J., Hiley, C. and Wilson, A., 1983, Comparison of the effects of the hypolipidaemic agents ICI53072 and clofibrate with those of phenobarbitone on liver size, blood flow and DNA content in the rat, *British Journal of Pharmacology*, **78**, 533–541

Beutler, B. and Cerami, A., 1988, Cachectin (tumor necrosis factor): a macrophage hormone governing cellular metabolism and inflammatory response, *Endocrinology Reviews*, **9**, 57–66

Bojes, H. and Thurman, R., 1996, Peroxisome proliferators activate Kupffer cells *in vivo*, *Cancer Research*, **56**, 1–4

Bojes, H., Germolec, D., Simeonova, P., Bruccoleri, A., Schoonhoven, R., Luster, M. and Thurman, R., 1997, Antibodies to tumor necrosis factor alpha prevent increases in cell replication in liver due to the potent peroxisome proliferator, WY-14,643, *Carcinogenesis*, **18**, 669–674

Braissant, O., Foufele, F., Scotto, C., Dauca, M. and Wahli, W., 1996, Differential expression of peroxisome proliferator-activated receptors (PPARs): tissue distribution of PPAR-alpha, -beta, and -gamma in the adult rat, *Endocrinology*, **137**, 354–366

BREMNER, R., KEMP, C. and BALMAIN, A., 1994, Induction of different genetic changes by different classes of chemical carcinogens during progression of mouse skin tumors, *Molecular Carcinogenesis*, **11**, 90–97

BURSCH, W., LAUER, B., TIMMERMANN-TROSIENER, I., BARTHEL, G., SCHUPPLER, J. and SCHULTE-HERMANN, R., 1984, Controlled death (apoptosis) of normal and putative preneoplastic cells in rat liver following withdrawal of tumor promoters, *Carcinogenesis*, **5**, 453–458

BURSCH, W., TAPER, H., LAUER, B. and SCHULTE-HERMANN, R., 1985, Quantitative histological and histochemical studies on the occurrence and stages of controlled cell death (apoptosis) during regression of rat liver hyperplasia, *Virchows Archives of B Cell Pathology Including Molecular Pathology*, **50**, 153–166

BUTLER, W., 1978, Long-term effects of phenobarbitone-Na on male Fischer rats, *British Journal of Cancer*, **37**, 418–423

CATTLEY, R., MARSMAN, D. and POPP, J., 1991, Age-related susceptibility to the carcinogenic effect of the peroxisome proliferator WY-14,643 in rat liver, *Carcinogenesis*, **12**, 469–473

CATTLEY, R.D.J., ELCOMBE, C.R., FENNER-CRISP, P., LAKE, B., MARSMAN, D., PASTOOR, T., *et al.*, 1998, Do peroxisome proliferating compounds pose a hepatocarcinogenic hazard to humans? *Regulatory Toxicology and Pharmacology*, **27**, 47–60

CERUTTI, P. and TRUMP, B., 1991, Inflammation and oxidative stress in carcinogenesis, *Cancer Cells*, **3**, 1–7

CHRISTOFORI, G., NAIK, P. and HANAHAN, D., 1994, A second signal supplied by insulin-like growth factor II in oncogene-induced tumorigenesis, *Nature*, **369**, 414–418

CHU, S., HUANG, Q., ALVARES, K., YELDANDI, A., RAO, M. and REDDY, J., 1995, Transformation of mammalian cells by overexpressing H2O2-generating peroxisomal fatty acyl-CoA oxidase, *Proceedings of the National Academy of Sciences of the United States of America*, **92**, 7080–7084

CRESSMAN, D., GREENBAUM, L., HABER, B. and TAUB, R., 1994, Rapid activation of post-hepatectomy factor/nuclear factor kappa B in hepatocytes, a primary response in the regenerating liver, *Journal of Biological Chemistry*, **269**, 30429–30435

CRESSMAN, D., GREENBAUM, L., DEANGELIS, R., CILIBERTO, G., FURTH, E., POLI, V. and TAUB, R., 1996, Liver failure and defective hepatocyte regeneration in interleukin-6-deficient mice, *Science*, **274**, 1379–1383

DAUGHADAY, W. and ROTWEIN, P., 1989, Insulin-like growth factors I and II. Peptide, messenger ribonucleic acid and gene structures, serum, and tissue concentrations, *Endocrinology Reviews*, **10**, 68–91

DEVCHAND, P., KELLER, H., PETERS, J., VAZQUEZ, M., GONZALEZ, F. and WAHLI, W., 1996, The PPAR α-leukotriene B4 pathway to inflammation control, *Nature*, **384**, 39–43

DEVIERE, J., CONTENT, J., DENYS, C., VANDENBUSSCHE, P., SCHANDENE, L., WYBRAN, J. and DUPONT, E., 1990, Excessive *in vitro* bacterial lipopolysaccharide-induced production of monokines in cirrhosis, *Hepatology*, **11**, 628–634

DIDONATO, J., HAYAKAWA, M., ROTHWARF, D., ZANDI, E. and KARIN, M., 1997, A cytokine-responsive IkappaB kinase that activates the transcription factor NF-kappaB, *Nature*, **388**, 548–554

EDWARDS, A. and LUCAS, C., 1985, Phenobarbital and some other liver tumor promoters stimulate DNA synthesis in cultured rat hepatocytes, *Biochemical and Biophysical Research Communications*, **131**, 103–108

ELCOMBE, C.R. and STYLES, J.A., 1989, Species differences in peroxisome proliferation, *The Toxicologist*, **9**, 249 only

FRANKE, T., KAPLAN, D. and CANTLEY, L., 1997, PI3K: downstream AKTion blocks apoptosis, *Cell*, **88**, 435–437

GILL, J., JAMES, N., ROBERTS, R. and DIVE, C., 1998, The non-genotoxic hepatocarcinogen nafenopin suppresses rodent hepatocyte apoptosis induced by TGF beta 1, DNA damage and Fas, *Carcinogenesis*, **19**, 299–304

GONZALEZ, F., 1997a, Recent update on the PPAR alpha-null mouse, *Biochimie*, **79**, 139–144

GONZALEZ, F., 1997b, The role of peroxisome proliferator activated receptor alpha in peroxisome proliferation, physiological homeostasis, and chemical carcinogenesis, *Advances in Experimental Medical Biology*, **422**, 109–125

GREEN, S., 1992, Peroxisome proliferators: a model for receptor mediated carcinogenesis, *Cancer Surveys*, **14**, 221–232

HARRINGTON, E., BENNETT, M., FANIDI, A. and EVAN, G., 1994, c-Myc-induced apoptosis in fibroblasts is inhibited by specific cytokines, *EMBO Journal*, **13**, 3286–3295

HELLER, R. and KRONKE, M., 1994, Tumor necrosis factor receptor-mediated signalling pathways, *Journal of Cell Biology*, **126**, 5–9

HELLERBRAND, C., JOBIN, C., IIMURO, Y., LICATO, L., SARTOR, R. and BRENNER, D., 1998, Inhibition of NFkappaB in activated rat hepatic stellate cells by proteasome inhibitors and an IkappaB super-repressor, *Hepatology*, **27**, 1285–1295

HERREN-FREUND, S., PEREIRA, M., KHOURY, M. and OLSON, G., 1987, The carcinogenicity of trichloroethylene and its metabolites, trichloroacetic acid and dichloroacetic acid, in mouse liver, *Toxicology and Applied Pharmacology*, **90**, 183–189

HUBER, W., GRASL-KRAUPP, B., STEKEL, H., GSCHWENTNER, C., LANG, H. and SCHULTE-HERMANN, R., 1997, Inhibition instead of enhancement of lipid peroxidation by pretreatment with the carcinogenic peroxisome proliferator nafenopin in rat liver exposed to a high single dose of corn oil, *Archives in Toxicology*, **71**, 575–581

Iimuro, Y., Nishiura, T., Hellerbrand, C., Behrns, K., Schoonhoven, R., Grisham, J. and Brenner, D., 1998, NFkappaB prevents apoptosis and liver dysfunction during liver regeneration, *Journal of Clinical Investigation*, **101**, 802–811

Issemann, I. and Green, S., 1990, Activation of a member of the steroid hormone receptor superfamily by peroxisome proliferators, *Nature*, **347**, 645–650

James, N. and Roberts, R., 1995, Species differences in the clonal expansion of hepatocytes in response to the coaction of epidermal growth factor and nafenopin, a rodent hepatocarcinogenic peroxisome proliferator, *Fundamental Applied Toxicology*, **26**, 143–149

James, N.H. and Roberts, R.A., 1996, Species differences in response to peroxisome proliferators correlate *in vitro* with induction of DNA synthesis rather than suppression of apoptosis, *Carcinogenesis*, **17**, 1623–1632

James, N.H., Gill, J.H., Brindle, R., Woodyatt, N.J., Macdonald, N., Rolfe, M., *et al.*, 1998, PPAR-alpha mediated responses and their importance in carcinogenesis, *Toxicology Letters*, **102–103**, 91–96

Kelekar, A., Chang, B., Harlan, J., Fesik, S. and Thompson, C., 1997, Bad is a BH3 domain-containing protein that forms an inactivating dimer with Bcl-xl, *Molecular and Cellular Biology*, **17**, 7040–7046

Kennedy, S., Wagner, A., Conzen, S., Jordan, J., Bellacosa, A., Tsichlis, P. and Hay, N., 1997, The PI 3-kinase/Akt signaling pathway delivers an anti-apoptotic signal, *Genes and Development*, **11**, 701–713

Kren, B., Trembley, J., Krajewski, S., Behrens, T., Reed, J. and Steer, C., 1996, Modulation of apoptosis-associated genes Bcl-2, Bcl-x, and bax during rat liver regeneration, *Cell Growth and Differentiation*, **7**, 1633–1642

Lake, B., Evans, J., Gray, T., Korosi, S. and North, C., 1989, Comparative studies on nafenopin-induced hepatic peroxisome proliferation in the rat, Syrian hamster, guinea pig, and marmoset, *Toxicology and Applied Pharmacology*, **99**, 148–160

Latruffe, N., 1997, Is hepatocarcinogenesis mediated by peroxisome proliferators restricted to rodents? *The Cancer Journal*, **10**, 162–165

Latruffe, N. and Vamecq, J., 1997, Peroxisome proliferators and peroxisome proliferator activated receptors (PPARs) as regulators of lipid metabolism, *Biochimie*, **79**, 81–94

Lee, S., Pineau, T., Drago, J., Lee, E., Owens, J., Kroetz, D., *et al.*, 1995, Targeted disruption of the alpha isoform of the peroxisome proliferator-activated receptor gene in mice results in abolishment of the pleiotropic effects of peroxisome proliferators, *Molecular and Cellular Biology*, **15**, 3012–3022

Liang, Q., He, J. and Fulco, A., 1995, The role of Barbie box sequences as cis-acting elements involved in the barbiturate-mediated induction of cytochromes P450BM-1 and P450BM-3 in Bacillus megaterium, *Journal of Biological Chemistry*, **270**, 4438–4450

Lock, E., 1990, Chronic nephrotoxicity of 2,2,4-trimethylpentane and other branched-chain hydrocarbons, *Toxicology Letters*, **53**, 75–80

Malinin, N., Boldin, M., Kovalenko, A. and Wallach, D., 1997, MAP3K-related kinase involved in NF-kappaB induction by TNF, CD95 and IL-1, *Nature*, **385**, 540–544

Mansbach, J., Mills, J., Boyer, I., De Souza, A., Hankins, G. and Jirtle, R., 1996, Phenobarbital selectively promotes initiated cells with reduced TGF beta receptor levels, *Carcinogenesis*, **17**, 171–174

Marsman, D., Cattley, R., Conway, J. and Popp, J., 1988, Relationship of hepatic peroxisome proliferation and replicative DNA synthesis to the hepatocarcinogenicity of the peroxisome proliferators di(2-ethylhexyl)phthalate and [4-chloro-6-(2,3-xylidino)-2-pyrimidinylthio]acetic acid (Wy-14,643) in rats, *Cancer Research*, **48**, 6739–6744

Marte, B. and Downward, J., 1997, PKB/Akt: connecting phosphoinositide 3-kinase to cell survival and beyond, *Trends in Biochemical Sciences*, **22**, 355–358

May, M., and Ghosh, S., 1998, Signal transduction through NF-kappa B, *Immunology Today*, **19**, 80–88

Mills, K. and Smart, R., 1989, Comparison of epidermal protein kinase C activity, ornithine decarboxylase induction and DNA synthesis stimulated by TPA or dioctanoylglycerol in mouse strains with differing susceptibility to TPA-induced tumor promotion, *Carcinogenesis*, **10**, 833–838

Moody, D., Rao, M. and Reddy, J., 1977, Mitogenic effect in mouse liver induced by a hypolipidemic drug, nafenopin, *Virchows Archives of B Cell Pathology*, **23**, 291–296

Moody, D., Reddy, J., Lake, B., Popp, J. and Reese, D., 1991, Peroxisome proliferation and nongenotoxic carcinogenesis: commentary on a symposium, *Fundamental Applied Toxicology*, **16**, 233–248

National Toxicology Progam, 1983, Carcinogenesis bioassays of di(2-ethylhexyl)phthalate (CAS No 117-81-7) in F344 rats and B6C3F1 mice, National Toxicology Program, National Institute of Health, Bethesda

Nebert, D., Puga, A. and Vasiliou, V., 1993, Role of the Ah receptor and the dioxin-inducible [Ah] gene battery in toxicity, cancer, and signal transduction, *Annals of the New York Academy of Sciences*, **685**, 624–640

Neveu, M., Hully, J., Paul, D. and Pitot, H., 1990, Reversible alteration in the expression of the gap junctional protein connexin 32 during tumor promotion in rat liver and its role during cell proliferation, *Cancer Communications*, **2**, 21–31

Nilakantan, V., Spear, B. and Glauert, H., 1998, Liver-specific catalase expression in transgenic mice inhibits NF-kappaB activation and DNA synthesis induced by the peroxisome proliferator ciprofibrate, *Carcinogenesis*, **19**, 631–637

OBERHAMMER, F. and QIN, H., 1995, Effect of three tumour promoters on the stability of hepatocyte cultures and apoptosis after transforming growth factor-beta 1, *Carcinogenesis*, **16**, 1363–1371

PERAINO, C., FRY, R., STAFFELDT, E. and CHRISTOPHER, J., 1975, Comparative enhancing effects of phenobarbital, amobarbital, diphenylhydantoin, and dichlorodiphenyltrichloroethane on 2-acetylaminofluorene-induced hepatic tumorigenesis in the rat, *Cancer Research*, **35**, 2884–2890

PERRONE, C.S. and WILLIAMS, G.M., 1998, Effect of rodent hepatocarcinogenic peroxisome proliferators on fatty acyl coA oxidase, DNA synthesis and apoptosis in cultured human and rat hepatocytes, *Toxicology and Applied Pharmacology*, **150**, 277–286

PETERS, J., CATTLEY, R. and GONZALEZ, F., 1997, Role of PPAR alpha in the mechanism of action of the nongenotoxic carcinogen and peroxisome proliferator Wy-14,643, *Carcinogenesis*, **18**, 2029–2033

PHILLIPS, J., PRICE, R., CUNNINGHAME, M., OSIMITZ, T., COCKBURN, A., GABRIEL, K., *et al.*, 1997, Effect of piperonyl butoxide on cell replication and xenobiotic metabolism in the livers of CD-1 mice and F344 rats, *Fundamental Applied Toxicology*, **38**, 64–74

PINKUS, R., BERGELSON, S. and DANIEL, V., 1993, Phenobarbital induction of AP-1 binding activity mediates activation of glutathione S-transferase and quinone reductase gene expression, *Biochemical Journal*, **290**, 637–640

POLAND, A. and KNUTSON, J., 1982, 2,3,7,8-tetrachlorodibenzo-p-dioxin and related halogenated aromatic hydrocarbons: examination of the mechanism of toxicity, *Annual Reviews in Pharmacology and Toxicology*, **22**, 517–554

REDDY, J.K. and RAO, M.S., 1989, Oxidative DNA damage caused by persistent peroxisome proliferation: its role in hepatocarcinogenesis, *Mutation Research*, **214**, 63–68

REDDY, J.K., AZARNOFF, D. and HIGNITE, C., 1980, Hypolipidaemic hepatic peroxisome proliferators form a novel class of chemical carcinogens, *Nature*, **283**, 397–398

REED, J., 1997, Double identity for proteins of the Bcl-2 family, *Nature*, **387**, 773–776

RESNICOFF, M., BURGAUD, J., ROTMAN, H., ABRAHAM, D. and BASERGA, R., 1995b, Correlation between apoptosis, tumorigenesis, and levels of insulin-like growth factor I receptors, *Cancer Research*, **55**, 3739–3741

RESNICOFF, M., ABRAHAM, D., YUTANAWIBOONCHAI, W., ROTMAN, H., KAJSTURA, J., RUBIN, R., *et al.*, 1995a, The insulin-like growth factor I receptor protects tumor cells from apoptosis *in vivo*, *Cancer Research*, **55**, 2463–2469

RICHBURG, J. and BOEKELHEIDE, K., 1996, Mono-(2-ethylhexyl) phthalate rapidly alters both Sertoli cell vimentin filaments and germ cell apoptosis in young rat testes, *Toxicology and Applied Pharmacology*, **137**, 42–50

ROBERTS, R.A. and KIMBER, I., 1999, Cytokines in non-genotoxic hepatocarcinogenesis, *Carcinogenesis*, **20**, 1397–1401

ROBERTS, R., NEBERT, D., HICKMAN, J., RICHBURG, J. and GOLDSWORTHY, T., 1997, Perturbation of the mitosis/apoptosis balance: a fundamental mechanism in toxicology, *Fundamental Applied Toxicology*, **38**, 107–115

ROBERTS, R., SOAMES, A., GILL, J., JAMES, N. and WHEELDON, E., 1995, Non-genotoxic hepatocarcinogens stimulate DNA synthesis and their withdrawal induces apoptosis, but in different hepatocyte populations, *Carcinogenesis*, **16**, 1693–1698

ROLFE, M., JAMES, N. and ROBERTS, R., 1997, Tumour necrosis factor alpha (TNF alpha) suppresses apoptosis and induces DNA synthesis in rodent hepatocytes: a mediator of the hepatocarcinogenicity of peroxisome proliferators? *Carcinogenesis*, **18**, 2277–2280

ROSE, M., GERMOLEC, D., SCHOONHOVEN, R. and THURMAN, R., 1997, Kupffer cells are causally responsible for the mitogenic effect of peroxisome proliferators, *Carcinogenesis*, **18**, 1453–1456

ROSSI, L., RAVERA, M., REPETTI, G. and SANTI, L., 1977, Long-term administration of DDT or phenobarbital-Na in Wistar rats, *International Journal of Cancer*, **19**, 179–185

RUSYN, I., TSUKAMATO, H. and THURMAN, R.G., 1998, WY-14, 643 rapidly activates nuclear factor kB in Kupffer cells before hepatocarcinogenesis, *Carcinogenesis*, **19**, 1217–1222

SASSONE-CORSI, P., RANSONE, L. and VERMA, I., 1990, Cross-talk in signal transduction: TPA-inducible factor jun/AP-1 activates cAMP-responsive enhancer elements, *Oncogene*, **5**, 427–431

SCHOONJANS, K., STAELS, B. and AUWERX, J., 1996b, Role of the peroxisome proliferator-activated receptor (PPAR) in mediating the effects of fibrates and fatty acids on gene expression, *Journal of Lipid Research*, **37**, 907–925

SCHOONJANS, K., PEINADO-ONSURBE, J., LEFEBVRE, A., HEYMAN, R., BRIGGS, M., DEEB, S., *et al.*, 1996a, PPARalpha and PPARgamma activators direct a distinct tissue-specific transcriptional response via a PPRE in the lipoprotein lipase gene, *EMBO Journal*, **15**, 5336–5348

SCHULTE-HERMANN, R., TIMMERMANN-TROSIENER, I. and SCHUPPLER, J., 1983, Promotion of spontaneous pre-neoplastic cells in rat liver as a possible explanation of tumour promotion by non-mutagenic compounds, *Cancer Research*, **43**, 839–844

SCHULTE-HERMANN, R., TIMMERMANN-TROSIENER, I., BARTHEL, G. and BURSCH, W., 1990, DNA synthesis, apoptosis, and phenotypic expression as determinants of growth of altered foci in rat liver during phenobarbital promotion, *Cancer Research*, **50**, 5127–5135

SCHULTE-HERMANN, R., HOFFMAN, V., PARZEFALL, W., KALLENBACH, M., GERHARDT, A. and SCHUPPLER, J., 1980, Adaptive responses of rat liver to the gestagen and anti-androgen cyproterone acetate and other inducers. II. Induction of growth, *Chemico-Biological Interactions*, **31**, 287–300

SMITH, C., FARRAH, T. and GOODWIN, R., 1994, The TNF receptor superfamily of cellular and viral proteins: activation, costimulation, and death, *Cell*, **76**, 959–962

SOLIMAN, M., CUNNINGHAM, M., MORROW, J., ROBERTS II, L.J., and Badr, M., 1997, Evidence against peroxisome proliferation-induced hepatic oxidative damage, *Biochemical Pharmacology*, **53**, 1369–1374

SONGYANG, Z., BALTIMORE, D., CANTLEY, L., KAPLAN, D. and FRANKE, T., 1997, Interleukin 3-dependent survival by the Akt protein kinase, *Proceedings of the National Academy of Sciences of the United States of America*, **94**, 11345–11350

TAUB, R., 1996, Liver regeneration in health and disease, *Clinical and Laboratory Medicine*, **16**, 341–360

TUGWOOD, J., ISSEMANN, I., ANDERSON, R., BUNDELL, K., MCPHEAT, W. and GREEN, S., 1992, The mouse peroxisome proliferator activated receptor recognizes a response element in the 5′ flanking sequence of the rat acyl CoA oxidase gene, *EMBO Journal*, **11**, 433–439

TZUNG, S., FAUSTO, N. and HOCKENBERY, D., 1997, Expression of Bcl-2 family during liver regeneration and identification of Bcl-x as a delayed early response gene, *American Journal of Pathology*, **150**, 1985–1995

UEDA, K. and GANEM, D., 1996, Apoptosis is induced by N-myc expression in hepatocytes, a frequent event in hepadnavirus oncogenesis, and is blocked by insulin-like growth factor II, *Journal of Virology*, **70**, 1375–1383

WALLACH, D., 1997, Apoptosis. Placing death under control, *Nature*, **388**, 123–126

WEST, D.R., JAMES, N.H., COSULICH, S.C., HOLDEN, P.R., BRINDLE, R., ROLFE, M. and ROBERTS, R.A., 1999, Role for tumor necrosis factor $\alpha$ receptor 1 (TNFR1) and interleukin 1 receptor (IL1R) in the suppression of mouse hepatocyte apoptosis by the peroxisome proliferator, nafenopin, *Hepatology*, in press

WILLIAMS, G.P.C., 1995, *Mechanism-based risk assessment of peroxisome proliferating rodent hepatocarcinogens*, New York: The New York Academy of Sciences

WORNER, W.S.D., 1996, Influence of liver tumour promoters on apoptosis in rat hepatocytes induced by 2-AAF, ultraviolet light or transforming growth factor beta, *Cancer Research*, **56**, 1272–1278

YAMADA, Y., KIRILLOVA, I., PESCHON, J. and FAUSTO, N., 1997, Initiation of liver growth by tumor necrosis factor: deficient liver regeneration in mice lacking type I tumor necrosis factor receptor, *Proceedings of the National Academy of Sciences of the United States of America*, **94**, 1441–1446

CHAPTER 9

# Toxic Mechanisms Mediated by Gene Expression

**PETER R. HOLDEN**

AstraZeneca Central Toxicology Laboratory, Alderley Park, Macclesfield, SK10 4TJ, UK

## Contents

## 9.1 INTRODUCTION

The transcriptional activation of genes that regulate apoptosis plays a pivitol role in interpreting cellular damage by toxicants and determining the response of a cell or tissue (Figure 9.1). A toxicant can elicit changes in a cell directly by activating gene expression or may act indirectly to perturb pathways associated with cell survival.

Gene transcription is a complex process that requires DNA-dependent RNA polymerase II to bind at the transcription start site within specific promoter sequences. The promoter is located 'upstream' of the coding region of a gene and contains a number of short and specific DNA sequences. The efficiency of transcription is dictated both by the efficiency of polymerase binding and by the rate of progress along the gene to be transcribed. The former is affected both by the accessibility of the DNA and by a complex array of co-factors. Some of these are transcription factors that have their own specific binding sites within the promoter region. These may enhance or repress transcription by promoting or blocking, respectively, the binding to DNA of the key components of the transcription initiation complex. Transcription factors frequently require activation by a specific ligand. This chapter will consider how toxicants modulate apoptosis by altering gene expression and will cover both those toxicants with adverse effects and those claimed to have benefits.

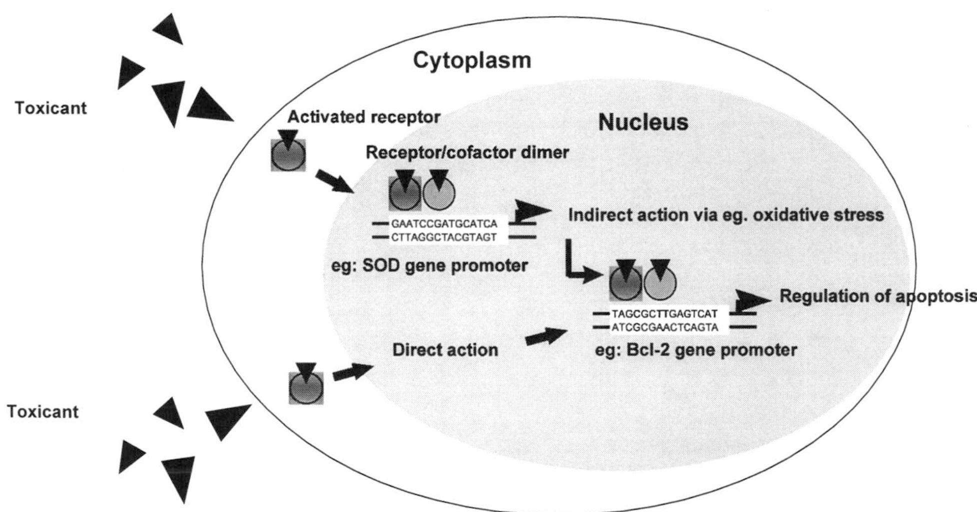


**Figure 9.1** Perturbation of apoptosis by toxicants: relevance of gene expression. A toxicant can activate specific receptors, triggering entry to the nucleus and binding to promoter regions of target genes. The promoter is located 'upstream' of the coding region of a gene and contains a number of short, specific DNA sequences that recognize and bind to gene regulatory proteins or transcription factors. Activated genes may be involved directly in the regulation of apoptosis such as the survival gene, Bcl-2, or indirectly involved via a controlling mechanism such as superoxide dismutase (SOD).

## 9.2 TOXIC MECHANISMS MEDIATED BY GENE EXPRESSION

### 9.2.1 Oxidative damage

Oxidative damage is known to be a key regulator in chemical-induced apoptosis. It also causes DNA damage which may lead to the development of a number of disease states including cancer, diabetes, atherosclerosis and Alzheimers disease (Breimer and Schellens, 1990). Although there is considerable evidence to implicate oxidative damage in the regulation of apoptosis, little is known of its mode of action. Cells and tissues are subjected to oxidative stress when the concentration of reactive oxygen species (ROS) exceeds the antioxidant capability of the cell (Sies, 1997). This occurs following either an increase in free radicals or peroxides such as $H_2O_2$, or a decrease in antioxidant levels in the cell (Trush and Kensler, 1991). ROS levels can be modulated by changes in endogenous cytochrome P450s or exogenously following the administration of drugs, hormones, xenobiotics and toxicants (Trush and Kensler, 1991). Oxidative damage can induce apoptosis following an increase in ROS (Buttke and Sandstrom, 1994, Suzuki *et al.*, 1997). Examples of this have been demonstrated in lung epithelial (Janssen *et al.*, 1997) and in human fibroblast cell lines (Gansauge *et al.*, 1997). In addition, antioxidants have been shown to inhibit dexamethasone- and etoposide-induced apoptosis in rat thymocytes (Wolfe *et al.*, 1994).

The molecular pathways that regulate apoptosis provide a possible target for oxidative damage, suggesting a role for transcriptional regulation of genes implicated in cell survival. One such group of genes are members of the family of mitogen activated protein kinases (MAPKs) which play a pivitol role in regulating changes in cell survival and proliferation. This family includes the extracellular signal-regulated protein kinases (ERKs) and the c-Jun N-terminal kinase/stress-activated protein kinase (JNK/SAPK) (Robinson and Cobb, 1997). Evidence shows that oxidative damage can activate and suppress the MAPKs since $H_2O_2$-induced apoptosis of HeLa cells is preceded by the inhibition of ERK and the activation of JNK/SAPK (Wang *et al.*, 1998). ERK activation also has an important role in protecting cardiac myocytes from apoptotic death following oxidative damage (Aikawa *et al.*, 1997). However, in cultured mouse keratinocytes, ERKs can be induced by treatment with $H_2O_2$ and butylated hydroxytoluene hydroperoxide indicating that cell type is an important factor in determining which signalling pathways are modulated (Guyton *et al.*, 1996a,b). It is thought that the dynamic balance between the ERKs and JNK/SAPK pathways is important in determining whether a cell survives or undergoes apoptosis (Xia *et al.*, 1995).

A number of other signalling proteins are involved in ROS-induced apoptosis. The anti-apoptotic protein Bcl-2 is normally down-regulated by oxidative damage and its activation can protect cells against apoptosis following treatment with $H_2O_2$ (Wang *et al.*, 1998). The pro-apoptotic genes p53 and p21WAF1/CIP1 accumulate during ROS-induced apoptosis in human fibroblast cells (Gansauge *et al.*, 1997). More recently the protein p85, a regulator of the signalling protein phosphatidyl-3-OH kinase (PI3kinase), has been shown to participate in p53-mediated apoptosis following oxidative damage (Yin *et al.*, 1998). The transcription factor NF-κB, which has been implicated in suppression of apoptosis and survival in some cell types, is also activated by ROS (Schreck *et al.*, 1991, 1992, reviewed in Sen and Packer, 1996). Cells overexpressing the $H_2O_2$-detoxifying enzyme catalase are unable to activate NF-κB whereas overexpression of Cu/Zn-dependent superoxide dismutase (SOD), which enhances the production of $H_2O_2$ from superoxide, increases NF-κB activation (Schmidt *et al.*, 1995). In summary, oxidative damage can elevate apoptosis, accompanied by changes in the expression or activation of pro- and anti-apoptotic genes. However, a clear role has been demonstrated with only a limited number of genes to date.

### 9.2.2 Detoxification by glutathione (GSH)

Glutathione (γ-glutamylcysteinylglycine, GSH) has multiple functions in detoxification and its depletion has been associated with an increased risk of chemical toxicity (Kearns and Hall, 1998). GSH can protect cells against toxic insults such as xenobiotics, radiation and infection and has been implicated in drug resistance. GSH also protects against oxidative stress by acting as an antioxidant and many reports suggest that GSH plays a role in apoptosis. For example, depletion of antioxidants such as GSH have been shown to induce apoptosis (Slater *et al.*, 1995, Burdon *et al.*, 1996) and elevation in the levels of antioxidants or overexpression of GSH-dependent enzymes can inhibit apoptosis (Hockenbery *et al.*, 1993, Iwata *et al.*, 1997). There is some debate as to whether apoptosis is triggered by a fall in GSH levels, or an increase in ROS, or both. A decrease in intracellular GSH concentration precedes the onset of apoptosis in murine thymocytes (Beaver and Waring, 1995) whereas GSH depletion is not sufficient to induce apoptosis in U937 human monocytic cells (Ghibelli *et al.*, 1995). In addition, Fas/APO-1 induced apoptosis in Jurkatt cells is associated with the removal of GSH from the cell, which suggests that in this case apoptosis results from a loss in GSH protection rather than an increase in ROS itself (Van den Dobbelsteen *et al.*, 1996). The search for a definitive role for GSH in apoptosis is gaining more momentum particularly with the suggestion that it might have a direct and fundamental role in the activation or inhibition of apoptosis regulatory proteins, such as Bcl-2 (Kearns and Hall, 1998).

### 9.2.3 DNA methylation

Changes in the methylation status of DNA have been linked with the regulation of apoptosis. The methylation of the C5 position of cytosine (5MC) in 5′-CpG-3′ dinucleotides of mammalian DNA is involved in tissue specific gene expression, cellular differentiation and the developmental expression of numerous genes (Cedar and Razin, 1990). The methylation status of a gene usually shows an inverse correlation with gene expression as hypermethylation usually results in gene inactivation (Baylin, 1997, Baylin *et al.*, 1998). In rat glioma C6.9 cells, 1,25-dihydroxyvitamin D3-induced apoptosis is suppressed by exposure to the DNA demethylating agents 5-azacytidine (5-AzaC) and 5-aza-2'-deoxycytidine (Canova *et al.*, 1998). The authors suggest that this change in DNA methylation could suppress apoptosis through the expression of previously hypermethylated genes with apoptotic suppressing activity. Likewise, in primary murine B cells, unmethylated CpG motifs protect CD40L-stimulated B cells from apoptosis mediated by a Fas-specific antibody (Wang *et al.*, 1997a). However other findings have suggested the opposite since rapid apoptosis is seen in human breast cancer cell lines following administration of 5-aza-2'-deoxycytidine (Ferguson *et al.*, 1997) and in PC12 cells following 5-Azacytidine (Hossain *et al.*, 1997). This would suggest that toxicant-induced changes in DNA methylation might alter the activity of apoptosis related genes.

## 9.3 PERTURBATION OF APOPTOSIS BY TOXICANTS

A number of toxicants can alter the rate of apoptosis (Table 9.1). For example, the carcinogenic transition metals, chromium, cadmium and nickel are capable of causing an increase in apoptosis along with DNA base modifications, strand breaks and rearrangements. The polycyclic aromatic hydrocarbons (PAHs) also cause apoptosis as well as impairments in carcinogenicity and development. In addition, nicotine and alcohol perturb apoptosis and are closely tied to the development of lung and liver cancers. Although apoptosis is modulated by these compounds, the molecular mechanisms that underpin the changes remain unknown.



### 9.3.1 Toxic metals

Evidence shows that the transition metals can cause changes in apoptosis. Chromium consistently induces apoptosis in many cell types including chinese hamster ovary (CHO) cells (Blankenship *et al.*, 1994, Shimada *et al.*, 1998) and cadmium can induce apoptosis in mouse liver cells (Habeebu *et al.*, 1998). However the effects of cadmium-induced changes in apoptosis are inconsistent among various culture systems (Ishido *et al.*, 1995) and has been recently shown to block chromium-induced apoptosis (Shimada *et al.*, 1998). The effects of cadmium or chromium on apoptosis-related gene expression are not well documented although cadmium has been shown to regulate expression of a number of cytokines and immediate-early genes (Beyersmann and Hechtenberg, 1997). Other toxic metals that perturb apoptosis include nickel (II) compounds that are associated with renal and lung cancers in humans and rodents (Ottolenghi *et al.*, 1975, Kasprzak *et al.*, 1994). Recent findings suggest that nickel can arrest growth and induce a dose-dependent increase in apoptosis in CHO cells without affecting the levels of p53 (Shiao *et al.*, 1998).

**TABLE 9.1**
Summary of changes in apoptosis, oxidative damage, glutathione (GSH), DNA methylation and gene expression/activation caused by toxic or protective compounds

| Compound | Apoptosis | Oxidative damage | GSH | DNA methylation | Gene expression/activation: Increase | Gene expression/activation: Decrease |
|---|---|---|---|---|---|---|
| **Toxic** | | | | | | |
| Chromium | ↑[1] | + | + | ?[2] | catalase GSH | |
| Cadmium | ↓↑ | + | + | ? | c-myc, *c-fos*, p53 | |
| Nickel | ↑ | + | + | + | | |
| BAP | ↓↑ | + | + | ? | caspase-3, Bcl-2, JNK | c-myc |
| TCDD | ↓↑ | + | + | + | | |
| DMBA | ↑ | + | + | ? | | |
| Nicotine | ↓ | + | + | ? | ERK-2, Bcl-2, PKC | |
| Alcohol | ↑ | ++ | + | + | TNF-α, IL-6, IL-1 | IGF, IGFR |
| **Preventative** | | | | | | |
| Butyrate | ↑ | + | + | ? | BAK, c-myc, p21, Bcl-2 | Bcl-2, Bcl-XL, JAK-1 |
| Selenium | ↑ | + | + | + | | GSH |
| GTPs | ↑ | + | + | ? | ERK, JNK, *c-jun*, *c-fos* | |
| Omega-3 PUFAs | ↑ | + | + | ? | | TNF-β, IL-2 |
| Omega-6 PUFAs | | + | + | ? | | |

1 Arrows represent an increase or decrease in apoptosis. + represent reported activation.
2 Question mark implies not reported. IGF: insulin-like growth factor; IGFR: insulin-like growth factor receptor; PKC: protein kinase C; IL-1: interleukin-1; IL-6: interleukin-6; JNK: *c-jun* N-terminal kinase; ERK: extracellular signal-regulated protein kinase; TNF-α: tumour necrosis factor alpha; JAK: Janus kinase.

Conversely, mutations in p53 have been found in human cells exposed to nickel (Fritsche *et al.*, 1993).

One mechanism by which these metals may cause toxicity is by oxidative damage following an increase in ROS. This follows the observation that the chemistry and type of mutation that results from metal-induced damage is similar to oxidative attack on DNA (Kasprzak *et al.*, 1995). Both nickel- (Robison *et al.*, 1982, Kasprzak *et al.*, 1990) and cadmium-induced oxidative damage (Snyder, 1988) have been demonstrated in cultured cells and experimental animals. More recently, using the frequency of DNA strand breaks as a marker, both nickel and cadmium were found to cause oxidative damage in human HeLa cells (Dally and Hartwig, 1997). In rat, nickel acetate causes oxidative damage in the kidney and liver although only renal tumours develop; this may be because the liver can support faster DNA repair compared with the kidney (Kasprzak *et al.*, 1992, Kasprzak *et al.*, 1997).

Other toxic metals such as lead and iron have been shown to possess carcinogenic activity and cause oxidative stress and damage. Iron is hepatocarcinogenic in rats and has been shown to modulate ROS production in hepatocytes (Abalea *et al.*, 1998). Lead causes generation of reactive oxygen species and functional impairment in rat sperm (Hsu *et al.*, 1997) and inhibits NO production *in vitro* by murine splenic macrophages (Tian and Lawrence, 1995, 1996). These authors have also shown that cadmium and mercury can inhibit NO production similar to lead but nickel and cobalt had a stimulatory effect (Tian and Lawrence, 1996). Copper, not iron, enhances the apoptosis-inducing activity of antioxidants in human promyelocytic leukemic cells (Satoh *et al.*, 1997). To conclude, an increase in oxidative damage (ROS) results from metal treatment in a number of human and rodent models. It follows that an increase in ROS could mediate the changes associated with the induction of apoptosis by these metals.

Metal toxicity has also been associated with GSH production. GSH modifies the toxicity of a number of metals (for review see Gochfeld, 1997) such as cadmium (Tang *et al.*, 1998), nickel (Lynn *et al.*, 1994) and chromium (Little *et al.*, 1996). It is possible that metals might suppress or remove the protective effect of GSH, causing cells to become more susceptible to apoptosis-inducing agents.

Like GSH, metallothionein (MT), a cysteine-rich metal binding protein could also be involved in co-ordinating the signals from metal toxicants to trigger apoptosis. Although the physiological function of MT is not entirely understood, induction of MT has been observed to be associated with protection from heavy metal toxicity and also cellular resistance to cytotoxic anti-cancer drugs (Klaassen and Liu, 1998). It is believed that it acts as a free radical scavenger and protects against oxidative damage. Indeed, MT-1 transgenic mice, which have a 10-fold higher concentration of MT in their liver than control mice, are resistant to cadmium-induced lethality and hepatotoxicity (Klaassen and Liu, 1998). Enhanced expression of MT in cells induces anti-apoptotic effects but a there is increased susceptibility to apoptotic cell death in MT null cells (Adbel-Mageed and Agrawal, 1997, Kondo *et al.*, 1997). Subsequent work has indicated that the activation of NF-κB may play a role in mediating the anti-apoptotic effects of MT (Adbel-Mageed and Agrawal, 1998). In addition, MT may be linked with the suppression of apoptosis via its antioxidant properties or via a more direct route by interaction with Bcl-2.

Finally, there is increasing evidence for alterations in DNA methylation in metal toxicity, particularly with nickel. Nickel-induced DNA condensation and hypermethylation lead to epigenetic silencing of a transgene in G12 cells (Lee *et al.*, 1995a) and several other gene sequences, including those for thrombospondin and a heme-containing peroxidase, are silenced in nickel-transformed cells (Salnikow *et al.*, 1994). More recently, nickel has been shown to inhibit cytosine 5-methyltransferase activity *in vivo* and *in vitro* as well as elevate total genomic DNA methylation levels (Lee *et al.*, 1998). Therefore it is possible that nickel, and other toxic metals, might perturb apoptosis by direct modulation of genes that regulate cell survival. Although the effects of metals on the methylation status of apoptotic genes has to be determined, it would present informative data on the mode of action with regard to apoptosis.

### 9.3.2 Polycyclic aromatic hydrocarbons (PAHs)

Benzo(a)pyrene (BaP) induces apoptosis and up-regulates the cysteine protease caspase-3 (ICE/Ced-3) in mouse Hepa cells (Lei *et al.*, 1998). BaP also caused a time and dose-dependent activation in JNK1 activity indicating that stress-activated kinase activation is involved in BaP-induced apoptosis (Lei *et al.*, 1998). BaP also induces cell death in mouse epidermis (Miller *et al.*, 1996) but, in contrast, can inhibit apoptosis in Syrian hamster embryo cells (Dhalluin *et al.*, 1997). In this latter experiment, Bcl-2 was up-regulated and levels of the Bcl-XL protein did not change, suggesting that inhibition of apoptosis via Bcl-2 might contribute to the early stages of the carcinogenic process.

Dioxin (2,3,7,8-tetrachlorodibenzo[p]dioxin; TCDD) can cause changes in the rate of apoptosis although the nature of these changes depends much on the experimental system. For example, TCDD can induce apoptosis in mouse thymocytes *in vivo* but not *in vitro* (Kamath *et al.*, 1997). Other reports in immature rat and mouse thymocytes (McConkey *et al.*, 1988, Rhile *et al.*, 1996) show that TCDD can induce apoptosis *in vitro*. Another PAH, dimethylbenz[a]anthracene (DMBA) has also been shown to induce apoptosis in an A20.1 murine B-cell line by a mechanism that is dependent on cellular $Ca^{2+}$ (Burchiel *et al.*, 1993).

The underlying mechanisms of action of PAHs on apoptosis are not clear. It was originally believed that all biological effects of TCDD were mediated through the aryl hydrocarbon (Ah) receptor (DeVito and Birnbaum, 1995), a ligand-activated member of the bHLH-PAS family of transcription factors (Burbach *et al.*, 1992). However, it has been shown recently that TCDD-induced apoptosis in a human leukemic lymphoblastic T cell line was not dependent on the Ah receptor (Hossain *et al.*, 1998). In this case, apoptosis was inhibited by caspase and tyrosine kinase inhibitors and was linked to a rapid increase in JNK and a decrease in Bcl-2 protein levels. The ability of single doses of TCDD to induce oxidative stress in hepatic and some extrahepatic tissues of animals is also well documented (Stohs *et al.*, 1990, Alsharif *et al.*, 1994). TCDD induces oxidative stress in mice as measured by production of superoxide anion, an effect that is controlled in part by the Ah receptor complex (Alsharif *et al.*, 1994). Recent data suggests that BaP also increases oxidative stress in human epidermoid carcinoma cells (Liu *et al.*, 1998). The effects of TCDD on glutathione synthesis has also been documented (reviewed in Schrenk, 1998).

### 9.3.3 Nicotine and alcohol

Nicotine blocks apoptosis induced by a variety of stimuli such as TNF-$\alpha$, UV light and a number of chemotherapeutic drugs (Wright *et al.*, 1993). The mechanisms involved in nicotine-induced suppression of apoptosis remain unclear. Nicotine has no effect on JNK and p38 MAPK activity but can activate ERK-2 which results in an increase in Bcl-2 and a suppression of apoptosis in human lung cancer cells (Heusch and Maneckjee, 1998). The same workers have shown that nicotine can stimulate protein kinase C (PKC) activity in lung cancer cell lines (Maneckjee and Minna, 1994) and can block the inhibition of PKC and ERK-2 activity by anti-cancer agents, such as therapeutic opioid drugs (Heusch and Maneckjee, 1998). It has been suggested that the activation of the nicotine receptors C6 or C10 can lead to a suppression of apoptosis which, in some cases, is mediated via the PKC pathway. The effect of nicotine on oxidative damage and DNA methylation have also been investigated although reports are few. The effects of snuff and nicotine on DNA methylation by 4-(methylnitrosamino)-1-(3-pyridyl)-1-butanone (NNK), a powerful tobacco-specific N-nitrosamine, have been investigated. NNK induced methylation in rat liver, nasal mucosae and oral cavity, snuff suppressed methylation whereas nicotine had no effect, indicating that nicotine has no effect on the genotoxicity of NKK (Prokopczyk *et al.*, 1987).

The mechanism by which chronic alcohol abuse induces widespread cell and tissue damage is unknown. Some reports suggest that alcohol toxicity is associated with increased levels of endotoxin in plasma (Su *et al.*, 1998). Others report that alcoholic liver injury results from a decrease in oxygen supply which produces hypoxia (Yuki and Thurman, 1980, Nanji *et al.*, 1995). There are also many examples of ethanol-induced apoptosis in hepatocytes and other cell types (Benedetti *et al.*, 1988, Aroor and Baker, 1997, Kurose *et al.*, 1997). In hepatocytes, ethanol administration induces an increase in lipid peroxidation either by enhancing the production of ROS and/or by decreasing the level of endogenous antioxidants (Sergent *et al.*, 1995, Higuchi *et al.*, 1996). In addition, agents that inhibit ethanol-induced oxidative stress effectively attenuate hepatocyte apoptosis and/or necrosis. Therefore, it has been suggested that ethanol-induced hepatocyte apoptosis is driven by an oxidant-dependent mechanism since oxidative stress is recognized to be a key step in the pathogenesis of ethanol-associated liver injury (Higuchi *et al.*, 1996). In support of this hypothesis other findings have shown that ethanol-induced ROS in hepatocytes mediates the DNA fragmentation step of apoptosis and that intracellular antioxidants such as glutathione protect hepatocytes against ethanol-induced apoptosis (Kurose *et al.*, 1997).

In addition to oxidative damage, alcohol causes changes in the expression of apoptosis-related genes. The effects of alcohol on insulin-like growth factor (IGF) and the IGF-1 receptor has been well documented (Resnicoff *et al.*, 1993, 1996). It has been shown that IGF-1 receptor controls the extent of cell proliferation in a variety of cell types and protects cells from apoptosis. Ethanol at low concentrations markedly inhibits IGF-1 receptor autophosphorylation and IGF-1-mediated cell growth. This receptor is likely to play an important role in ethanol-induced apoptosis as it can also act as a strong inhibitor of etoposide-induced apoptosis in BALB/c 3T3 cells (Sell *et al.*, 1995). Up-regulation in cytokine expression has also been implicated in ethanol induced apoptosis. In sublingual salivary gland cells chronic alcohol ingestion causes enhancement of TNF-α expression and apoptosis (Slomiany *et al.*, 1997). In HepG2 cells ethanol-induced cytotoxicity is accompanied by an increase in TNF-α, interleukin-1α (IL-1α) and interleukin-6 (IL-6) and a reduction in glutathione reductase levels (Neuman *et al.*, 1998).

In summary, both nicotine and alcohol are important toxic regulators of apoptosis. Nicotine suppression of apoptosis appears to involve Bcl-2 and the activation of the PKC pathway whereas alcohol induced apoptosis is regulated primarily through oxidative damage. However the observation that alcohol can inhibit IGF signalling suggests that apoptosis-inducing toxicants may operate by inhibiting survival pathways.

## 9.4 PERTURBATION OF APOPTOSIS BY PROTECTIVE COMPOUNDS

Compounds with beneficial health effects may act by regulating apoptosis. Not surprisingly, many of these are known dietary supplements. This includes butyric acid or butyrate, selenium, green tea polyphenols (GTPs) and the polyunsaturated fatty acids (PUFAs). Most of these compounds are described as having a health protective role although they can both induce or suppress apoptosis.

### 9.4.1 Butyrate

Butyrate, in the form of dietary fibre supplementation, leads to reduced colonic cell proliferation and a lower tumour mass (Boffa *et al.*, 1992, McIntyre *et al.*, 1993). It inhibits growth in human hepatocellular carcinoma (HCC) cells (Yamamoto *et al.*, 1998) and induces apoptosis in human colorectal carcinoma cells (Hague *et al.*, 1993, Mandal and Kumar, 1996). In human colonic adenoma cell lines butyrate-induced apoptosis can be associated with either an increase in the levels of the pro-apoptotic gene, Bak, or a reduction of the anti-apoptotic gene Bcl-2 (Hague *et al.*, 1997). Likewise in colorectal carcinoma cells an inverse relationship was found between the levels of Bcl-2 and sensitivity of the cells to undergo apoptosis in response to butyrate (Mandal *et al.*, 1997). Over expression of Bcl-2 results in the suppression of butyrate-induced apoptosis and enhanced cell survival (Mandal *et al.*, 1997). Butyrate has also been shown to down-regulate the levels of Bcl-2 in other cell types including human breast MCF-7 cells (Mandal and Kumar, 1996) and in myeloid leukemic cells (Calabresse *et al.*, 1994). However, more recent evidence has shown that sodium butyrate has no affect on apoptosis in HCC cells where it up-regulates Bcl-2; thus, the mechanism of butyrate-induced apoptosis is not fully understood (Saito *et al.*, 1998).

Butyrate targets a number of other apoptotic genes. In addition to Bcl-2 and Bak, butyrate significantly lowers the levels of the apoptosis antagonist Bcl-XL in human fibroblasts (Chung *et al.*, 1998). In a chronic leukemia cell line the butyrate analogue, arginine butyrate, inhibits the expression of the anti-apoptotic factor p210 Bcr-abl and other phosphoproteins including JAK-1 kinase (Urbano *et al.*, 1998). Furthermore, both c-myc and p21 WAF/Cip1 are induced by butyrate, independently of p53 (Hague *et al.*, 1993, Janson *et al.*, 1997, Heerdt *et al.*, 1998).

Butyrate has been shown to be a potential inhibitor of apoptosis in other tissues since it markedly suppresses c-myc-mediated apoptosis in murine plasmacytomas and human Burkitt lymphomas (Alexandrov *et al.*, 1998). Also, butyrate can induce the synthesis of MT in ROS 17/2.8 cells thus protecting against the possible pro-apoptotic effects of oxidative damage (Thomas *et al.*, 1991). In summary, the protective effects of butyrate may be based upon an induction of apoptosis, possibly due to the down-regulation of Bcl-2. However consideration should be given to conflicting reports that show butyrate-induced suppression of apoptosis.

### 9.4.2 Selenium

Selenium markedly reduces the incidence of lung, colon and prostate cancers (Clark *et al.*, 1996, Harrison *et al.*, 1997). Sodium selenite is one of the most effective selenium compounds and has been shown, like butyrate, to induce apoptosis in human colonic carcinoma cells (Stewart *et al.*, 1997). Sodium selenite is metabolized to hydrogen selenide through its reduction to glutathione and this is thought to be an important step in exerting its anti-proliferative and apoptotic effect. Selenodiglutathione (SDG), the primary metabolite of selenite reduction to glutathione, also has potent anti-carcinogenic and cytotoxic properties (Lanfear *et al.*, 1994). Like selenite, SDG has been shown to induce apoptosis in a number of mammary carcinoma cell lines (Lanfear *et al.*, 1994, Thompson *et al.*, 1994) although the biological mechanisms underlying these effects are unknown. Apoptosis may result from $H_2O_2$-induced oxidative stress since $H_2O_2$ is a by-product of selenite reduction to SDG and the role of oxidative damage as a pro-apoptotic factor has been well documented. Selenomethione, another selenium compound and primary constituent of dietary selenium supplement, has anti-carcinogenic properties that have been demonstrated in both lung and colon tumour cell lines (Redman *et al.*, 1997). Selenomethione induced apoptosis and decreased the polyamine content in these cancer cell lines suggesting that the anti-carcinogenic effects of selenium supplementation might be due to depletion of polyamine levels. Polyamines, such as spermine and spermidine, are required for normal cell proliferation and development but it appears that imbalances in the metabolism of polyamines may constitute a signal for apoptosis.

### 9.4.3 Tea polyphenols

Green tea polyphenols (GTPs), major constituents of green tea, are potent chemopreventative agents in a number of animal experimental models of cancer. Feeding of GTP has been shown to suppress skin tumourigenesis induced by both aromatic hydrocarbons and UV light and inhibit tobacco induced lung tumourigenesis in mice (Wang *et al.*, 1991, Katiyar *et al.*, 1993a, b, Shi *et al.*, 1994).

Around 10% of the dry weight of green tea is made up of catechins, of which (-)-epigallocatechin-3-gallate (EGCG) is thought to be an active anti-cancer component (Balentine, 1992). However, during the manufacture of black tea the catechins are oxidized and polymerized to form the black tea polyphenols, theaflavins and thearubigins (Balentine, 1992). These compounds have antioxidative activities, due to their radical scavenging and metal chelating functions, and are also thought to have tumour inhibitory activities (Serafini *et al.*, 1996).

Several lines of evidence suggest that the induction of apoptosis may explain the growth inhibitory properties of tea extracts. An induction in apoptosis and concomitant fall in DNA synthesis has been observed in human stomach cancer KATO III cells following exposure to green tea catechin extract and EGCG (Hibasami *et al.*, 1998). Likewise green tea has been shown to cause apoptosis and cell cycle arrest in a number of other human carcinoma cell lines (Ahmad *et al.*, 1997, Zhao *et al.*, 1997, Katdare *et al.*, 1998). Also exposure of a lung tumour cell line H661 to either EGCG or theaflavins led to an increase in apoptosis as well as a growth inhibition (Yang *et al.*, 1998). These workers have also shown that tea polyphenols induce production of $H_2O_2$ which may mediate this apoptosis.

Other evidence suggests that the protective effects of GTP are due to the inhibition of enzymes, such as cytochrome P450, involved in the bioactivation of carcinogens (Bu-Abbas *et al.*, 1994a,b). Reports indicate that GTP causes increases in the activities of the glutathione S-transferases (Khan *et al.*, 1992, Lee *et al.*, 1995b). Tea and tea polyphenol compounds have been shown to inhibit the activities of a number of other important regulatory enzymes including protein kinase C, ornithine decarboxylase and DNA topoisomerase II (Austin *et al.*, 1992, Jankun *et al.*, 1997). Further studies have shown that the mitogen-activated protein kinases ERK-2 and JNK-1 are strongly activated by GTPs as are the immediate-early genes c-*jun* and c-*fos* (Yu *et al.*, 1997). These workers have suggested that the stimulation of MAPKs by GTP may provide the early signalling events that lead to the eventual activation of Phase II detoxifying enzymes. However despite an increasing list of GTP targets, a mode of action is still to be elucidated.

### 9.4.4 Dietary fats

Epidemiological and experimental studies indicate a strong relationship between dietary fats such as polyunsaturated fatty acids (PUFAs) and cancer. Dietary fat can be both beneficial or detrimental to animals. For example, corn oil which contains high levels of omega-6 PUFAs, such as linoleic acid, enhances chemically-induced colon tumourigenesis in rodents, whereas fish oil, which is rich in omega-3 PUFAs such as eicosapentaenoic acid

(EPA) and docosahexaenoic acid (DHA) reduces colon carcinogenesis (Torosian *et al.*, 1995).

In rat, the protective effects of fish oil against experimentally induced colon tumorigenesis results from an increase in apoptosis and differentiation, rather than a decrease in proliferation (Chang *et al.*, 1998). In addition, the the anti-tumour effect of DHA corresponds with its induction of apoptosis (Calviello *et al.*, 1998). In contrast the anti-tumoural effects of EPA appears related mainly to inhibition of cell proliferation (Calviello *et al.*, 1998). Other Omega-3 PUFAs inhibited growth by inducing apoptosis in three pancreatic cancer cell lines and in a HL-60 leukaemic cell line (Hawkins *et al.*, 1998). In this latter study apoptosis was preceded by lipid peroxidation indicating that apoptosis might be caused by an oxidative mechanism.

Conversely, the omega-6 PUFAs, such as linoleic acid, enhance tumour growth not only by promoting cell proliferation but also by suppressing apoptosis (Tang *et al.*, 1997). Although NO production is inhibited by the anti-tumourigenic omega-3 PUFAs, probably by suppression of NO synthetase, no inhibition is observed with omega-6 PUFAs such as linoleic and oleic acid or the saturated fatty acid, stearic acid (Ohata *et al.*, 1997).

To conclude, these protective compounds share the ability of detrimental toxicants to induce apoptosis. However, the apoptosis induced by the protective compounds may remove unwanted or deleterious cells from the target tissue. Evidence for changes in apoptosis related gene expression is inconsistent, although in many cases, down-regulation of *bcl-2* precedes induction of apoptosis. Again, oxidative damage, GSH and MT are also potential mediators of changes in apoptosis caused by these compounds.



## 9.5 SUMMARY

Toxicants may have adverse or beneficial effects on health in humans and in animal models. Some toxicants can suppress apoptosis (see Chapter 8) although examples of an induction of apoptosis are more numerous. There are a variety of mechanisms implicated in the regulation of apoptosis by toxicants, including changes in the expression or activation of specific genes that regulate cell survival and apoptosis (Table 9.1). However, increases in ROS (oxidative damage) and a decrease in GST production are frequently observed, suggesting that the effects on gene expression may be indirect. Toxicant-induced changes in the methylation status of apoptosis-related genes may present some new leads. In addition, alcohol toxicity suggests that toxicants may cause the suppression of survival pathways, mediated by IGFs. The study of toxicant-induced gene expression should provide a mechanistic explanation for the regulation of apoptosis by toxic and protective compounds.

## ACKNOWLEDGEMENTS

The author would like to thank Susan Hasmall for critical appraisal of this manuscript. The editor would like to thank Zana Dye for invaluable editorial assistance.

LIVERPOOL
JOHN MOORES UNIVERSITY
AVRIL ROBARTS LRC
TITHEBARN STREET
LIVERPOOL L2 2ER
TEL. 0151 231 4022

## REFERENCES

ABALEA, V., CILLARD, J., DUBOS, M.P., ANGER, J.P., CILLARD, P. and MOREL, I., 1998, Iron-induced oxidative DNA damage and its repair in primary rat hepatocyte culture, *Carcinogenesis*, **19**, 1053–1059

ABDEL-MAGEED, A. and AGRAWAL, K.C., 1997, Antisense down-regulation of metallothionein induces growth arrest and apoptosis in human breast carcinoma cells, *Cancer Gene Therapy*, **4**, 199–207

ABDEL-MAGEED, A.B. and AGRAWAL, K.C., 1998, Activation of nuclear factor kappaB: potential role in metallothionein-mediated mitogenic response, *Cancer Research*, **58**, 2335–2338

AHMAD, N., FEYESM, D.K., NIEMINEN, A.L., AGARWAL, R. and MUKHTAR, H., 1997, Green tea constituent epigallocatechin-3-gallate and induction of apoptosis and cell cycle arrest in human carcinoma cells, *Journal of the National Cancer Institute*, **89**, 1881–1886

AIKAWA, R., KOMURO, I., YAMAZAKI, T., ZOU, Y., KUDOH, S., TANAKA, M., *et al.*, 1997, Oxidative stress activates extracellular signal-regulated kinases through Src and Ras in cultured cardiac myocytes of neonatal rats, *Journal of Clinical Investigation*, **100**, 1813–1821

ALEXANDROV, I., ROMANOVA, L., MUSHINSKI, F. and NORDAN, R., 1998, Sodium butyrate suppresses apoptosis in human Burkitt lymphomas and murine plasmacytomas bearing c-myc translocations, *FEBS Letters*, **434**, 209–214

ALSHARIF, N.Z., LAWSON, T. and STOHS, S.J., 1994, Oxidative stress induced by 2,3,7,8-tetrachlorodibenzo-p-dioxin is mediated by the aryl hydrocarbon (Ah) receptor complex, *Toxicology*, **92**, 39–51

AROOR, A.R. and BAKER, R.C., 1997, Ethanol-induced apoptosis in human HL-60 cells, *Life Sciences*, **61**, 2345–2350

AUSTIN, C.A., PATEL, S., ONO, K., NAKANE, H. and FISHER, L.M., 1992, Site-specific DNA cleavage by mammalian DNA topoisomerase II induced by novel flavone and catechin derivatives, *Biochemistry Journal*, **282**, 883–889

BALENTINE, D.A., 1992, Manufacturing and chemistry of tea, in Ho, C.-T., Huang, M.-T. and Lee, C.Y. (eds), *Phenolic compounds in food and their effects on health* (1st edn). Washington DC: American Chemical Society, pp. 102–117

BAYLIN, S.B., 1997, Tying it all together: epigenetics, genetics, cell cycle, and cancer, *Science*, **277**, 1948–1949

BAYLIN, S.B., HERMAN, J.G., GRAFF, J.R., VERTINO, P.M. and ISSA, J.P., 1998, Alterations in DNA methylation: a fundamental aspect of neoplasia, *Advances in Cancer Research*, **72**, 141–196

BEAVER, J.P. and WARING, P., 1995, A decrease in intracellular glutathione concentration precedes the onset of apoptosis in murine thymocytes, *European Journal of Cell Biology*, **68**, 47–54

BENEDETTI, A., BRUNELLI, E., RISICATO, R., CILLUFFO, T., JEZEQUEL, A.M. and ORLANDI, F., 1988, Subcellular changes and apoptosis induced by ethanol in rat liver, *Journal of Hepatology*, **6**, 137–143

BEYERSMANN, D. and HECHTENBERG, S., 1997, Cadmium, gene regulation, and cellular signalling in mammalian cells, *Toxicology and Applied Pharmacology*, **144**, 247–261

BLANKENSHIP, L.J., MANNING, F.C., ORENSTEIN, J.M. and PATIERNO, S.R., 1994, Apoptosis is the mode of cell death caused by carcinogenic chromium, *Toxicology and Applied Pharmacology*, **126**, 75–83

BOFFA, L.C., LUPTON, J.R., MARIANI, M.R., CEPPI, M., NEWMARK, H.L., SCALMATI, A. and LIPKIN, M., 1992, Modulation of colonic epithelial cell proliferation, histone acetylation, and luminal short chain fatty acids by variation of dietary fiber (wheat bran) in rats, *Cancer Research*, **52**, 5906–5912

BREIMER, D.D. and SCHCLLENS, J.H.M., 1990, A cocktail strategy to assess *in vivo* oxidative drug metabolism in humans, *Trends in Pharmacological Sciences*, **11**, 223–225

BU-ABBAS, A., IOANNIDES, C. and WALKER, R., 1994a, Evaluation of the antimutagenic potential of anthracene: *in vitro* and *ex vivo* studies, *Mutation Research*, **309**, 101–107

BU-ABBAS, A., CLIFFORD, M.N., WALKER, R. and IOANNIDES, C., 1994b, Selective induction of rat hepatic CYP1 and CYP4 proteins and of peroxisomal proliferation by green tea, *Carcinogenesis*, **15**, 2575–2579

BURBACH, K.M., POLAND, A. and BRADFIELD, C.A., 1992, Cloning of the Ah-receptor cDNA reveals a distinctive ligand-activated transcription factor, *Proceedings of the National Academy of Sciences of the United States of America*, **89**, 8185–8189

BURCHIEL, S.W., DAVIS, D.A., RAY, S.D. and BARTON, S.L., 1993, DMBA induces programmed cell death (apoptosis) in the A20.1 murine B cell lymphoma, *Fundamental and Applied Toxicology*, **21**, 120–124

BURDON, R.H., GILL, V. and ALLIANGANA, D., 1996, Hydrogen peroxide in relation to proliferation and apoptosis in BHK-21 hamster fibroblasts, *Free Radical Research*, **24**, 81–93

BUTTKE, T.M. and SANDSTROM, P.A., 1994, Oxidative stress as a mediator of apoptosis, *Immunology Today*, **15**, 7–10

CALABRESSE, C., VENTURINI, L., RONCO, G., VILLA, P., DEGOS, L., BELPOMME, D. and CHOMIENNE, C., 1994, Selective induction of apoptosis in myeloid leukemic cell lines by monoacetone glucose-3 butyrate, *Biochemical Biophysical Research Communications*, **201**, 266–283

CALVIELLO, G., PALOZZA, P., PICCIONI, E., MAGGIANO, N., FRATTUCCI, A., FRANCESCHELLI, P. and BARTOLI, G.M., 1998, Dietary supplementation with eicosapentaenoic and docosahexaenoic acid inhibits growth of Morris hepatocarcinoma 3924A in rats: effects on proliferation and apoptosis, *International Journal of Cancer*, **75**, 699–705

CANOVA, C., CHEVALIER, G., REMY, S., BRACHET, P. and WION, D., 1998, Epigenetic control of programmed cell death: inhibition by 5-azacytidine of 1,25-dihydroxyvitamin D3-induced programmed cell death in C6.9 glioma cells, *Mechanisms of Ageing and Development*, **101**, 153–166

CEDAR, H. and RAZIN, A., 1990, DNA methylation and development, *Biochimica Biophysia Acta*, **1049**, 1–8

CHANG, W.L., CHAPKIN, R.S. and LUPTON, J.R., 1998, Fish oil blocks azoxymethane-induced rat colon tumorigenesis by increasing cell differentiation and apoptosis rather than decreasing cell proliferation, *Journal of Nutrition*, **128**, 491–497

CHUNG, D.H., ZHANG, F., CHEN, F., MCLAUGHLIN, W.P. and LJUNGMAN, M., 1998, Butyrate attenuates BCLX(L) expression in human fibroblasts and acts in synergy with ionizing radiation to induce apoptosis, *Radiation Research*, **149**, 187–194

CLARK, L.C., COMBS JR, G.F., TURNBULL, B.W., SLATE, E.H., CHALKER, D.K., CHOW, J., *et al.*, 1996, Effects of selenium supplementation for cancer prevention in patients with carcinoma of the skin. A randomized controlled trial. Nutritional Prevention of Cancer Study Group, *Journal of the American Medical Association*, **276**, 1957–1960

DALLY, H. and HARTWIG, A., 1997, Induction and repair inhibition of oxidative DNA damage by nickel(II) and cadmium(II) in mammalian cells, *Carcinogenesis*, **18**, 1021–1026

DEVITO, M.J. and BIRNBAUM, L.S., 1995, Dioxins: model chemicals for assessing receptor-mediated toxicity, *Toxicology*, **102**, 115–123

DHALLUIN, S., GATE, L., VASSEUR, P., TAPIERO, H. and NGUYEN-BA, G., 1997, Dysregulation of ornithine decarboxylase activity, apoptosis and Bcl-2 oncoprotein in Syrian hamster embryo cells stage-exposed to di(2-ethylhexyl)phthalate and tetradecanoylphorbol acetate, *Carcinogenesis*, **18**, 2217–2223

FERGUSON, A.T., VERTINO, P.M., SPITZNER, J.R., BAYLIN, S.B., MULLER, M.T. and DAVIDSON, N.E., 1997, Role of estrogen receptor gene demethylation and DNA methyltransferase. DNA adduct formation in 5-aza-2'deoxycytidine-induced cytotoxicity in human breast cancer cells, *Journal of Biological Chemistry*, **19**, 32260–32267

FRITSCHE, M., HAESSLER, C. and BRANDNER, G., 1993, Induction of nuclear accumulation of the tumor-suppressor protein p53 by DNA-damaging agents, *Oncogene*, **8**, 307–318

GANSAUGE, S., GANSAUGE, F., GAUSE, H., POCH, B., SCHOENBERG, M.H. and BEGER, H.G., 1997, The induction of apoptosis in proliferating human fibroblasts by oxygen radicals is associated with a p53- and p21WAF1CIP1 induction, *FEBS Letters*, **404**, 6–10

GHIBELLI, L., COPPOLA, S., ROTILIO, G., LAFAVIA, E., MARESCA, V. and CIRIOLO, M.R., 1995, Non-oxidative loss of glutathione in apoptosis via GSH extrusion, *Biochemical and Biophysical Research Communications*, **216**, 313–320

GOCHFELD, M., 1997, Factors influencing susceptibility to metals, *Environmental Health Perspectives*, **105**, 817–822

GUYTON, K.Z., GOROSPE, M., KENSLER, T.W. and HOLBROOK, N.J., 1996b, Mitogen-activated protein kinase (MAPK) activation by butylated hydroxytoluene hydroperoxide: implications for cellular survival and tumor promotion, *Cancer Research*, **56**, 3480–3485

GUYTON, K.Z., LIU, Y., GOROSPE, M., XU, Q. and HOLBROOK, N.J., 1996a, Activation of mitogen-activated protein kinase by H2O2. Role in cell survival following oxidant injury, *Journal of Biological Chemistry*, **271**, 4138–4142

HABEEBU, S.S., LIU, J. and KLAASSEN, C.D., 1998, Cadmium-induced apoptosis in mouse liver, *Toxicology and Applied Pharmacology*, **149**, 203–209

HAGUE, A., DIAZ, G.D., HICKS, D.J., KRAJEWSKI, S., REED, J.C. and PARASKEVA, C., 1997, Bcl-2 and bak may play a pivotal role in sodium butyrate-induced apoptosis in colonic epithelial cells; however overexpression of bcl-2 does not protect against bak-mediated apoptosis, *International Journal of Cancer*, **72**, 898–905

HAGUE, A., MANNING, A.M., HANLON, K.A., HUSCHTSCHA, L.I., HART, D. and PARASKEVA, C., 1993, Sodium butyrate induces apoptosis in human colonic tumour cell lines in a p53-independent pathway: implications for the possible role of dietary fibre in the prevention of large-bowel cancer, *International Journal of Cancer*, **55**, 498–500

HARRISON, P.R., LANFEAR, J., WU, L., FLEMING, J., MCGARRY, L. and BLOWER, L., 1997, Chemopreventive and growth inhibitory effects of selenium, *Biomedical and Environmental Sciences*, **10**, 235–245

HAWKINS, R.A., SANGSTER, K. and ARENDS, M.J., 1998, Apoptotic death of pancreatic cancer cells induced by polyunsaturated fatty acids varies with double bond number and involves an oxidative mechanism, *Journal of Pathology*, **185**, 61–70

HEERDT, B.G., HOUSTON, M.A., ANTHONY, G.M. and AUGENLICHT, L.H., 1998, Mitochondrial membrane potential (delta psi(mt)) in the coordination of p53-independent proliferation and apoptosis pathways in human colonic carcinoma cells, *Cancer Research*, **58**, 2869–2875

HEUSCH, W.L. and MANECKJEE, R., 1998, Signalling pathways involved in nicotine regulation of apoptosis of human lung cancer cells, *Carcinogenesis*, **19**, 551–556

HIBASAMI, H., KOMIYA, T., ACHIWA, Y., OHNISHI, K., KOJIMA, T., NAKANISHI, K., *et al.*, 1998, Induction of apoptosis in human stomach cancer cells by green tea catechins, *Oncology Reports*, **5**, 527–529

HIGUCHI, H., KUROSE, I., KATO, S., MIURA, S. and ISHII, H., 1996, Ethanol-induced apoptosis and oxidative stress in hepatocytes, *Alcoholism-Clinical and Experimental Research*, **20** (Suppl. 9), A340–A346

HOCKENBERY, D.M., OLTVAI, Z.N., YIN, X.M., MILLIMAN, C.L. and KORSMEYER, S.J., 1993, Bcl-2 functions in an antioxidant pathway to prevent apoptosis, *Cell*, **75**, 241–251

HOSSAIN, A., TSUCHIYA, S., MINEGISHI, M., OSADA, M., IKAWA, S., TEZUKA, F.A., *et al.*, 1998, The Ah receptor is not involved in 2,3,7,8-tetrachlorodibenzo-p-dioxin-mediated apoptosis in human leukemic T cell lines, *Journal of Biological Chemistry*, **273**, 19853–19858

HOSSAIN, M.M., TAKASHIMA, A., NAKAYAMA, H. and DOI, K., 1997, 5-Azacytidine induces toxicity in PC12 cells by apoptosis, *Experimental and Toxicological Pathology*, **49**, 201–206

HSU, P.C., LIU, M.Y., HSU, C.C., CHEN, L.Y. and LEON GUO, Y., 1997, Lead exposure causes generation of reactive oxygen species and functional impairment in rat sperm, *Toxicology*, **122**, 133–143

ISHIDO, M., HOMMA, S.T., LEUNG, P.S. and TOHYAMA, C., 1995, Cadmium-induced DNA fragmentation is inhibitable by zinc in porcine kidney LLC-PK1 cells, *Life Sciences*, **56**, 351–356

IWATA, S., HORI, T., SATO, N., HIROTA, K., SASADA, T., MITSUI, A., *et al.*, 1997, Adult T cell leukemia (ATL)-derived factor/human thioredoxin prevents apoptosis of lymphoid cells induced by L-cystine and glutathione depletion: possible involvement of thiol-mediated redox regulation in apoptosis caused by pro-oxidant state, *Journal of Immunology*, **158**, 3108–3117

JANKUN, J., SELMAN, S.H., SWIERCZ, R. and SKRZYPCZAK-JANKUN, E., 1997, Why drinking green tea could prevent cancer, *Nature*, **387**, 561 only

JANSON, W., BRANDNER, G. and SIEGEL, J., 1997, Butyrate modulates DNA-damage-induced p53 response by induction of p53-independent differentiation and apoptosis, *Oncogene*, **15**, 1395–1406

JANSSEN, Y.M., MATALON, S. and MOSSMAN, B.T., 1997, Differential induction of c-fos, c-jun, and apoptosis in lung epithelial cells exposed to ROS or RNS, *American Journal of Physiology*, **273**, 789–796

KAMATH, A.B., NAGARKATTI, P.S. and NAGARKATTI, M., 1997, Characterization of phenotypic alterations induced by 2,3,7,8-tetrachlorodibenzo-p-dioxin on thymocytes *in vivo* and its effect on apoptosis, *Toxicology and Applied Pharmacology*, **150**, 117–124

KASPRZAK, K.S., 1995, Possible role of oxidative damage in metal-induced carcinogenesis, *Cancer Investigations*, **13**, 411–430

KASPRZAK, K.S., DIWAN, B.A. and RICE, J.M., 1994, Iron accelerates while magnesium inhibits nickel-induced carcinogenesis in the rat kidney, *Toxicology*, **90**, 129–140

KASPRZAK, K.S., DIWAN, B.A., KONISHI, N., MISRA, M. and RICE, J.M., 1990, Initiation by nickel acetate and promotion by sodium barbital of renal cortical epithelial tumours in male F344 rats, *Carcinogenesis*, **11**, 647–652

KASPRZAK, K.S., DIWAN, B.A., RICE, J.M., MISRA, M., RIGGS, C.W., OLINSKI, R. and DIZDAROGLU, M., 1992, Nickel(II)-mediated oxidative DNA base damage in renal and hepatic chromatin of pregnant rats and their fetuses. Possible relevance to carcinogenesis, *Chemical Research in Toxicology*, **5**, 809–815

KASPRZAK, K.S., JARUGA, P., ZASTAWNY, T.H., NORTH, S.L., RIGGS, C.W., OLINSKI, R. and DIZDAROGLU, M., 1997, Oxidative DNA base damage and its repair in kidneys and livers of nickel(II)-treated male F344 rats, *Carcinogenesis*, **18**, 271–277

KATDARE, M., OSBORNE, M.P. and TELANG, N.T., 1998, Inhibition of aberrant proliferation and induction of apoptosis in pre-neoplastic human mammary epithelial cells by natural phytochemicals, *Oncology Reports*, **5**, 311–315

KATIYAR, S.K., AGARWAL, R. and MUKHTAR, H., 1993a, Inhibition of both stage I and stage II skin tumor promotion in SENCAR mice by a polyphenolic fraction isolated from green tea: inhibition depends on the duration of polyphenol treatment, *Carcinogenesis*, **14**, 2641–2643

KATIYAR, S.K., AGARWAL, R. and MUKHTAR, H., 1993b, Protection against malignant conversion of chemically induced benign skin papillomas to squamous cell carcinomas in SENCAR mice by a polyphenolic fraction isolated from green tea, *Cancer Research*, **53**, 5409–5412

KEARNS, P.R. and HALL, A.G., 1998, Glutathione and the response of malignant cells to chemotherapy, *Drug Discovery Today*, **3**, 113–121

KHAN, S.G., KATIYAR, S.K., AGARWAL, R. and MUKHTAR, H., 1992, Enhancement of antioxidant and phase II enzymes by oral feeding of green tea polyphenols in drinking water to SKH-1 hairless mice: possible role in cancer chemoprevention, *Cancer Research*, **52**, 4050–4052

KLAASSEN, C.D. and LIU, J., 1998, Induction of metallothionein as an adaptive mechanism affecting the magnitude and progression of toxicological injury, *Environmental Health Perspectives*, **106**, 297–300

KONDO, Y., RUSNAK, J.M., HOYT, D.G., SETTINERI, C.E., PITT, B.R. and LAZO, J.S., 1997, Enhanced apoptosis in metallothionein null cells, *Molecular Pharmacology*, **52**, 195–201

KUROSE, I., HIGUCHI, H., MIURA, S., SAITO, H., WATANABE, N., HOKARI, R., *et al.*, 1997, Oxidative stress-mediated apoptosis of hepatocytes exposed to acute ethanol intoxication, *Hepatology*, **25**, 368–378

LANFEAR, J., FLEMING, J., WU, L., WEBSTER, G. and HARRISON, P.R., 1994, The selenium metabolite selenodiglutathione induces p53 and apoptosis: relevance to the chemopreventive effects of selenium? *Carcinogenesis*, **15**, 1387–1392

LEE, S.F., LIANG, Y.C. and LIN, J.K., 1995b, Inhibition of 1,2,4-benzenetriol-generated active oxygen species and induction of phase II enzymes by green tea polyphenols, *Chemico-Biological Interactions*, **98**, 283–301

LEE, Y.W., BRODAY, L. and COSTA, M., 1998, Effects of nickel on DNA methyltransferase activity and genomic DNA methylation levels, *Mutation Research*, **415**, 213–218

LEE, Y.W., KLEIN, C.B., KARGACIN, B., SALNIKOW, K., KITAHARA, J., DOWJAT, K., *et al.*, 1995a, Carcinogenic nickel silences gene expression by chromatin condensation and DNA methylation: a new model for epigenetic carcinogens, *Molecular and Cell Biology*, **15**, 2547–2557

LEI, W., YU, R., MANDLEKAR, S. and KONG, A.N., 1998, Induction of apoptosis and activation of interleukin 1beta-converting enzyme/Ced-3 protease (caspase-3) and c-Jun NH2-terminal kinase 1 by benzo(a)pyrene, *Cancer Research*, **15**, 2102–2106

LITTLE, M.C., GAWKRODGER, D.J. and MACNEIL, S., 1996, Chromium- and nickel-induced cytotoxicity in normal and transformed human keratinocytes: an investigation of pharmacological approaches to the prevention of Cr(VI)-induced cytotoxicity, *British Journal of Dermatology*, **134**, 199–207

LIU, Z., LU, Y., ROSENSTEIN, B., LEBWOHL, M. and WEI, H., 1998, Benzo[a]pyrene enhances the formation of 8-hydroxy-2'-deoxyguanosine by ultraviolet A radiation in calf thymus DNA and human epidermoid carcinoma cells, *Biochemistry*, **14**, 10307–10312

LYNN, S., YEW, F.H., HWANG, J.W., TSENG, M.J. and JAN, K.Y., 1994, Glutathione can rescue the inhibitory effects of nickel on DNA ligation and repair synthesis, *Carcinogenesis*, **15**, 2811–2816

MANDAL, M. and KUMAR, R., 1996, Bcl-2 expression regulates sodium butyrate-induced apoptosis in human MCF-7 breast cancer cells, *Cell Growth and Differentiation*, **7**, 311–318

MANDAL, M., WU, X. and KUMAR, R., 1997, Bcl-2 deregulation leads to inhibition of sodium butyrate-induced apoptosis in human colorectal carcinoma cells, *Carcinogenesis*, **18**, 229–232

MANECKJEE, R. and MINNA, J.D., 1994, Opioids induce while nicotine suppresses apoptosis in human lung cancer cells, *Cell Growth and Differentiation*, **10**, 1033–1040

MCCONKEY, D.J., HARTZELL, P., DUDDY, S.K., HAKANSSON, H. and ORRENIUS, S., 1988, 2,3,7,8-Tetrachlorodibenzo-p-dioxin kills immature thymocytes by $Ca^{2+}$-mediated endonuclease activation, *Science*, **242**, 256–259

MCINTYRE, A., GIBSON, P.R. and YOUNG, G.P., 1993, Butyrate production from dietary fibre and protection against large bowel cancer in a rat model, *Gut*, **34**, 386–391

MILLER, M.L., ANDRINGA, A., CODY, T., DIXON, K. and ALBERT, R.E., 1996, Cell proliferation and nuclear abnormalities are increased and apoptosis is decreased in the epidermis of the p53 null mouse after topical application of benzo[a]pyrene, *Cell Proliferation*, **29**, 561–576

NANJI, A.A., GREENBERG, S.S., TAHAN, S.R., FOGT, F., LOSCALZO, J., SADRZADEH, S.M., *et al.*, 1995, Nitric oxide production in experimental alcoholic liver disease in the rat: role in protection from injury, *Gastroenterology*, **109**, 899–907

NEUMAN, M.G., SHEAR, N.H., BELLENTANI, S. and TIRIBELLI, C., 1998, Role of cytokines in ethanol-induced cytotoxicity *in vitro* in Hep G2 cells, *Gastroenterology*, **115**, 157–166

OHATA, T., FUKUDAM, K., TAKAHASHI, M., SUGIMURA, T. and WAKABAYASHI, K., 1997, Suppression of nitric oxide production in lipopolysaccharide-stimulated macrophage cells by omega 3 polyunsaturated fatty acids, *Japanese Journal of Cancer Research*, **88**, 234–237

OTTOLENGHI, A.D., HASEMAN, J.K., PAYNE, W.W., FALK, H.L. and MACFARLAND, H.N., 1975, Inhalation studies of nickel sulfide in pulmonary carcinogenesis of rats, *Journal of the National Cancer Institute*, **54**, 1165–1172

PROKOPCZYK, G., ADAMS, J.D., LAVOIE, E.J. and HOFFMANN, D., 1987, Effect of snuff and nicotine on DNA methylation by 4-(methylnitrosamino)-1-(3-pyridyl)-1-butanone, *Carcinogenesis*, **8**, 1395–1397

REDMAN, C., XU, M.J., PENG, Y.M., SCOTT, J.A., PAYNE, C., CLARK, L.C. and NELSON, M.A., 1997, Involvement of polyamines in selenomethionine induced apoptosis and mitotic alterations in human tumor cells, *Carcinogenesis*, **18**, 1195–1202

RESNICOFF, M., CUI, S., COPPOLA, D., HOEK, J.B. and RUBIN, R., 1996, Ethanol-induced inhibition of cell proliferation is modulated by insulin-like growth factor-I receptor levels, *Alcoholism and Clinical and Experimental Research*, **20**, 961–966

RESNICOFF, M., SELL, C., AMBROSE, D., BASERGA, R. and RUBIN, R., 1993, Ethanol inhibits the autophosphorylation of the insulin-like growth factor 1 (IGF-1) receptor and IGF-1-mediated proliferation of 3T3 cells, *Journal of Biological Chemistry*, **268**, 21777–21782

RHILE, M.J., NAGARKATTI, M. and NAGARKATTI, P.S., 1996, Role of Fas apoptosis and MHC genes in 2,3,7,8-tetrachlorodibenzo-p-dioxin (TCDD)-induced immunotoxicity of T cells, *Toxicology*, **110**, 153–167

ROBISON, S.H., CANTONI, O. and COSTA, M., 1982, Strand breakage and decreased molecular weight of DNA induced by specific metal compounds, *Carcinogenesis*, **3**, 657–662

ROBINSON, M.J. and COBB, M.H., 1997, Mitogen-activated protein kinase pathways, *Current Opinions in Cell Biology*, **9**, 180–186

SAITO, H., EBINUMA, H., TAKAHASHI, M., KANEKO, F., WAKABAYASHI, K., NAKAMURA, M. and ISHII, H., 1988, Loss of butyrate-induced apoptosis in human hepatoma cell lines HCC-M and HCC-T having substantial Bcl-2 expression, *Hepatology*, **27**, 1233–1240

SALNIKOW, K., COSENTINO, S., KLEIN, C. and COSTA, M., 1994, Loss of thrombospondin transcriptional activity in nickel-transformed cells, *Molecular and Cell Biology*, **14**, 851–858

SATOH, K., KADOFUKU, T. and SAKAGAMI, H., 1997, Copper, but not iron, enhances apoptosis-inducing activity of antioxidants, *Anticancer Research*, **17**, 2487–2490

SCHMIDT, K.N., AMSTAD, P., CERUTTI, P. and BAEUERLE, P.A., 1995, The roles of hydrogen peroxide and superoxide as messengers in the activation of transcription factor NF-κB, *Chemistry and Biology*, **2**, 13 only

SCHRECK, R., RIEBER, P., BAEUERLE, P.A., 1991, Reactive oxygen intermediates as apparently widely used messengers in the activation of the NF-kappa B transcription factor and HIV-1, *EMBO Journal*, **10**, 2247–2258

SCHRECK, R., MEIER, B., MANNEL, D.N., DROGE, W. and BAEUERLE, P.A., 1992, Dithiocarbamates as potent inhibitors of nuclear factor kappa B activation in intact cells, *Journal of Experimental Medicine*, **175**, 1181–1194

SCHRENK, D., 1998, Impact of dioxin-type induction of drug-metabolizing enzymes on the metabolism of endo- and xenobiotics, *Biochemical Pharmacology*, **55**, 1155–1162

SELL, C., BASERGA, R. and RUBIN, R., 1995, Insulin-like growth factor I (IGF-I) and the IGF-I receptor prevent etoposide-induced apoptosis, *Cancer Research*, **55**, 303–306

SEN, C.K. and PACKER, L., 1996, Antioxidant and redox regulation of gene transcription, *FASEB Journal*, **10**, 709–720

SERAFINI, M., GHISELLI, A. and FERRO-LUZZI, A., 1996, In vivo antioxidant effect of green and black tea in man, *European Journal of Clinical Nutrition*, **50**, 28–32

SERGENT, O., MOREL, I., CHEVANNE, M., CILLARD, P. and CILLARD, J., 1995, Oxidative stress induced by ethanol in rat hepatocyte cultures, *Biochemical and Molecular Biology Interactions*, **35**, 575–583

SHI, S.T., WANG, Z.Y., SMITH, T.J., HONG, J.Y., CHEN, W.F., HO, C.T. and YANG, C.S., 1994, Effects of green tea and black tea on 4-(methylnitrosamino)-1-(3-pyridyl)-1-butanone bioactivation, DNA methylation, and lung tumorigenesis in A/J mice, *Cancer Research*, **54**, 4641–4647

SHIAO, Y.H., LEE, S.H. and KASPRZAK, K.S., 1998, Cell cycle arrest, apoptosis and p53 expression in nickel(II) acetate-treated Chinese hamster ovary cells, *Carcinogenesis*, **19**, 1203–1207

SHIMADA, H., SHIAO, Y.H., SHIBATA, M.A., WAALKES, M.P., 1998, Cadmium suppresses apoptosis induced by chromium, *Journal of Toxicology and Enviromental Health*, Pt. A, **54**, 159–168

SIES, H., 1997, Oxidative stress: oxidants and antioxidants, *Experimental Physiology*, **82**, 291–295

SLATER, A.F., STEFAN, C., NOBEL, I., VAN DEN DOBBELSTEEN, D.J. and ORRENIUS, S., 1995, Signalling mechanisms and oxidative stress in apoptosis, *Toxicology Letters*, **82–83**, 149–153

SLOMIANY, B.L., PIOTROWSKI, J. and SLOMIANY, A., 1997, Chronic alcohol ingestion enhances tumor necrosis factor-alpha expression and salivary gland apoptosis, *Alcoholism and Clinical and Experimental Research*, **21**, 1530–1533

SNYDER, R.D., 1988, Role of active oxygen species in metal-induced DNA strand breakage in human diploid fibroblasts, *Mutation Research*, **193**, 237–246

STEWART, M.S., DAVIS, R.L., WALSH, L.P. and PENCE, B.C., 1997, Induction of differentiation and apoptosis by sodium selenite in human colonic carcinoma cells (HT29), *Cancer Letters*, **117**, 35–40

STOHS, S.J., SHARA, M.A., ALSHARIF, N.Z., WAHBA, Z.Z. and AL-BAYATI, Z.A., 1990, 2,3,7,8-Tetrachlorodibenzo-p-dioxin-induced oxidative stress in female rats, *Toxicology and Applied Pharmacology*, **106**, 126–130

SU, G.L., RAHEMTULLA, A., THOMAS, P., KLEIN, R.D., WANG, S.C. and NANJI, A.A., 1998, CD14 and lipopolysaccharide binding protein expression in a rat model of alcoholic liver disease, *American Journal of Pathology*, **152**, 841–849

SUZUKI, Y.J., FORMAN, H.J. and SEVANIAN, A., 1997, Oxidants as stimulators of signal transduction, *Free Radical Biology and Medicine*, **22**, 269–285

TANG, D.G., GUAN, K.L., LI, L., HONN, K.V., CHEN, Y.Q., RICE, R.L., *et al.*, 1997, Suppression of W256 carcinosarcoma cell apoptosis by arachidonic acid and other polyunsaturated fatty acids, *International Journal of Cancer*, **72**, 1078–1087

TANG, W., SADOVIC, S. and SHAIKH, Z.A., 1998, Nephrotoxicity of cadmium-metallothionein: protection by zinc and role of glutathione, *Toxicology and Applied Pharmacology*, **151**, 276–282

THOMAS, D.J., ANGLE, C.R., SWANSON, S.A. and CAFFREY, T.C., 1991, Effect of sodium butyrate on metallothionein induction and cadmium cytotoxicity in ROS 17/2.8 cells, *Toxicology*, **66**, 35–46

THOMPSON, H.J., WILSON, A., LU, J., SINGH, M., JIANG, C., UPADHYAYA, P., *et al.*, 1994, Comparison of the effects of an organic and an inorganic form of selenium on a mammary carcinoma cell line, *Carcinogenesis*, **15**, 183–186

TIAN, L. and LAWRENCE, D.A., 1995, Lead inhibits nitric oxide production *in vitro* by murine splenic macrophages, *Toxicology and Applied Pharmacology*, **132**, 156–163

TIAN, L. and LAWRENCE, D.A., 1996, Metal-induced modulation of nitric oxide production *in vitro* by murine macrophages: lead, nickel, and cobalt utilize different mechanisms, *Toxicology and Applied Pharmacology*, **141**, 540–547

TOROSIAN, M.H., CHARLAND, S.L. and LAPPIN, J.A., 1995, Differential effects of .omega.-3 and .omega.-6 fatty acids on primary tumor growth and metastasis, *International Journal of Oncology*, **7**, 667–672

TRUSH, M.A. and KENSLER, T.W., 1991, An overview of the relationship between oxidative stress and chemical carcinogenesis, *Free Radical Biology and Medicine*, **10**, 201–209

URBANO, A., KOC, Y. and FOSS, F.M., 1998, Arginine butyrate downregulates p210 bcr-abl expression and induces apoptosis in chronic myelogenous leukemia cells, *Leukemia*, **12**, 930–936

VAN DEN DOBBELSTEEN, D.J., NOBEL, C.S.I., SCHLEGEL, J., COTGREAVE, I.A., ORRENIUS, S. and SLATER, A.F., 1996, Rapid and specific efflux of reduced glutathione during apoptosis induced by anti-Fas/APO-1 antibody, *Journal of Biological Chemistry*, **271**, 15420–15427

WANG, Z.Y., AGARWAL, R., BICKERS, D.R. and MUKHTAR, H., 1991, Protection against ultraviolet B radiation-induced photocarcinogenesis in hairless mice by green tea polyphenols, *Carcinogenesis*, **12**, 1527–1530

WANG, Z., KARRAS, J.G., COLARUSSO, T.P., FOOTE, L.C. and ROTHSTEIN, T.L., 1997a, Unmethylated CpG motifs protect murine B lymphocytes against Fas-mediated apoptosis, *Cellular Immunology*, **180**, 162–167

WANG, X., MARTINDALE, J.L., LIU, Y. and HOLBROOK, N.J., 1998, The cellular response to oxidative stress: influences of mitogen-activated protein kinase signalling pathways on cell survival, *Biochemical Journal*, **333**, 291–300

WOLFE, J.T., ROSS, D. and COHEN, G.M., 1994, A role for metals and free radicals in the induction of apoptosis in thymocytes, *FEBS Letters*, **352**, 58–62

WRIGHT, S.C., ZHONG, J., ZHENG, H. and LARRICK, J.W., 1993, Nicotine inhibition of apoptosis suggests a role in tumor promotion, *FASEB Journal*, **7**, 1045–1051

XIA, Z., DICKENS, M., RAINGEAUD, J., DAVIS, R.J. and GREENBERG, M.E., 1995, Opposing effects of ERK and JNK-p38 MAP kinases on apoptosis, *Science*, **270**,1326–1331

YAMAMOTO, H., FUJIMOTO, J., OKAMOTO, E., FURUYAMA, J., TAMAOKI, T. and HASHIMOTO-TAMAOKI, T., 1998, Suppression of growth of hepatocellular carcinoma by sodium butyrate *in vitro* and *in vivo*, *International Journal of Cancer*, **76**, 897–902

YANG, G.Y., LIAO, J., KIM, K., YURKOW, E.J. and YANG, C.S., 1998, Inhibition of growth and induction of apoptosis in human cancer cell lines by tea polyphenols, *Carcinogenesis*, **19**, 611–616

YIN, Y., TERAUCHI, Y., SOLOMON, G.G., AIZAWA, S., RANGARANJAN, P.N., YASAKI, Y., *et al.*, 1998, Involvement of p85 in p53-dependent apoptotic response to oxidative stress, *Nature*, **391**, 701–711

YU, R., JIAO, J.J., DUH, J.L., GUDEHITHLU, K., TAN, T.H. and KONG, A.N., 1997, Activation of mitogen-activated protein kinases by green tea polyphenols: potential signalling pathways in the regulation of antioxidant-responsive element-mediated phase II enzyme gene expression, *Carcinogenesis*, **18**, 451–456

YUKI, T. and THURMAN, R.G., 1980, The swift increase in alcohol metabolism. Time course for the increase in hepatic oxygen uptake and the involvement of glycolysis, *Biochemical Journal*, **186**, 119–126

ZHAO, Y., CAO, J., MA, H. and LIU, J., 1997, Apoptosis induced by tea polyphenols in HL-60 cells, *Cancer Letters*, **121**, 163–167

CHAPTER

# 10

# Detection and Biomarkers of Apoptosis

**JOHN R. FOSTER**

AstraZeneca Central Toxicology Laboratory, Alderley Park, Macclesfield, SK10 4TJ, UK

## Contents

## 10.1 INTRODUCTION

Apoptosis, as a mediator of the action of toxicants, has been described in the preceding chapters. In addition, the pathways that regulate apoptotic cell death have been considered in detail both at the cellular and at the molecular level. However, a clearer understanding of the role of apoptosis in mediating any biological process depends upon accurate quantitation. This chapter will describe the morphology and biomarkers of apoptosis and will consider the application of such knowledge to the quantitation of cell death *in vivo* and *in vitro*.

The first description of the morphology of apoptosis involved stimulation of thymocytes with glucocorticoids (Arends and Wyllie, 1991). Subsequently, it became apparent that the morphological features described were common to many other tissues such as the kidney (Chapter 6) and liver. In whole tissues or for adherent cells in culture, the cell assumes a rounded appearance and a clear 'halo' appears as the cell becomes separated from its neighbours (Figure 10.1A). Using classical histological stains (Kerr *et al.*, 1972), the nucleus becomes deeply stained with haematoxylin as the chromatin condenses and marginates and the cytoplasm becomes deeply eosinophilic. This series of morphological changes is represented diagramatically in Figure 10.1B.

## 10.2 PARAMETERS AFFECTING THE DETECTION OF APOPTOSIS

As discussed in Chapter 1, the biochemical stages of apoptosis can be described as imposition of damage, sensing, coupling and execution. The majority of techniques available detect the cellular changes associated with the final, execution phase of apoptosis rather than any of the latent phases of signalling and commitment. As such, all methodologies underestimate the number of apoptotic cells present. Other factors affecting success in quantitating apoptosis include the tissue to be studied and the marker chosen (see Table 10.1). In addition, some stages within execution are more readily detected than others; for example, the crescentic, marginated chromatin is far more easily recognized than the following stages where nuclear disintegration has occurred (Clarke, 1990). Different tissues in the vertebrate body display different rates of apoptosis depending upon their function. Those tissues which display high rates of cell replication such as the haemopoietic tissue and GI tract generally show higher rates of apoptosis than those tissues, like the liver, where cell replication rates tend to be low. Tissues with a small proportion of apoptotic cells will present the researcher with additional problems from those with high rates of apoptosis. Techniques involving homogenization of the whole tissue, such as ELISA assays, may dilute out the small proportion of apoptotic cells to such an extent that they fall below detection limits (Naruse *et al.*, 1994, Kawarada *et al.*, 1998).

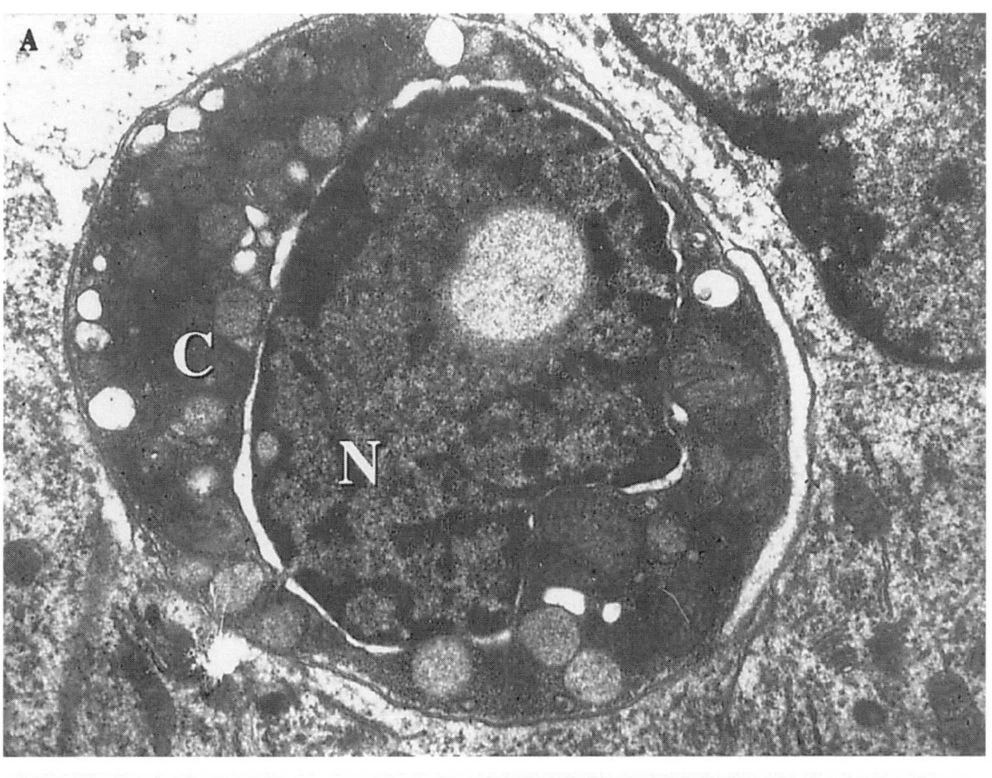


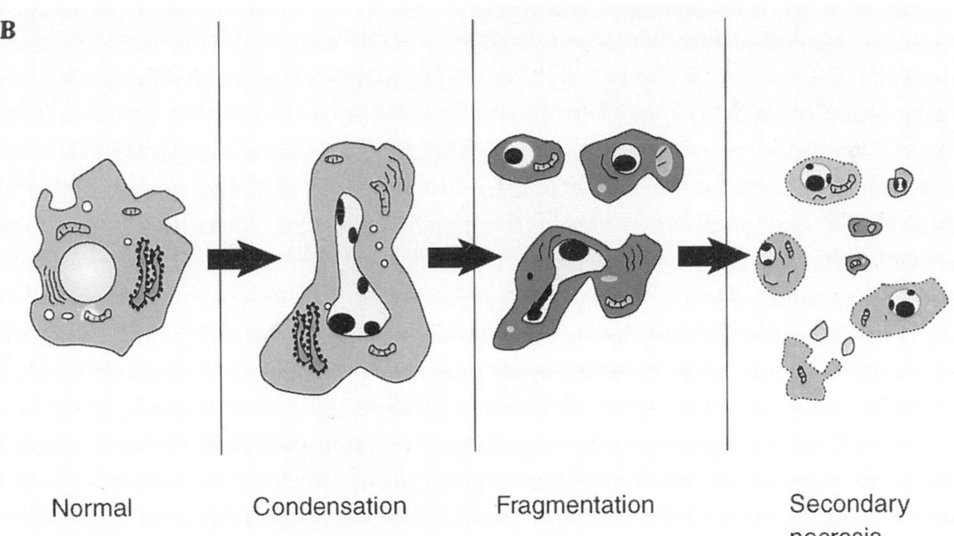


**Figure 10.1** (A) Electron micrograph of an apoptotic heptocyte. The cell has rounded off and lost contact with its neighbours. It shows condensed cytoplasm (C) and nucleus (N) but still retains recognizable organelles such as mitochondria. Magnification X 4,750. (B) Sequence of morphological events that occur as a cell becomes apoptotic. The end stage of secondary necrosis normally takes place after the cell fragments have been phagocytosed by other cells.

**TABLE 10.1**
Potential marker molecules for apoptotic studies

| Marker | Specificity for apoptosis | Change | Reference |
|---|---|---|---|
| c-myc/fos | Not specific | ↑ | Buttyan *et al.*, 1988 |
| Bcl-2/Bax family | Good | ↓/↑ | Vaux *et al.*, 1988 |
| $Ca^{2+}$ | Not specific | ↑ | Kyprianou and Isaacs, 1988 |
| DNA fragmentation | Good | ↑ | Gavrieli *et al.*, 1992 |
| p53 | Limited | ↑ | Yonish-Rouach, 1996 |
| Morphology | Very good | Condensation/ fragmentation | Bursch *et al.*, 1986 |
| Hoechst 33342 | Good? | Condensation of DNA | James and Roberts, 1996 |
| Transglutaminase | Good | ↑ | Fesus *et al.*, 1987 |
| Caspases | Good | ↑ (Enzymatic activity) | Gagliardini *et al.*, 1994 |

Similarly, histological techniques using haematoxylin and eosin on tissue sections may detect a single apoptotic cell present within a tissue section but quantitation will require considerable effort and operator skill.

ELISA assays provide quantitative data following tissue homogenization and differential centrifugation of the cellular components. However it is difficult to relate figures obtained by ELISA assay back to changes in cell populations since a knowledge of the cellular content of the marker in question is required. Similarly, DNA laddering has been regarded as a definitive marker of apoptosis but there is a question regarding whether the phenomenon occurs in all instances of apoptosis (Cohen *et al.*, 1994), and in situations where only a small proportion of the cells undergo apoptosis its lower level of sensitivity may preclude it detecting the increase due to the dilution factor of a large amount of DNA from non-apoptotic cells.

Histological methods can be sensitive, depending upon the marker employed (see Table 10.1), and are applicable to almost all tissues requiring study. However the effort involved in assessing histological preparations of tissues in a quantitative way is always going to be considerable even though their accuracy may be great (Bursch *et al.*, 1986).

Frequently, there may be a compromise between specificity and ease of detection. A marker which is not specifically expressed by apoptotic cells may still be useful if it allows an investigator to distinguish the apoptotic cell from its neighbours (McCarthy and Evan, 1998). An example of such a marker is haematoxylin and eosin, in routine histological preparations, and Hoechst 33342 on cells in culture. Neither stain specifically localizes apoptotic cells but, because they both differentially stain the apoptotic cells, they are of real use.

## 10.3 METHODS FOR DETECTING APOPTOTIC CELLS

Methods for detecting apoptosis can be divided into three broad categories based upon the methodology for tissue preparation. The first involves processing of intact tissue for histology and then utilizing specific or non-specific markers of apoptotic cells. The second involves the preparation of individual cells from the intact tissue and the third method involves disruption of the tissue and cells and analysis of the released cellular proteins and/or DNA.

### 10.3.1 Methods using intact tissue

Tissues can be fixed either by chemical fixatives or by freezing. By definition, these methods involve minimal disruption of the tissue and are able to give precise spatial information regarding the distribution of apoptotic cells within a tissue or organ. Following fixation and sectioning, the choice of histological methodology falls broadly into morphological recognition of the cells by microscopy, immunocytochemical detection using antibodies raised against a molecule expressed by cells undergoing apoptosis or detection of the DNA fragmentation occurring during the apoptotic process. Each of these will be discussed separately.

Morphological recognition of apoptotic cells by either light or electron microscopy remains a widespread technique. With training, the cells can be recognized accurately and may be counted using morphometric techniques on a sufficient number of tissue sections to provide a representative sample. Advantages include ease of use since these morphological techniques are compatible with routine histological or ultrastructural examination. The main disadvantages arise from the subjectivity of the recognition of apoptotic cells which may depend upon the experience of the operator. The studies of Bursch *et al.* (1986) have incorporated counts of apoptotic cells both with and without nuclear fragments, with the identification of the latter being purely on the basis of round eosinophilic bodies, most often contained within other unaffected cells (Bursch *et al.*, 1994). Thus, inter-operator and inter-laboratory variability must be managed to avoid inaccuracies. In addition, judgement of sampling size both within the section and between sections of the same biological sample must be determined statistically to limit variability.

Light microscopy of apoptotic cells (Figure 10.2) has advantages in that the preparation is rapid and inexpensive and requires minimal technical input. In addition, the numbers of apoptotic cells in each section are sufficient for quantitation; this can be automated using similar technology to that described previously for routine evaluation of cell proliferation indices (Soames *et al.*, 1994).

In addition to light microscopy, electron microscopy can be applied to identify apoptotic cells by their morphology. This methodology probably represents the 'gold standard' as far as apoptosis in intact tissue is concerned since the apoptotic bodies may be

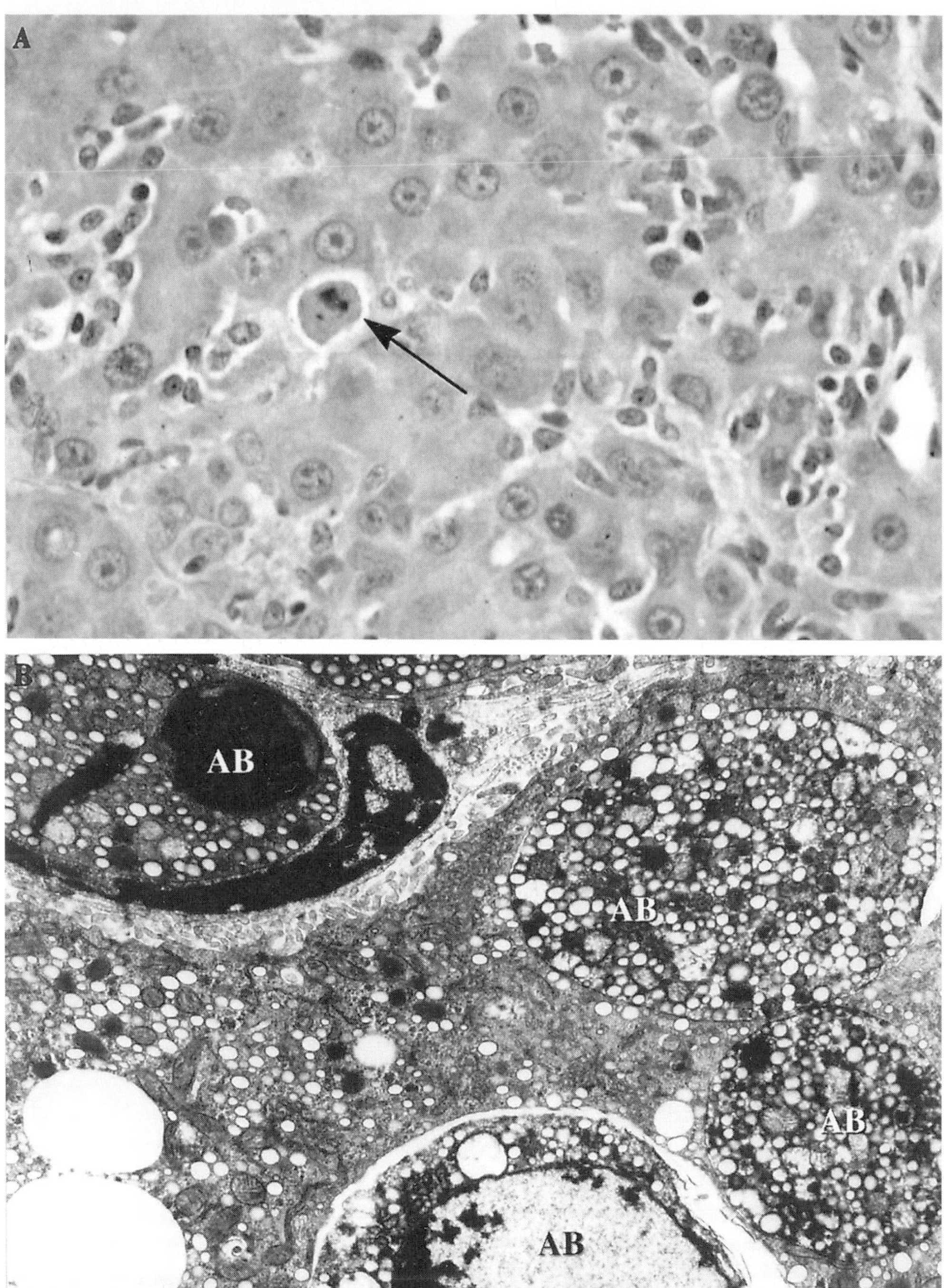


**Figure 10.2** (A) Histological appearance of an apoptotic cell in the liver of a rat. Haematoxylin and eosin stained, X 380 magnification. (B) Electron micrograph of a number of apoptotic bodies (AB), some with and some without nuclei. These are present within both hepatocytes and within Kupffer cells in the liver of a mouse given a chlorinated solvent. In all cases recognizable cytoplasmic organelles are present with a similar morphology to those present within the non-apoptotic hepatocytes that have engulfed the cells. X 2,375 magnification.

identified in all of their stages of progression (Clarke *et al.*, 1990), with or without nuclear fragments (Figure 10.2B).

Immunocytochemistry can be used to detect apoptosis since many proteins are expressed specifically or preferentially by cells undergoing apoptosis. There are several candidate proteins listed in Table 10.1. Transglutaminase is an enzyme involved in the reaction resulting in the condensation of tubular proteins which enables the apoptotic cells to be recognized in tissue sections stained with routine histological stains (Folk 1980, Hager *et al.*, 1997). The transglutaminase-catalyzed cross-linking reaction leads to protein polymerization which is extremely strong (Folk and Finlayson, 1977). The over-expression of tissue or type II transglutaminase is thought to be specific for apoptosis (Piacentini *et al.*, 1994) although low level expression was noted in viable endothelial cells, smooth muscle cells and mesangial cells (Thomazy and Fesus, 1989).

Although marker proteins such as c-myc, p34, SGP-2 and vitronectin have been advocated for use in identifying apoptotic cells (Buttyan *et al.*, 1989, Savill *et al.*, 1990, Dini *et al.*, 1992, Evan *et al.*, 1992, Shi *et al.*, 1994) they are certainly not specific to the process and their use has to be combined with morphology to positively identify the cell as apoptotic. One recent development is the use of caspases (see Chapter 2) as biomarkers of apoptosis. Caspases constitute a family of proteases that are activated at the final, execution stage of apoptosis by cleavage from a pro-form. Activation of the caspase proteolytic cascade results in apoptotic cell death (see Chapter 2). Although antibodies to the caspases are available that can be used for immunocytochemistry, caspase protein levels may appear unchanged since activation results from enzymatic cleavage. Thus, caspase assays based on activity are very useful as biomarkers of apoptosis and will be discussed in section 10.4.3.



With all immunocytochemistry procedures, the fixation methodology must be tailored to suit the tissue and the epitopes (Swanson, 1997). Aldehyde fixatives such as paraformaldehyde or formaldehyde, or alcohol-based fixatives such as methacarn (Oyaizu *et al.*, 1996) are all suitable for immunocytochemistry. In general, the alcohol-based fixatives are less destructive to the antigenic determinants since the aldehyde fixatives introduce extensive cross linking that may render the epitope unrecognizable to an antibody raised against the native protein. In some cases, fixatives may be counter-indicated and tissues would have to be fresh-frozen prior to sectioning. In the case of tissues already fixed, antigen retrieval methodology may have to be applied (Werner *et al.*, 1996). In general, a delicate balance is required between fixation which preserves morphology at the expense of antigenicity; these factors must be balanced for successful identification of apoptotic cells by immunocytochemical techniques.

During apoptosis, DNA fragmentation occurs simultaneously with the morphological changes of chromatin condensation. Several systems have been developed that permit the localization of these DNA breaks in tissue sections (Gorczyca *et al.*, 1993) and in isolated cells in culture (reviewed in Roberts *et al.*, 1995). Overall, the methods fall into three functional categories dependent on the enzyme used for the tagging procedure. *In situ* end labelling (ISEL) uses DNA polymerase I (Kornberg polymerase) which has dual functionality; it can add digoxigenin (DIG)-labelled bases to recessed free 3′ hydroxyl groups but can also nick translate from single stranded breaks, hence the alternative acronym ISNT (*in situ* nick translation). Alternatively, the Klenow fragment of DNA polymerase I (which lacks the nick translation activity of Kornberg) can be used to carry out end labelling of DNA on tissue sections. Alternatively, the terminal deoxynucleotidyl transferase enzyme (Tdt) has been used for Tdt-mediated dUTP nick end labelling (TUNEL) or tailing. This enzyme can add DIG-labelled bases to both double stranded or single stranded 3′-hydroxyl groups. The bound markers on newly synthesized DNA can then be detected by binding to avidin-conjugated peroxidase or other suitable enzyme marker.

The methodology has been criticized as being non-specific since it also stains necrotic cells (Grasl-Kraupp *et al.*, 1995) and gives false positives (Stahelin et al., 1998). However, this would be anticipated since DNA strand breaks are not unique to apoptosis; as with all immunocytochemical techniques, care is needed in interpretation and coincidence of the stain with apoptotic morphology should be used as an additional criterion.

### 10.3.2 Methods using isolated cells

In many applications the necessity for maintaining tissue organization is secondary to the many advantages offered by isolated cells. For example, cells can be made relatively homogeneous and can be generated in large numbers. In some cases, cells can be synchronized to enter apoptosis in large numbers (Sen *et al.*, 1990) and the temporal aspects of apoptosis can be studied in the same cell as it progresses from normality to secondary necrosis (Pulkkinen *et al.*, 1996). Importantly, the effects of exogenous factors such as calcium (Story *et al.*, 1992), xenobiotics (James and Roberts, 1996) and growth factors such as transforming growth factor-β (James and Roberts, 1996) can be studied under defined conditions (Wang *et al.*, 1995, Tong *et al.*, 1996).

McCarthy and Evan (1998) utilized time-lapse video microscopy to study the effects of cell density and cell contact on the apoptotic rate and the subsequent phagocytosis of apoptotic cell fragments. Using this system, they demonstrated that survival factors such as Bcl-2 and insulin-like growth factor I suppressed the onset of apoptosis in cells but did not affect the kinetics of individual events. In contrast, inhibitors of caspases had no effect on the onset of apoptosis but prevented each apoptotic death from going to completion (McCarthy *et al.*, 1997). Thus, time-lapse video microscopy can provide unique information about the fates and rates of apoptosis in individual cells. However, the need for specialist equipment can be prohibitive.

The process of apoptosis can be studied by immunocytochemistry in isolated cells using antibodies against a range of apoptosis-associated antigens (Table 10.1) as described earlier for tissue sections (section 10.3.1). Again, operator expertise is essential for accurate

**TABLE 10.2**
Commonly used DNA stains for morphology and flow cytometry (from Fraker *et al.*, 1995)

| DNA binding dye | | Laser excitation line (nm) | Emission range (nm) |
|---|---|---|---|
| Propidium iodide | Intercalative | 488 | 550–640 |
| Ethidium bromide | Intercalative | 488 | 550–640 |
| Acridine orange | Intercalative | 488 | 510–530 |
| 7-amino-actinomycin D | Intercalative | 488, 541 | 640–680 |
| Daunomycin | Intercalative | 488 | 560–600 |
| Mithramycin A | External | 457 | 510–530 |
| Chromomycin A | External | 457 | 550–580 |
| Olivomycin | External | 457, 480–500 | 530–560 |
| Hoechst 33258 | External | 351–364 | 380–420 |
| Hoechst 33342 | External | 351–364 | 380–420 |
| DAPI | External | 351–364 | 380–420 |

identification of apoptotic cells and appropriate fixation methodology is required to achieve optimal immunostaining. In addition, cells in culture have an intact cell membrane which has to be made permeable to allow the entry of the relatively large antibody molecules. This is usually achieved by alcohol fixation or by a combination of exposure to hypotonic saline and freeze-thawing (Farahat *et al.*, 1994). The choice of final reaction product to visualize the apoptotic cells can be a chromogen or a fluorochrome (Heidenreich *et al.*, 1996).

As an alternative to immunocytochemistry, DNA intercalating dyes can be used for detection of apoptosis in isolated primary cells or in cell lines *in vitro*. There are several available such as propidium iodide, acridine orange or Hoescht 33258. These dyes allow detection of condensed DNA within the nuclei of apoptotic cells and provide an important tool in the study of apoptosis *in vitro* (Juan *et al.*, 1996). Under the right conditions very accurate estimations of the number of apoptotic cells present following the treatment of the culture with, for example, growth factors or cytotoxic drugs, can be made (James and Roberts, 1996, McGahon *et al.*, 1998). The TUNEL technology, described for tissue sections in 10.3.1, is also applicable to cells in culture (Gold *et al.*, 1993).

Flow cytometry is probably one of the most efficient methods for the study and quantification of apoptosis (Dive *et al.*, 1992, Telford *et al.*, 1994, van Engeland *et al.*, 1998). As cells pass through a laser beam, they can be differentiated on the basis of their size and density. Using propidium iodide, the DNA content per cell can also be determined to permit analysis of the proportion of cells undergoing apoptosis (Dive *et al.*, 1992). Table 10.2 shows a list of the common DNA dyes used in flow cytometry (taken from Fraker *et al.*, 1995). Flow cytometry can be especially useful when mixed cell populations need to be analyzed since the technique can differentiate more than one parameter simultaneously. For example, thymocytes can be analyzed simultaneously for DNA content using propidium iodide and cell typed by their expression of CD4/CD8 using antibodies (McCarthy and Evan, 1998). Alternatively, propidium iodide can be used alongside other markers associated with apoptosis such as transglutaminase, annexin V (see 10.3.3), TUNEL (see 10.3.3) or Bcl-2 (Al-Rubeai, 1998).

### 10.3.3 Methods using subcellular fractions

Many assays use other properties of apoptotic cells such as protein, membrane and DNA changes. There are many kits available commercially that exploit these changes to provide rapid and cost-effective measurement of cell death. Some assays utilize an ELISA (Nilsson, 1989) to detect specific proteins or look for the presence of DNA in inappropriate cellular fractions (Chen *et al.*, 1997, Jaeschke *et al.*, 1998, Kawarada *et al.*, 1998, Wassermann *et al.*, 1998). One method used in the study of apoptosis uses primary antibodies directed either against single-stranded DNA itself (Naruse *et al.*, 1994, Kawarada *et al.*, 1998) or raised against histone protein (Suenaga and Abdou, 1996).

A combination of the use of antibodies raised against both DNA and histone proteins was the basis of a commercial technique marketed as a 'Cell death ELISA' (Figure 10.3). During the initial stages of apoptosis, the plasma membrane becomes relatively impermeable (Piacentini *et al.*, 1994) but the nuclear membrane remains unchanged; as the nuclear DNA breaks down, these fragments can be detected in the cytoplasm (Duke and Cohen, 1986). Recently, Ayala *et al.* (1998) used the technique to study the onset of hepatocellular injury induced by sepsis in mice. The disadvantages of this and any other ELISA-based technique have been reviewed by Allen *et al.* (1997). The major drawback is the inability to relate data to actual cell numbers and the subsequent inability to compare with previous complementary data.

Another popular technique used for apoptosis is the detection of phosphatidyl serine (PS), a negatively charged phospholipid normally restricted to the inner surface of the plasma membrane bilayer (Tilcock *et al.*, 1984). During apoptosis, PS is externalized, acting as a trigger for phagocytosis (Fadok *et al.*, 1992). This process can be monitored using the phospholipid binding protein, annexin V or anti-PS antibodies. Annexin V can be used in live cells and is useful to study the dynamics and rate of apoptosis in cell culture (Ernst *et al.*, 1998, Gatti *et al.*, 1998).

Caspases constitute a family of proteases that are activated at the execution stage of apoptosis (see Chapter 1) by cleavage from a pro-form. Activation of the caspase proteolytic cascade results in apoptotic cell death (see Chapter 2). The activation of caspases during apoptosis together with the discovery of several caspase substrates has led to the development of caspase assays as biomarkers for apoptosis (reviewed in Cohen, 1997, Villa

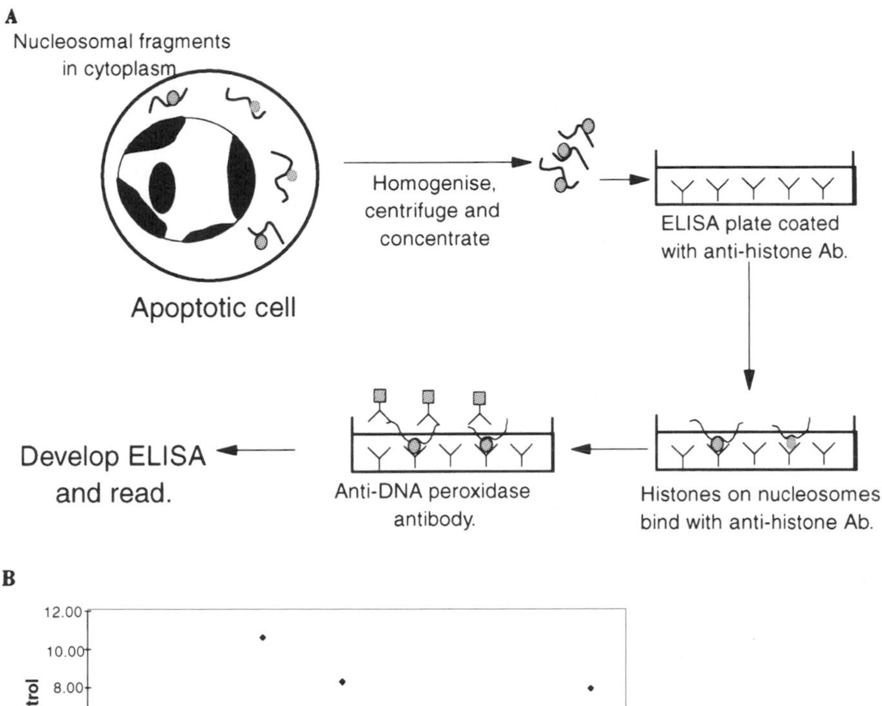


**Figure 10.3** (A) Diagramatic representation of the cell death ELISA technology. Anti-histone antibody is adsorbed onto the wall of the titre plate then the nucleosomes contained in the sample bind, via their histone proteins, to the immobilized anti-histone antibodies. A peroxidase-conjugated anti-DNA antibody reacts with the DNA in the bound sample and the amount of antibody bound is directly proportional to the amount of cytoplasmic mono- and oligonucleosomes present in the original tissue (Allen *et al.*, 1997). (B) A scatter diagram of the apoptotic index derived from the cell death ELISA (shown as optical density, OD) compared with the apoptotic index derived from histological sections of TUNEL-stained liver sections (AB% control).

*et al.*, 1997, McCarthy and Evan, 1998). These assays fall roughly into two categories, those that monitor the cleavage of caspases themselves by western blotting and those that measure substrate cleavage either by western blotting or colourimetric/fluorogenic assays. Anti-caspase antibodies are now widely available commercially and can be used to monitor the proteolytic cleavage of key members of the caspase family such as caspase 3, 8 and 10. These types of study provide great detail on the progression of the caspase cascade with time and have been used to determine the roles played be different family members in different tissues. The disadvantages of this approach are antibody specificity and complexity. For most applications, assays based on the cleavage of caspase substrates are more flexible and practical. Kits that assay cleavage by western blot of the natural caspase-3 substrate, poly (ADP-ribose) polymerase (PARP; see Chapter 2) are now available. In addition, there is a wide range of artificial peptide substrates based on the known caspase cleavage sites linked to a fluorophore or colouriphore (see Chapter 2). Some peptide substrates are highly specific whereas others are cleaved by several caspase family members. Inhibitors are also available, providing additional demonstration of specificity.

### 10.3.4 Quantitation

The techniques reviewed above are varied but all are aimed at providing a meaningful measurement of the quantity of cell death under different experimental conditions. Converting the raw numbers obtained into such meaningful results presents varied challenges. For methods that count the proportion of apoptotic and viable cells either in cell culture or *in vivo*, the apoptotic index can be calculated directly (James and Roberts, 1996). For *in vivo* estimations, morphological assessment and manual counting of tissue sections is most common by analogy with the labelling index methodology used to count cells that have undergone DNA synthesis (Foster, 1997). Typical methodologies count 2000 cells and evaluate the apoptotic index as a ratio of apoptotic cells to the total (Roberts *et al.*, 1995, Wheeldon *et al.*, 1995, Goldsworthy *et al.*, 1996). However, this may be insufficient due to low basal rates of apoptosis (8–15 apoptotic cells per 10,000 cells in rat liver). Thus, care must be taken to count sufficient cells for meaningful data to be generated. At Central Toxicology Laboratory (CTL) we have developed a TUNEL technique that combines automated image analysis with manual counting (Figure 10.4A). The total number of cells in each lobe of the liver can be estimated by image analysis of the number of nuclei in each section (taken from a mean of 5 random fields) and an estimation of the section area. The selection of 5 fields is based upon the minimum number required to give acceptable standard deviations of within 10% of the mean value. Following this automated assessment of total cell numbers, the total number of apoptotic cells is counted manually by scanning the section systematically. An example of the effect of four chemicals that cause liver growth in rodents on the apoptotic rate present within the liver is given in Figure 10.4B. Despite very low levels of apoptosis in untreated liver, a statistically significant inhibition in the apoptotic index could be detected.

Data obtained from flow cytometry lends itself to quantitation since the apoptotic index is readily derived from the flow profile (Gorman *et al.*, 1997, Al-Rubeai, 1998). For cell suspensions or in systems using lymphocytes, for example, the preparation for flow cytometry is relatively simple and the readout readily understandable (Fraker *et al.*, 1995, Sherwood and Schimke, 1995). Despite this, there are reports where only a limited correlation was seen between estimates of apoptosis by flow cytometry and morphological discrimination (Bryson *et al.*, 1994).

A

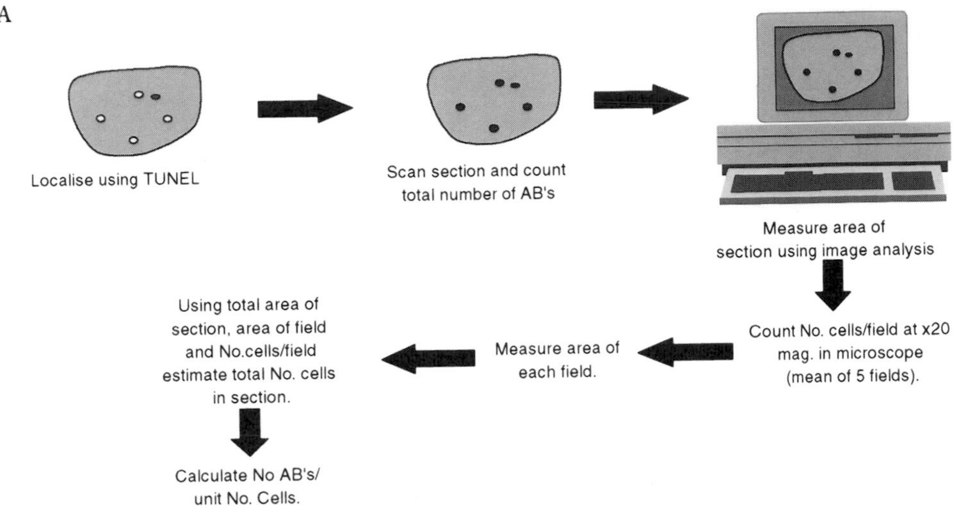


B

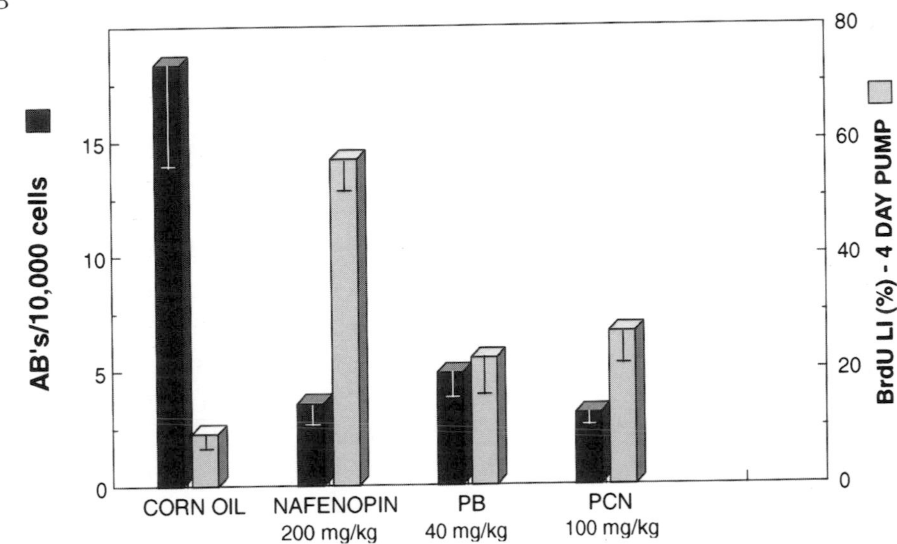


**Figure 10.4** (A) Diagram of the methodology used to calculate the apoptotic index in sections of liver stained with the TUNEL technique. (B) The effect of nafenopin, sodium phenobarbitone (PB) and pregnelolone carbonitrile (PCN) on the apoptotic index (AB/10,000 cells) and the S-phase labelling index (BrdU LI) in the livers of rats given the chemicals at the doses shown for 4 days. It can be seen that as the labelling index increases with treatment, the apoptotic index decreases and the livers of the treated rats grow in size.

For ELISA methods, an optical density measurement is provided that requires intepretation from a set of standard curves. As described in section 10.3.3, it may be difficult to refer optical density measurements back to actual cell numbers. Figure 10.3C illustrates a comparison between the apoptotic index derived using the cell death ELISA methodology, described in section 10.3.3, with that derived from a TUNEL procedure.

TABLE 10.3
Comparison of the available techniques for estimating apoptotic index

| Technique | Reference |
| --- | --- |
| **Morphology** | |
| 1. Light microscopy | Bursch *et al.*, 1986 |
| 2. Electron microscopy | |
| **DNA fragmentation** | |
| 1. Light microscopy | Wheeldon *et al.*, 1995 |
| 2. Flow cytometry | Gold *et al.*, 1993 |
| **Antibody techniques** | |
| 1. Light microscopy | Pyrah *et al.*, 1998 |
| 2. 'Cell death' ELISA | Ayala *et al.*, 1998 |
| 3. Flow cytometry | Dive *et al.*, 1992 |

## 10.4 SUMMARY

There are many varied methodologies for evaluating changes in the apoptotic index, each with their own merits (Table 10.3). These will continue to be developed to increase the ease and accuracy with which the relatively rare events of apoptosis can be monitored. Modifications in preparative procedures and increased specificity of the applied techniques, together with the discovery of new, more specific markers promises to make the field ever more productive in the future.

## ACKNOWLEDGEMENTS

J.R. Foster would like to acknowledge the help of the students and technical staff of CTL Research Pathology in providing some of the material and in evaluating some of the techniques described for the detection and quantitation of apoptosis. Particular thanks go to Ms C. Eagleton, Mr N. Monks, Mr A.R. Soames, Mrs S.M. Williams and Mr N.J. McArdle.

The editor would like to thank Dr Sabina Cosulich for critical appraisal of this chapter and Zana Dye for invaluable editorial assistance.

## REFERENCES

ALLEN, R.T., HUNTER III, W.J. and AGRAWAL, D.K., 1997, Morphological and biochemical characterisation and analysis of apoptosis, *Journal of Pharmacological and Toxicological Methods*, **37**, 215–228

AL-RUBEAI, M., 1998, Apoptosis and cell culture technology, *Advances in Biochemistry Engineering and Biotechnology*, **59**, 225–249

Arends, M.J. and Wyllie, A.H., 1991, Apoptosis: mechanisms and roles in pathology, *International Review of Experimental Pathology*, **32**, 223–254

Ayala, A., Evans, T.A. and Chaudry, I.H., 1998, Does hepatocellular injury in sepsis involve apoptosis? *Journal of Surgical Research*, **76**, 165–173

Bryson, G.J., Harmon, B.V. and Collins, R.J., 1994, A flow cytometric study of cell death: failure of some models to correlate with morphological assessment, *Immunology and Cell Biology*, **72**, 35–41

Bursch, W., Dusterberg, B. and Schulte-Hermann, R., 1986, Growth, regression and cell death in rat liver as related to tissue levels of the hepatomitogen cyproterone acetate, *Archives of Toxicology*, **59**, 221–227

Bursch, W., Grasl-Kraupp, B., Ellinger, A., Torok, L., Kienzl, H., Mullauer, L. and Schulte-Hermann, R., 1994, Active cell death: role in hepatocarcinogenesis and subtypes, *Biochemistry and Cell Biology*, **72**, 669–675

Buttyan, R., Zakeri, Z., Lockshin, R. and Wolgemuth, D., 1988, Cascade induction of c-fos, c-myc and heat shock 70K transcripts during regression of the rat ventral prostate gland, *Molecular Endocrinology*, **2**, 650–657

Buttyan, R., Olsson, C.A., Pintar, J., Chang, C., Bandyk, M., Poying, N.G. and Sawczuk, I.S., 1989, Induction of TRPM-2 gene in cells undergoing programmed cell death, *Molecular and Cell Biology*, **9**, 3473–3481

Chen, G., Sordillo, E.M., Ramey, W.G., Reidy, J., Holt, P.R., Krajewski, S., *et al.*, 1997, Apoptosis in gastric epithelial cells is induced by Helicobacter pylori and accompanied by increased expression of BAK, *Biochemistry Biophysics Research Communications*, **239**, 626–632

Clarke, P.G.H., 1990, Developmental cell death: morphological diversity and multiple mechanisms, *Anatomy and Embryology*, **181**, 195–213

Cohen, G.M., 1997, Caspases: the executioners of apoptosis, *Biochemical Journal*, **326**, 1–16

Cohen, G.M., Sun, X., Fearnhead, H., MacFarlane, M., Brown, D.G., Snowden, R.T. and Dinsdale, D., 1994, Formation of large molecular weight fragments of DNA is a key committed step of apoptosis in thymocytes, *Journal of Immunology*, **153**, 507–516

Dini, L., Autuori, F., Lentini, A., Oliverio, S. and Piacentini, M., 1992, The clearance of apoptotic cells in the liver is mediated by the asiologlycoprotein receptor, *FEBS Letters*, **296**, 174–178

Dive, C., Gregory, C.D., Phipps, D.J., Evans, D.L., Milner, A.E. and Wyllie, A.H., 1992, Analysis and discrimination of necrosis and apoptosis (programmed cell death) by multi-parameter flow cytometry, *Biochimica et Biophysica Acta*, **1133**, 275–285

Duke, R.C. and Cohen, J.J., 1986, IL-2 addiction: withdrawal of growth factor activates a suicide program in dependent T cells, *Lymphokine Research*, **5**, 289–299

ERNST, J.D., YANG, L., ROSALES, J.L. and BROADDUS, V.C., 1998, Preparation and characterization of an endogenously fluorescent annexin for detection of apoptotic cells, *Analytical Biochemistry*, **260**, 18–23

EVAN, G.I., WYLLIE, A.H., GILBERT, C.S., LITTLEWOOD, T.D., LAND, H., BROOKS, M., *et al.*, 1992, Induction of apoptosis in fibroblasts by c-myc protein, *Cell*, **69**, 119–128

FADOK, V.A., VOELKER, D.R., CAMPBELL, P.A., COHEN, J.J., BRATTON, D.L. and HENSON, P., 1992, Exposure of phophatidyl serine on the surface of apoptotic lymphocytes triggers their specific removal by macrophages, *Journal of Immunology*, **148**, 2207–2216

FARAHAT, N., MORILLA, A., OWUSU-ANKOMAH, K., MORILLA, R., PINKERTON, C.R., TRELEAVEN, J.G., *et al.*, 1994, Demonstration of cytoplasmic and nuclear antigens in acute leukaemia using flow cytometry, *Journal of Clinical Pathology*, **47**, 843–849

FESUS, L., THOMAZY, V. and FALUS, A., 1987, Induction and activation of tissue transglutaminase during programmed cell death, *FEBS Letters*, **224**, 104–108

FOLK, J.E., 1980, Transglutaminases, *Annual Reviews of Biochemistry*, **49**, 517–531

FOLK, J.E. and FINLAYSON, S., 1977, The ε(γ-glutamyl)lysine cross-links and the catalytic role of transglutaminase, *Advances in Protein Chemistry*, **31**, 1–133

FOSTER, J.R., 1997, The role of cell proliferation in chemically induced carcinogenesis, *Journal of Comparative Pathology*, **116**, 113–114

FRAKER, P.J., KING, L.E., LILL-ELGHANIAN, D. and TELFORD, W.G., 1995, Quantification of apoptotic events in pure and heterogeneous populations of cells using the flow cytometer, *Methods in Cell Biology*, **46**, 57–76

GAGLIARDINI, V., FERNANDEZ, P.-A., LEE, R.K.K., DREXLER, H.C.A., ROTELLO, R.J., FISHMAN, M.C. and YUAN, J., 1994, Prevention of vertebrate neuronal cell death by the crumb A gene, *Science*, **263**, 826–828

GATTI, R., BELLETTI, S., ORLANDINI, G., BUSSOLATI, O., DALL'ASTA, V. and GAZZOLA, G.C., 1998, Comparison of annexin V and calcein-AM as early vital markers of apoptosis in adherent cells by confocal laser microscopy, *Journal of Histochemistry and Cytochemistry*, **46**, 895–900

GAVRIELLI, Y., SHERMAN, Y. and BEN-SASSON, S.A., 1992, Identification of programmed cell death in situ via specific labelling of nuclear DNA fragmentation, *Journal of Cell Biology*, **119**, 493–501

GOLD, R., SCHMIED, M., ROTHE, G., ZISCHLER, H., BREITSCHOPF, H., WEKERLE, H., LASSMANN, H., 1993, Detection of DNA fragmentation in apoptosis: application of in situ nick translation to cell culture systems and tissue sections, *Journal of Histochemistry and Cytochemistry*, **41**, 1023–1030

GOLDSWORTHY, T.L., FRANSSON-STEEN, R., MARONPOT, R.R., 1996, Importance of and approaches to quantification of hepatocyte apoptosis, *Toxicological Pathology*, **24**, 24–35

GORCZYCA, W., GONG, J. and DARZYNKIEWICZ, Z., 1993, Detection of DNA strand breaks in individual apoptotic cells by the in situ terminal deoxynucleotidyl transferase and nick translation assays, *Cancer Research*, **53**, 1945–1951

GORMAN, A.M., SAMALI, A., MCGOWAN, A.J., COTTER, T.G., 1997, Use of flow cytometry techniques in studying mechanisms of apoptosis in leukemic cells, *Cytometry*, **29**, 97–105

GRASL-KRAUPP, B., RUTTKAY-NEDECKY, B., KOUDELKA, H., BUKOWSKA, K., BURSCH, W. and SCHULTE-HERMANN, R., 1995, In situ detection of fragmented DNA (TUNEL assay) fails to discriminate among apoptosis, necrosis, and autolytic cell death: a cautionary note, *Hepatology*, **21**, 1465–1468

HAGER, H., JENSEN, P.H., HAMILTON-DUTOIT, S., NEILSEN, M.S., BIRCKBICHLER, P. and GLIEMANN, J., 1997, Expression of tissue transglutaminase in human bladder carcinoma, *Journal of Pathology*, **183**, 398–403

HAWORTH, R.A., 1990, Use of isolated adult myocytes to evaluate cardiotoxicity. II. Preparation and properties, *Toxicological Pathology*, **18**, 521–530

HEIDENREICH, S., SCHMIDT, M., BACHMANN, J. and HARRACH, B., 1996, Apoptosis of monocytes cultured from long-term hemodialysis patients, *Kidney International*, **49**, 792–799

JAESCHKE, H., FISHER, M.A., LAWSON, J.A., SIMMONS, C.A., FARHOOD, A. and JONES, D.A., 1998, Activation of caspase 3 (CPP32)-like proteases is essential for TNF-alpha-induced hepatic parenchymal cell apoptosis and neutrophil-mediated necrosis in a murine endotoxin shock model, *Journal of Immunology*, **160**, 3480–3486

JAMES, N.H. and ROBERTS, R.A., 1996, Species differences in response to peroxisome proliferators correlate *in vitro* with induction of DNA synthesis rather than with suppression of apoptosis, *Carcinogenesis*, **17**, 1623–1632

JUAN, G., PAN, W. and DARZYNKIEWICZ, Z., 1996, DNA segments sensitive to single-strand-specific nucleases are present in chromatin of mitotic cells, *Experimental Cell Research*, **227**, 197–202

KAWARADA, Y., MIURA, N. and SUGIYAMA, T., 1998, Antibody against single-stranded DNA useful for detecting apoptotic cells recognises hexadeoxynucleotides with various base sequences, *Journal of Biochemistry (Tokyo)*, **123**, 492–498

KERR, J.F.R., WYLLIE, A.H. and CURRIE, A.R., 1972, Apoptosis: a basic biological phenomenon with wide-ranging implications in tissue kinetics, *British Journal of Cancer*, **26**, 239–357

KYPRIANOU, N. and ISAACS, J.T., 1988, Activation of programmed cell death in the rat ventral prostate after castration, *Endocrinology*, **122**, 552–562

MCCARTHY, N.J. and EVAN, G.I., 1998, Methods for detecting and quantifying apoptosis, *Current Topics in Developmental Biology*, **36**, 259–278

McCarthy, N.J., Whyte, M., Gilbert, C. and Evan, G.I., 1997, Inhibition of Ced-3/ICE-related proteases does not prevent cell death induced by oncogenes, DNA damage or the Bcl-2 homologue Bak, *Journal of Cell Biology*, **136**, 215–227

McGahon, A.J., Costa Pereira, A.P., Daly, L., Cotter, T.G., 1998, Chemotherapeutic drug-induced apoptosis in human leukaemic cells is independent of the Fas (APO-1/CD95) receptor/ligand system, *British Journal of Haematology*, **101**, 539–547

Naruse, I., Keino, H. and Kawarada, Y., 1994, Antibody against single-stranded DNA detects both programmed cell death and drug-induced apoptosis, *Histochemistry*, **101**, 73–78

Nilsson, K.G.I., 1989, Preparation of nanoparticles conjugated with enzyme and antibody and their use in heterogeneous enzyme Immonoassays. *Journal of Immunological Methods*, **122**, 273–277

Oyaizu, T., Arita, S., Hatano, T. and Tsubura, A., 1996, Immunohistochemical detection of estrogen and progesterone receptors performed with an antigen-retrieval technique on methacarn-fixed paraffin-embedded breast cancer tissues, *Journal of Surgical Research*, **60**, 69–73

Piacentini, M., Davies, P.J.A. and Fesus, L., 1994, Transglutaminase in cells undergoing apoptosis, in Tomei, L.O. and Cope, F.O. (eds) *Apoptosis II. The molecular basis of apoptosis in disease.* Cold Spring Harbor: Cold Spring Harbor Laboratory Press, 143–163

Pulkkinen, J.O., Elomaa, L., Joensuu, H., Martikainen, P., Servomaa, K. and Grenman, R., 1996, Paclitaxel-induced apoptotic changes followed by time-lapse video microscopy in cell lines established from head and neck cancer, *Journal of Cancer Research and Clinical Oncology*, **122**, 214–218

Pyrah, I.T.G., Stehr, J.E. and Foster, J.R., 1998, Transglutaminase expression in the adult rat testis as a marker of apoptotic cell death, *European Journal of Veterinary Pathology*, **4**, 1–4

Roberts, R.A., Soames, A.R., James, N.H., Gill, J.H. and Wheeldon, E.B., 1995, Dosing-induced stress causes hepatocyte apoptosis in rats primed by the rodent non-genotoxic hepatocarcinogen cyproterone acetate, *Toxicology and Applied Pharmacology*, **135**, 192–199

Savill, J.S., Dransfield, I., Hogg, A. and Haslett, C., 1990, Vitronectin receptor-mediated phagocytosis of cells undergoing apoptosis, *Nature*, **343**, 170–173

Sen, S., Erba, E. and D'Incalci, M., 1990, Synchronisation of cancer cell lines of human origin using methotrexate, *Cytometry*, **11**, 595–602

Sherwood, S.W. and Schimke, R.T., 1995, Cell cycle analysis of apoptosis using flow cytometry, *Methods in Cell Biology*, **46**, 77–97

SHI, L., NISHIOKA, W.K., TH'NG, J., BRADBURY, E.M., LITCHFIELD, D.W. and GRENBERG, A.H., 1994, Premature p34cdc2 activation required for apoptosis, *Science*, **263**, 1143–1145

SOAMES, A.R., LAVENDER, D., FOSTER, J.R., WILLIAMS, S.M. and WHEELDON, E.B., 1994, Image analysis of bromodeoxyuridine (BrdU) staining for measurement of S-phase in rat and mouse liver, *Journal of Histochemistry and Cytochemistry*, **42**, 939–944

STAHELIN, B.J., MARTI, U., SOLIOZ, L., ZIMMERMANN, H. and REICHEN, J., 1998, False positive staining in the TUNEL assay to detect apoptosis in liver and intestine is caused by endogenous nucleases and inhibited by diethyl pyrocarbonate, *Journal of Clinical Pathology-Molecular Pathology*, **51**, 204–208

STORY, M.D., STEPHENS, L.C., TOMASOVIC, S.P. and MEYN, R.E., 1992, A role for calcium in regulating apoptosis in rat thymocytes irradiated *in vitro*, *International Journal of Radiation Biology*, **61**, 243–251

SUENAGA, R. and ABDOU, N.I., 1996, Anti-(DNA-histone) antibodies in active lupus nephritis, *Journal of Rheumatology*, **23**, 279–284

SWANSON, P.E., 1997, HIERanarchy: the state of the art in immunohistochemistry, *American Journal of Clinical Pathology*, **107**, 139–140

TELFORD, W.G., KING, L.E. and FRAKER, P.J., 1994, Rapid quantitation of apoptosis in pure and heterogeneous cell populations using flow cytometry, *Journal of Immunological Methods*, **172**, 1–16

THOMAZY, V. and FESUS, L., 1989, Differential expression of tissue transglutaminase in human cells, *Cell and Tissue Research*, **255**, 215–224

TILCOCK, C.P.S., BALLY, M.B., FARREN, S.B., CULLIS, P.R. and GRUNER, S.M., 1984, Cation dependent segregation phenomena and phase behaviour in model membrane systems containing PS: influence of cholesterol and acyl chain composition, *Biochemistry*, **23**, 2696–2703

TONG, J.X., EICHLER, M.E. and RICH, K.M., 1996, Intracellular calcium levels influence apoptosis in mature sensory neurons after trophic factor deprivation, *Experimental Neurology*, **138**, 45–52

VAN ENGELAND, M., NIELAND, L.J., RAMAEKERS, F.C., SCHUTTE, B. and REUTELINGSPERGER, C.P., 1998, Annexin V-affinity assay: a review on an apoptosis detection system based on phosphatidylserine exposure, *Cytometry*, **31**, 1–9

VAUX, D.L., CORY, S. and ADAMS, J.M., 1988, Bcl-2 gene promotes haemopoietic cell survival and co-operates with c-myc to immortalise pre-B cells, *Nature*, **335**, 440–442

VILLA, P., KAUFMANN, S.H. and EARNSHAW, W.C., 1997, Caspases and caspase inhibitors, *Trends in Biochemochemical Sciences*, **22**, 388–393

WANG, C.Y., ESHLEMAN, J.R., WILLSON, J.K. and MARKOWITZ, S., 1995, Both transforming growth factor-beta and substrate release are inducers of apoptosis in a human colon adenoma cell line, *Cancer Research*, **55**, 5101–5105

WASSERMANN, R.J., POLO, M., SMITH, P., WANG, X., KO, F. and ROBSON, M.C., 1998, Differential production of apoptosis-modulating proteins in patients with hypertrophic burn scar, *Journal of Surgical Research*, **75**, 74–80

WEI, S.J., CHAO, Y., HUNG, Y.M., LIN, W.C., YANG, D.M., SHIH, Y.L., *et al.*, 1998, S- and G2-phase cell cycle arrests and apoptosis induced by ganciclovir in murine melanoma cells transduced with herpes simplex virus thymidine kinase, *Experimental Cell Research*, **241**, 66–75

WERNER, M., VON-WASIELEWSKI, R. and KOMMINOTH, P., 1996, Antigen retrieval, signal amplification and intensification in immunohistochemistry, *Histochemistry and Cell Biology*, **105**, 253–260

WHEELDON, E.B., WILLIAMS, S.M., SOAMES, A.R., JAMES, N.H. and ROBERTS, R.A., 1995, Quantitation of apoptotic bodies in rat liver by in situ end labelling (ISEL): correlation with morphology, *Toxicological Pathology*, **23**, 410–415

YONISH-ROUACH, E., 1996, The p53 tumour suppressor gene: a mediator of a GI growth arrest and of apoptosis, *Experientia*, **52**, 1001–1007

# Index

Numerals in **bold** refer to Figures and those in *italic* refer to Tables.